药物合成实验指导

尹 伟
韩丽娟 等编著

Guidance for
Drug Synthesis Experiments

·北京·

内容简介

实验是高等学校化学相关专业最为基本的教学形式之一，在培养学生科学思维、科学方法、创新意识以及创新能力等方面，有着极其重要的作用。本书共整理、编写了 36 个基础有机化学实验、25 个药物合成实验，大部分的实验相关内容都经过了教学实践的检验，同时吸收了国内外不同化学实验课程书籍的优点而编写。

本书适合高职类院校化学、药学等专业师生使用。

图书在版编目（CIP）数据

药物合成实验指导：汉英对照 / 尹伟等编著. —北京：化学工业出版社，2024.8
ISBN 978-7-122-45757-8

Ⅰ. ①药… Ⅱ. ①尹… Ⅲ. ①药物化学-有机合成-化学实验-汉、英 Ⅳ. ①TQ460.31-33

中国国家版本馆 CIP 数据核字（2024）第 108104 号

责任编辑：张　蕾　　　　　装帧设计：史利平
责任校对：刘　一

出版发行：化学工业出版社
　　　　　（北京市东城区青年湖南街 13 号　邮政编码 100011）
印　　装：北京印刷集团有限责任公司
787mm×1092mm　1/16　印张 21　字数 495 千字
2025 年 6 月北京第 1 版第 1 次印刷

购书咨询：010-64518888　　　售后服务：010-64518899
网　　址：http://www.cip.com.cn
凡购买本书，如有缺损质量问题，本社销售中心负责调换。

定　　价：59.80 元　　　　　　　　　　　版权所有　违者必究

编著者名单

尹　伟（合肥职业技术学院）
韩丽娟（合肥京东方医院）
谢俊俊（江西省药品检查员中心）
王曙光（滁州城市职业学院）
杨培星（安徽省第二人民医院灵璧医院）
崔海奇（合肥高新区瑞慈瑞合综合门诊部）
陈晓新（合肥市第三人民医院）
胡　琪（南京中芯启恒生物技术有限公司）
李新章（南通凯恒生物科技发展有限公司）
刘　伟（合肥医工医药股份有限公司）
林　腾（铜陵市义安区市场监督管理局）
方丽波（合肥职业技术学院）
王　刚（安徽中医药大学）
张国升（安徽中医药大学）
卫　强（安徽新华学院）
戴　一（安徽新华学院）
张　爽（中国科学技术大学）
周青浩（中国科学技术大学）

前言

当今世界科学技术的快速发展，全面主导着社会的进步。全方位地推进、优化素质教育就显得尤为重要。高等学校应不断调整、更新课堂教学以及实验教学的相关内容，才能适应社会高速发展的需要。化学实验课程教学是高等学校化学相关专业最为基本的教学形式之一，在培养学生科学思维、科学方法、创新意识以及创新能力等方面，有着极其重要的作用。

《药物合成实验指导》共整理、编写了 36 个基础有机化学实验、25 个药物合成实验，大部分实验相关内容都经过了教学实践的检验，同时吸收了国内外不同化学实验书籍的优点而编写。本书具有以下特点：

一、本书将大量各种现行的化学实验课程教材以及理论课程教材中的实验内容，进行了汇总，对理论教学与实践教学进行了有机整合，体现了本门课程的系统性与科学性。

二、本书内容主要分为有机化学实验与药物合成实验两大部分，内容具有层次感，从基础实验开始，到药物合成的综合性实验，既满足了基础知识的学习，又迎合了知识拔高的需要，同时亦在一定程度上锻炼了学生分析问题、解决问题的能力。

三、本书中安排了在临床中常用药物以及行业前沿药物的合成，在一定程度上增加了学生对本书兴趣，在帮助学生掌握具体实验技能的同时，亦能掌握药物合成的新知识、新动向。

四、本书编写人员有高校教师、化学药品生产企业工程师、临床医生以及药品生产稽查专家，涉及药品研发生产及使用的全过程，应用面较广，覆盖面较全。

五、本书采用了双语体系编写，在掌握化学实验内容的同时，有利于提高高校学生的专业英语能力。

本著作通篇主要由尹伟、韩丽娟完成，在此感谢谢俊俊、王曙光、杨培星、崔海奇、陈晓新、胡琪、李新章、刘伟、林腾、方丽波、王刚、张国升、卫强、戴一、张爽、周青浩等人员给予的指导及帮助。

本著作在编写的过程中，参考了相关文献、书籍，谨此深表谢意。

由于本著作编写时间较短，受编者水平所限等原因，难免存在不足之处，敬请广大读者在使用过程中予以指正，以提高本书质量，感谢！

编 者
2025 年 4 月

目 录

第一部分　实验室基本知识

一、实验室管理制度 ……………………………………………………………… 002
二、实验室气体采购和气瓶安全管理规定 ……………………………………… 005
三、实验室化学危险品管理办法 ………………………………………………… 007
四、实验记录与实验报告 ………………………………………………………… 010

Part Ⅰ　Basic Knowledge of Laboratory

Ⅰ. Laboratory Management System ……………………………………………… 014
Ⅱ. Laboratory gas procurement and gas cylinder safety management regulations … 017
Ⅲ. Measures for the control of hazardous chemicals in laboratories …………… 020
Ⅳ. Experimental records and reports ……………………………………………… 025

第二部分　有机化学实验
Part Ⅱ　Organic chemistry experiment

实验一　1-溴丁烷的合成 ………………………………………………………… 029
Experiment 1　Synthesis of 1-bromobutane …………………………………… 031
实验二　对硝基苯乙酸的合成 …………………………………………………… 033
Experiment 2　Synthesis of p-nitrophenyl acetic acid ……………………… 035
实验三　乙酸松油酯的合成 ……………………………………………………… 037
Experiment 3　Synthesis of terpinyl acetate …………………………………… 039
实验四　氯代环己烷的合成 ……………………………………………………… 041
Experiment 4　Synthesis of chloro-cyclohexane ……………………………… 043
实验五　苄基三乙基氯化铵的合成 ……………………………………………… 045
Experiment 5　Synthesis of benzyl triethyl ammonium chloride ……………… 047
实验六　乳酸丁酯的合成 ………………………………………………………… 049
Experiment 6　Synthesis of butyl lactate ……………………………………… 051

实验七　桂皮酰哌啶的合成 053
Experiment 7　Synthesis of cinnamyl piperidine 055
实验八　3-苯甲酰基丙烯酸的合成 058
Experiment 8　Synthesis of 3-benzoyl acrylic acid 060
实验九　D-葡萄糖酸-δ-内脂的合成 062
Experiment 9　Synthesis of D-gluconic acid -δ-endolipin 064
实验十　N-苄基乙酰苯胺的合成 066
Experiment 10　Synthesis of N-benzyl acetanilide 068
实验十一　对硝基乙酰苯胺的合成 070
Experiment 11　Synthesis of p-nitro acetanilide 072
实验十二　4-氨基-1,2,4-三唑-5-酮的合成 074
Experiment 12　Synthesis of 4-amino-1, 2, 4-triazole-5-ketone 075
实验十三　丙酰氯的合成 077
Experiment 13　Synthesis of propionyl chloride 079
实验十四　紫罗兰酮的合成 081
Experiment 14　Synthesis of ionone 083
实验十五　α-呋喃丙烯酸的合成 085
Experiment 15　Synthesis of α-furanacrylic acid 087
实验十六　4-溴-2-萘酚的合成 089
Experiment 16　Synthesis of 4-bromo-2-naphthol 091
实验十七　对硝基苯甲醛的合成 093
Experiment 17　Synthesis of p-nitro benzaldehyde 095
实验十八　2-庚酮的合成 097
Experiment 18　Synthesis of 2-heptanone 099
实验十九　查尔酮的合成 101
Experiment 19　Synthesis of chalcone 102
实验二十　己酸异戊酯的合成 104
Experiment 20　Synthesis of isoamyl caproate 105
实验二十一　二苯甲醇的合成（1） 107
Experiment 21　Synthesis of diphenyl carbinol (1) 109
实验二十二　正丁醛的合成 111
Experiment 22　Synthesis of n-butyl aldehyde 113
实验二十三　2-氨基丙醇的合成 115
Experiment 23　Synthesis of 2-amino-propyl alcohol 117
实验二十四　3,4-二甲氧基硝基苯的合成 119
Experiment 24　Synthesis of 3, 4-dimethoxy-nitrobenzene 120
实验二十五　对氨基苯甲酰-β-丙氨酸的合成 122
Experiment 25　Synthesis of p-amino benzoyl-β-alanine 123
实验二十六　$KMnO_4$氧化法对硝基苯甲酸的合成 125

Experiment 26　Synthesis of *p*-nitrobenzoic acid by KMnO₄ oxidation ·················127
实验二十七　硝基苯的合成·················129
Experiment 27　Synthesis of nitrobenzene·················131
实验二十八　对硝基肉桂酸的合成·················133
Experiment 28　Synthesis of *p*-nitro cinnamic acid·················135
实验二十九　环己酮的合成·················137
Experiment 29　Synthesis of cyclohexanone·················139
实验三十　扁桃酸乙酯的合成·················141
Experiment 30　Synthesis of ethyl mandelate·················143
实验三十一　苯甲酸的合成·················145
Experiment 31　Synthesis of benzoic acid·················147
实验三十二　4-溴代丁酸乙酯的合成·················149
Experiment 32　Synthesis of 4-bromo-ethyl butyrate·················150
实验三十三　香豆素-3-羧酸的合成·················152
Experiment 33　Synthesis of coumarin-3-carboxylic acid·················154
实验三十四　对甲基苯乙酮的合成·················156
Experiment 34　Synthesis of *p*-methyl acetophenone·················158
实验三十五　Cannizzaro 反应合成苯甲酸和苯甲醇·················160
Experiment 35　Synthesis of benzoic acid and benzyl alcohol by Cannizzaro reaction·················162
实验三十六　二苯甲醇的合成（2）·················165
Experiment 36　Synthesis of diphenyl carbinol (2)·················167

第三部分　药物合成实验
Part Ⅲ　Drug synthesis experiment

实验一　青霉素 G 钾盐的氧化·················170
Experiment 1　Oxidation of penicillin G potassium salts·················172
实验二　盐酸胍法辛的合成·················174
Experiment 2　Synthesis of guanfacine hydrochloride·················176
实验三　诺氟沙星的合成·················179
Experiment 3　Synthesis of norfloxacin·················182
实验四　来曲唑的合成·················186
Experiment 4　Synthesis of letrozole·················188
实验五　贝诺酯的合成·················191
Experiment 5　Synthesis of benorilate·················193
实验六　醋酸胍那苄的合成·················196
Experiment 6　Synthesis of Guanabenz Acetate·················198

实验七 阿司匹林的合成 200
Experiment 7　Synthesis of aspirin 203
实验八 盐酸萘替芬的合成 206
Experiment 8　Synthesis of Naftifine hydrochloride 209
实验九 依达拉奉的合成 212
Experiment 9　Synthesis of Edaravone 214
实验十 盐酸苯海索的合成 216
Experiment 10　Synthesis of trihexyphenidyl hydrochloride 218
实验十一 葡甲胺的合成 221
Experiment 11　Synthesis of meglumine 224
实验十二 利巴韦林的合成 227
Experiment 12　Synthesis of ribavirin 229
实验十三 L-抗坏血酸棕榈酸酯的合成 232
Experiment 13　Synthesis of L-ascorbyl palmitate 234
实验十四 奥沙普秦的合成 236
Experiment 14　Synthesis of oxaprozin 238
实验十五 苯佐卡因的合成 240
Experiment 15　Synthesis of benzocaine 243
实验十六 对乙酰氨基酚（扑热息痛）的合成及鉴定 247
Experiment 16　Synthesis and identification of acetaminophen(paracetamol) 249
实验十七 亚胺-154 的合成 252
Experiment 17　Synthesis of Ethylimine 254
实验十八 丙戊酸钠的合成 256
Experiment 18　Synthesis of sodium valproate 259
实验十九 苯妥英钠的合成 263
Experiment 19　Synthesis of phenytoin sodium 265
实验二十 硝苯地平的合成 268
Experiment 20　Synthesis of Nifedipine 270
实验二十一 维生素 K_3 的合成 272
Experiment 21　Synthesis of vitamin K_3 274
实验二十二 巴比妥酸的合成 277
Experiment 22　Synthesis of barbiturates 279
实验二十三 乳酸米力农的合成 281
Experiment 23　Synthesis of Milrinone lactate 284
实验二十四 美沙拉秦的合成 287
Experiment 24　Synthesis of Mesalazine 289
实验二十五 曲尼司特的合成 292
Experiment 25　Synthesis of Tranilast 294

附 录

附录 01　常用的溶剂的纯化处理措施、方法一览表 ·········· 297
附录 02　国际原子量一览表 ·········· 302
附录 03　不同温度下水的蒸气压一览表 ·········· 303
附录 04　常见化合物的物理常数一览表 ·········· 304
附录 05　常见干燥剂使用一览表 ·········· 307
附录 06　常见试剂的除水剂一览表 ·········· 308
附录 07　常见溶剂与水形成的二元共沸物一览表 ·········· 309
附录 08　常见有机溶剂间的共沸物一览表 ·········· 310
附录 09　实验室常用酸碱的浓度一览表 ·········· 311
附录 10　常见冰盐浴冷却剂一览表 ·········· 312
附录 11　常见冷却剂一览表 ·········· 313
附录 12　常见干燥剂的分类以及使用方法一览表 ·········· 314
附录 13　常见溶剂的提纯、干燥以及贮藏方法一览表 ·········· 316
附录 14　标准缓冲液的配制方法一览表 ·········· 319
附录 15　常用的缓冲液的配制一览表 ·········· 320
附录 16　常用酸碱试剂的含量与密度一览表 ·········· 321
附录 17　常用酸碱指示剂以及配制方法一览表 ·········· 322
附录 18　常用试剂的配制方法一览表 ·········· 323
附录 19　常用二元体系展开剂的洗脱顺序一览表 ·········· 325

参考文献 ·········· 326

第一部分　实验室基本知识

一、实验室管理制度

（一）化学实验室规则

1. 进入实验室工作的人员，必须严格遵守实验室的规章制度，服从实验室人员的安排和管理，保持室内安静与整洁，做到文明实验。
2. 使用仪器设备必须严格遵守操作规程，认真填写使用记录，发生破碎或损坏的，应及时报告实验室管理人员。
3. 保证实验室账、物相符。对实验室仪器设备，要定期进行保养、维修、检验，保持仪器设备完好与实验数据的准确、可靠。提倡分工协作、专管专用，提高仪器设备的使用率。
4. 仪器设备的管理、维护、保养和档案材料的填写、整理、保管等工作需要有专人负责。
5. 实验室应保持整洁、安静，禁止吸烟，严禁存放个人物品，不得随意住宿，更不得将仪器设备、场地私自租借给他人使用。
6. 未经实验室负责人同意，非实验室人员不得在实验室内做实验。任何人不得以任何借口，长期占用实验室。
7. 注意安全，做好防火、防盗、防爆炸、防破坏工作，防止事故的发生。
8. 校外人员进入实验室做实验或参观学习，须经主管部门批准同意后方可进行。

（二）学生实验守则

1. 遵守实验室的规章制度，服从教师指导，保持实验室的整洁、安静，不准吸烟、随地吐痰、乱扔杂物等。
2. 实验前应认真预习，明确实验目的、要求，掌握所用仪器的性能及操作方法，按要求做好一切准备。经教师检查许可后，方可进行实验。
3. 实验课不得迟到，衣冠不整不得进入实验室，不准将与实验课无关的物品带入实验室。
4. 严格按操作规程进行实验，认真如实地记录各种实验数据，不得擅离操作岗位。
5. 实验完毕后，经教师检查仪器、工具、器皿及实验记录后，方可离开实验室。
6. 发现仪器设备损坏，及时报告，查明原因。凡违反操作规程造成事故的，按有关规定处理。
7. 注意安全，一旦发生事故，应立即切断电源、火源，并向教师如实报告，采取紧急措施。
8. 爱护实验室内一切设施，不得乱写乱画，禁止动用与本实验无关的仪器设备、器材与相关设施。
9. 要勤俭节约，不浪费水、电、材料。
10. 对不遵守纪律和实验不认真者，教师有权令其停止或重做。

（三）实验室安全管理规定

1. 实验室人员要严格执行实验室安全管理规定，根据各实验室具体情况，建立实验室的安全操作规定、防盗安全制度和防火公约，明确职责，任务落实到人。

2. 每个实验室、库房都要选派1名责任心强、熟悉情况的老师担任安全员，全面负责实验室的安全工作，发生事故应立即采取应急措施，及时处理和上报，重大事故要保护好现场等待处理。

3. 凡做带有危险性的实验，必须有安全防护措施，并要有教师或实验技术人员的监护，否则不得进行。

4. 易燃、易爆、高压、高温、有毒、有害等危险品，按规定设专用库房，并有专人妥善保管，要严格领用手续。

5. 安全用水、电、气，不得乱拉电线，水、电、气使用完毕后，立即关闭开关，对杂物要及时清扫。

6. 不得擅自安装和使用煤气炉、电炉、电暖气和大功率加热器，因教学、科研需要时，要经主管领导批准，人员离开时要及时切断电源。

7. 消防器材与设施，要放在明显的位置，经常检查，发现故障及时排除和维修，使之处于完好状态。

8. 各实验室的钥匙要加强管理，不得私自配备或转借他人。

（四）实验室工作人员守则

1. 加强精神文明建设，优化育人制度。
2. 严格遵守学校的各项规章制度，树立良好的职业道德。
3. 认真做好实验前的准备工作，严格履行岗位职责。
4. 严格要求学生，按实验程序和操作规程进行实验。
5. 要做好仪器设备、实验器材的账、物管理工作，做到日清月结，保持账、物一致。
6. 爱护仪器设备和实验设施，不得随意借用，更不允许私人占用。
7. 熟练掌握仪器设备的操作技能和技巧，认真做好仪器设备的维护保养工作，提高维修技术，保证仪器设备的完好率以及实验项目的开出率。
8. 认真学习实验教育理论，刻苦钻研业务技术，不断提高实验教学水平。

（五）药品领用、存储及操作制度

1. 操作危险性化学药品，请务必遵守操作守则，遵照规定的操作流程，勿自行更换实验流程。
2. 领取药品时，该确认容器上标示中文名称是否为需要的实验用药品。
3. 领取药品时，请看清楚药品危害标示与图样，确认是否有危害。
4. 使用挥发性有机溶剂、强酸强碱性、高腐蚀性、有毒性的药品，必须在排烟柜及桌上型抽烟管下进行操作。
5. 有机溶剂、固体化学药品、酸、碱等化合物均须分开存放，挥发性的化学药品必须放置于具有抽气装置的药品柜中。
6. 高挥发性或易于氧化的化学药品，必须存放于冰箱或冰柜之中。
7. 避免独自一人在实验室做危险实验。

8. 若须进行无人监督的实验，其实验装置必须防火、防爆、防水，保持实验室灯常开，并在门上留下紧急处理时联络人电话。

9. 进行危险性实验前，必须经实验室负责人批准，有 2 人以上在场方可进行，节假日与夜间严禁做危险性实验。

10. 做有危害性气体的实验，必须在通风橱里进行。

11. 做放射性、激光等对人体危害较重的实验，应制定严格的安全措施，做好个人防护。

12. 废弃、过期药液或废弃物，必须依照分类标示清楚，药品使用后的废（液）弃物，严禁倒入水槽或水沟，应倒入专用收集容器中，进行回收。

二、实验室气体采购和气瓶安全管理规定

（一）安全管理

学校是实验室气瓶安全管理的责任主体，各实验室责任人对所属实验室的气瓶负有安全管理责任。责任人必须随时关注检查气瓶的安全情况，掌握气瓶的使用状态，确保相关使用条件和应急处置措施符合规范要求，及时发现安全隐患，并对气瓶使用人员进行安全技术教育。

（二）档案管理

实验教学管理中心，负责建立气瓶管理档案，档案文件包括：
1. 实验室气瓶安全管理制度以及事故应急处理预案。
2. 实验室气瓶台账，包括气瓶数量、存放地点、充装气体成分、管理人员等。
3. 实验室气体供应商的资质材料，包括企业营业执照、危险化学品经营许可证、气瓶充装许可证等的复印件，并加盖公章。

（三）购置

实验室原则上不得购置气瓶。《气瓶安全监察规定》第二十六条规定，气瓶充装单位，有义务向气体消费者提供气瓶，并对气瓶的安全全面负责。实验室购买实验气体时，由供应商提供在检测合格期内的气瓶，并出具气瓶定期检测证书。使用人须先填写实验用气体购买申请表，经批准后方可采购，申请表由实验教学管理中心存档。

（四）验收

气瓶必须经检查以及验收合格后，方能投入正常使用。
1. 检查气瓶所充装气体是否正确，气体名称标注应清晰可见。
2. 检查气瓶有无定期检验，有无钢印，信息是否完整（制造厂、制造日期、气瓶型号、工作压力、气压实验压力、气压实验日期及下次送检日期、气体容积、气瓶重量）。
3. 检查气瓶有无出厂合格证。
4. 检查气瓶附件（防震圈、防护帽、瓶阀）是否齐全、完好，管路材质是否合适，有无破损或老化现象。
5. 检查气瓶外观是否正常、气嘴有无变形、瓶身是否存在腐蚀、变形、磨损、裂纹等缺陷。
6. 气瓶检查合格后，办理实验室气瓶验收登记表，登记表存放于实验室备查。

（五）储存

1. 实验室内存放气瓶数量合理，氧气与可燃气体均不得超过1瓶，且不可混放，其他气瓶的存放以不影响工作为准，控制在最小需求量。

2. 气瓶存放点必须通风、远离热源、避免暴晒和强烈震动，地面平整干燥。

3. 所有气瓶均应通过气瓶柜、气瓶防倒链、防倒栏栅等方式竖直固定。空瓶与实瓶应分开放置，并有明显的区分标志。

4. 严禁与易燃、易爆、有毒危险化学品、固废混存，并避开各种放射源。

5. 使用剧毒、易燃易爆气体的实验室，须配有通风设施和监控报警装置，张贴必要的安全警示标识。

（六）使用

使用人须严格按照气瓶及所属仪器的操作规程（说明书、注意事项等）进行操作，学生自行实验使用气瓶前，指导教师须向其告知潜在的危险因素、后果和应急措施，经指导教师批准、签字（实验室/仪器设备使用申请单）后，学生方可进行相关实验。实验室责任人必须对学生使用者，进行安全知识和使用资格确认。

1. 气瓶须加装相应的减压器，并进行正确设置。可燃性气体以及可能造成回流的使用场合，必须配置防倒灌的装置，如单向阀、止回阀、缓冲罐等，并随时检查其有效性。

2. 操作者应站在与气瓶接口相垂直的位置，操作减压器和开关阀的动作必须缓慢，使用气瓶时，先开阀门，后开减压阀，关闭气瓶时，先关阀门，放尽余气后再关减压阀。

3. 气瓶内气体不得用尽，必须留有剩余压力或重量，永久性气体气瓶剩余压力应不小于0.05MPa（表压），液化气体气瓶应留有不少于0.5%~1.0%规定充装量的剩余气体。

4. 气瓶必须定期检漏，有气体泄漏的气瓶严禁使用。

5. 气瓶必须在规定检验期限内，使用完毕或送检。

6. 乙炔气瓶不得放于绝缘体上，使用前，必须先直立20min后，再连接减压阀使用。

7. 氧气瓶或氢气瓶严禁与油类接触，操作人员不能穿戴有油脂或油污的工作服和手套等进行操作，以免引起燃烧或爆炸。

（七）搬运

搬运气瓶前，应关紧阀门，拆下减压器，检查各部件标牌是否完好。操作人员必须了解瓶内气体的名称、性质和搬运注意事项，并备齐相应的工具和防护用品。

1. 近距离（5m以内）移动气瓶，应手扶气瓶肩，转动瓶底，并且使用手套。移动距离较远时，应使用专用小车或吊篮。

2. 使用专用气瓶车，进行远距离搬运时，应旋紧防护瓶帽，装上防震垫圈，严禁使用叉车、铲车搬运，不得与化学品混装混运。

3. 装卸气瓶时，应轻装轻卸，禁止手执开关阀搬抬气瓶，禁止采用抛、滑、摔、滚、碰等方式搬运，卸车时，应在气瓶落地点铺上软垫或橡胶皮垫，逐个卸车，严禁溜放，严防因违规或不当操作引发事故。

三、实验室化学危险品管理办法

（一）总则

1. 为进一步规范和加强实验室化学危险品的安全管理，严防事故发生，保障师生员工生命财产的安全，更好地为教学、科研服务，根据《危险化学品安全管理条例》《易制毒化学品管理条例》的有关规定，制定本办法。

2. 本管理办法中，所指化学危险品包括国家标准《危险货物分类和品名编号》（GB 6944—2012）、《危险货物品名表》（GB 12268—2012），将危险化学品分为九大类：①爆炸品；②气体；③易燃液体；④易燃固体，易于自燃的物质，遇水放出易燃气体的物质；⑤氧化性物质和有机过氧化物；⑥毒害物质和感染性物质；⑦放射性物质；⑧腐蚀性物质；⑨杂项危险物质和物品，包括危害环境物质。

3. 凡学校购买、运输、储存、使用和销毁化学危险品的部门和个人必须遵守本办法。

（二）管理体制

1. 学校化学危险品的安全管理实行学校、实验室、使用者分级负责制，在学校化学危险品与放射源管理领导小组领导下，保卫处、实验管理中心负责化学危险品的归口管理，代表学校行使安全管理职能。

2. 管理职能划分

（1）保卫处负责全校化学危险品的安全监督管理工作，代表学校与各学院签订化学危险品安全管理责任书。

（2）实验管理中心负责化学危险品购置审批手续办理和国家管控化学危险品的集中采购、运输、储存保管工作。

（3）学校负责本单位的化学危险品安全管理工作。明确一位领导作为责任人，并确定一名专（兼）职管理人员，协助责任人开展此项工作，制定本单位化学危险品安全管理规章制度，落实安全责任制和责任人，负责本单位非国家管控化学危险品的采购工作，建立相应的化学危险品储存库房，负责本单位化学危险品的保管，督促各实验室加强化学危险品的安全管理，杜绝安全事故的发生。

（4）实验室负责本部门化学危险品的安全使用管理工作。建立健全本实验室化学危险品的安全管理责任制度和安全操作规程，制定相应的化学危险品事故处理应急措施，指定专人对化学危险品进行管理。对本实验室师生员工进行安全教育，组织必要的安全学习和技术培训，提高全体人员的安全管理意识和安全使用水平。

（三）化学危险品的申购与运输

1. 为了确保化学危险品的安全，化学危险品的购置，实行国家管控化学危险品和一般化学危险品分级购置。

（1）非国家管控的化学危险品，经学校主管领导批准同意后，由学校指派专人到有销售资质的化学危险品销售公司购买。

（2）国家管控的化学危险品采购，按照公安机关要求实行申购审批制度。学校首先向实验管理中心提出购买申请，实验管理中心汇总，并经学校保卫处审核同意后，报公安机关审批（爆炸品、剧毒化学品、麻醉品的购置需报公安分局审批，易制毒化学品的购置需报公安禁毒支队审批）。

（3）国家管控的化学危险品由实验管理中心指派专人负责集中采购，按照国家的有关规定，办理相关的购置手续，到指定地点购买，并负责运输。

2. 化学危险品的运输，必须符合国家有关化学危险品运输规定

（1）装运化学危险品必须小心谨慎，严防震荡、撞击、摩擦、重压和倾斜，装运气瓶时，必须旋紧瓶帽，轻装轻卸，防止碰撞。

（2）性质互相抵触的化学危险品严禁混装运输。

（3）易燃品、油脂或带有油污的物品，严禁与氧气瓶和强氧化剂同车装运。

（4）盛放危险品的容器，事先必须进行严格检查，确定安全后才能使用。

（5）易燃、易爆、有毒的化学危险品，必须使用具有专门许可证的车辆运输。

（6）运输危险品时，车辆上应按规定悬挂相应的警示标志，车上严禁烟火。

（7）运输危险品时，必须带有必要的消防和防护设施，夏、秋季运输危险品时，必须采取遮阳或防湿等安全措施。

（8）严禁随身携带化学危险品乘坐公共交通工具。

（四）化学危险品的储存与保管

1. 学校设立校、学院两级化学危险品储存保管库房，对化学危险品进行分级管理。校级库房负责储存、保管剧毒化学品等国家管控的化学危险品，学院库房负责储存、保管本部门购买的化学危险品和正在（或近期）使用的国家管控的化学危险品。

2. 化学危险品储存、保管必须符合国家有关规定，并指定专人进行保管。

（1）建立相应的化学危险品储存与管理规章制度，落实安全责任制。

（2）设立专（兼）职化学危险品库房保管人员，库房保管人员必须经过公安机关的上岗培训，具有相应的资质。

（3）库房保管人员对新购入的化学危险品，必须严格按采购计划及合同进行验收，并及时入库存放。

（4）爆炸品和剧毒化学品出、入库房时，必须有2人同时在场，进行严格检查和验收，并做好收、发登记工作。

（5）化学危险品库房内外30m范围内严禁烟火。进入化学危险品库房，必须交出随身携带的火种，杜绝一切可能产生火花的因素。

（6）化学危险品库房必须配备必要的、性能适用的消防器材及报警和防护设施。

（7）严禁将性质互相抵触或灭火方法不同的化学危险品混放在一起。

（8）蒸气有毒或蒸气与空气混合后，容易引起爆炸的物品，必须将瓶塞严密封闭，并放置在阴凉处，同时注意通风。

（9）严禁将遇木材着火的物品（如过氯酸等）直接放在木质架上。

（10）严禁将遇水燃烧、怕冻、怕晒的化学危险品存放在室外。

（11）严禁将盛放易燃或自燃气体的气瓶、油脂或带油污的物品与盛装压缩气体的钢瓶混放在一起，必须按规定定期进行技术检验。

（12）存放易燃、易爆物品的仓库，夏季必须采取防暑降温措施。

（13）爆炸品、剧毒化学品必须放在保险柜内，并且严格实行"双人双锁"保管。

（14）对储存的化学危险品，应定期进行检查，防止变质、自燃或爆炸事故，对变质过期的炸药、火工品和需要销毁的爆炸品，必须上报并送交有处理资质的单位进行回收处理。

（15）发现化学危险品丢失、被盗时，必须立即报告上级领导和保卫部门。

（五）化学危险品的领取及使用管理

1. 实验管理中心，按申购计划购买国家管控的化学危险品后，及时通知相关学院，学院指派2人到校级化学危险品库房，办理相关手续统一领取。

2. 学院要切实加强化学危险品的日常使用管理工作，建立相应的化学危险品领取、使用制度，对化学危险品使用过程，必须有可追溯的记录，确保化学危险品的使用安全。

（1）学院必须指定具有化学危险品业务知识的2人负责爆炸品、剧毒化学品的领用、发放，并认真做好记录。

（2）实验室需要领用化学危险品，必须有专人负责，按实验需求领取。

（3）实验室领用爆炸品、剧毒化学品，必须详细写明用途，领取最少用量，经实验室负责人签字同意后，由2人同时到院级化学危险品库房，办理领用登记手续后方可领取。

（4）学生在使用化学危险品时，教师必须详细指导，教授安全操作方法，并采取必要的安全防护措施。

（5）对每一瓶剧毒化学品，使用单位必须有完整的使用记录，其内容包括使用时间、使用地点、使用人、使用数量等。

（6）实验室领用的爆炸品、剧毒化学品有多余或当天使用不完的，必须立即退还院级化学危险品库房，严禁使用人自行保管。

（7）严禁转让或借用国家管控的化学危险品。

（六）废弃化学危险品处理

1. 对实验产生的废气、废液、废渣排放处理必须符合国家环保要求。

2. 对废弃化学危险品，各使用单位不得自行处置，应进行分类收集，妥善储存，容器外加贴标签，注明废弃物内容和品名，送交有相应资质的单位，进行回收处理。

3. 对已使用完的化学危险品容器，不得随意丢弃或另作他用，必须送交有相应资质的单位进行回收处理。

四、实验记录与实验报告

实验是在相关理论指导下的科学实践，其主要目的在于经过有效实践，掌握科学观察的基本方法和基本技能，以便培养学生的科学思维、分析判断以及解决实际问题的能力。也是培养探求真知、注重科学事实和真理的学风以及培养科学态度的重要环节。

因此要求学生在进行相关实验前，必须预习、理解基本原理和实验基本操作步骤、注意事项，列出所需试剂和仪器，实验中组织安排好时间，严肃认真地进行操作，细致观察变化，如实作好记录、书写实验报告等。

（一）实验记录

1. 实验记录整洁、字迹清楚。
2. 实验中观察到的现象、结果和数据，要及时记在记录本上，原始记录必须准确简练、详尽、清楚。
3. 记录时，应做到如实正确记录实验结果，不可夹杂主观因素。在实验条件下观察到的现象，也应如实仔细地记录下来。在定量实验中观测到的数据，如称量物的重量、滴定管的读数、分光光度计的数值等，都应设计一定的表格准确记录。并应根据仪器的精确度，准确记录有效数字。要求对实验的每个结果都应正确无遗漏地做好记录。
4. 实验中使用仪器的类型、试剂的规格、化学反应试剂名称、分子量、浓度等，都应记录清楚。完整的实验记录，应包括日期、实验题目、目的、操作、结果。

（二）实验报告

实验报告是在科学研究活动中，科研人员为了检验某一种科学理论或假设，通过实验中的观察、分析、综合、判断，如实地把实验的全过程和实验结果用文字形式记录下来的书面材料。实验报告具有情报交流的作用和保留资料的作用。

科技实验报告是描述、记录某个科研课题过程和结果的一种科技应用文体。撰写实验报告是科技实验工作不可缺少的重要环节。虽然实验报告与科技论文一样都以文字形式阐明了科学研究的成果，但二者在内容和表达方式上仍有所差别。科技论文一般是把成功的实验结果作为论证科学观点的根据。实验报告则客观地记录实验的过程和结果，着重告知一项科学事实，不夹带实验者的主观看法。

1. 实验结束后，应及时整理和总结实验结果及记录，写出实验报告。
2. 实验报告应包括实验名称、实验目的及要求、实验原理、试剂配制、仪器设备、操作方法、实验结果、讨论等内容。
3. 书写实验报告时，目的与要求、原理以及操作方法部分，进行简单扼要的叙述，但对实验条件和操作的关键环节必须写清楚。对于实验结果部分，应根据实验的具体要求，把所得的实验结果和数据进行整理、归纳、分析和对比，并尽量总结成各种图表，如标准曲线图以及实验组与对照组实验结果的比较表等。

同时对实验结果进行必要的说明和分析。讨论部分不是对结果的重述,而是对实验方法、实验结果和异常现象进行探讨和评论以及对于实验设计的认识、体会和建议。

一般来说,实验要有结论,结论要归纳总结,言简意赅,能够有效说明本次实验所获得的结果。

常见实验报告格式

实验题目：_____

实验小组成员：_____实验日期：_____天气：_____室温：_____

一、实验目的

二、实验原理

三、主要仪器和试剂

四、实验步骤

五、注意事项

六、思考题

Part I Basic Knowledge of Laboratory

Ⅰ. Laboratory Management System

1. Chemistry laboratory rules

(1) The staff who enter the laboratory must strictly abide by the laboratory regulations, obey the arrangement and management of laboratory staff, keep the room quiet and clean, and conduct experiments in a civilised.

(2) The use of instruments and equipment must strictly abide by operating procedures. Usage records must be filled cut carefully. In the event of breakage or damage, it should be reported to the laboratory management staff in time.

(3) Ensure that laboratory accounts and materials are consistent. The laboratory instruments and equipment should be maintained, repaired and inspected regularly to keep the equipment in good condition, and ensure the accurary and reliability of experimental data. Promote division of labor and cooperation, with dedicated management for specific use, to improve the utilization rate of instruments and equipment.

(4) The management, maintenance and maintenance of instruments and equipments, as well as the filling, sorting and safekeeping of archival materials need to be responsible for the work.

(5) The laboratory should be kept clean and quiet. No smoking. No personal belongings. No accommodation. No equipment and no private rental to others.

(6) Without the consent of the laboratory supervisor, non-laboratory personnel are not allowed to conduct experiments in the laboratory. No one can use the laboratory for a long time under any pretext.

(7) Pay attention to safety, And ensure fire prevention, theft prevention, explosion prevention, damage prevention, to prevent the occurrence of accidents.

(8) Off-campus personnel entering the laboratory for experiment or study visit can only be approved by the subjective department.

2. Student experiment code

(1) Comply with the rules and regulations of the laboratory, and obey the teacher's guidance. Keep the laboratory clean and quiet, no smoking, spitting, and mess still sundry.

(2) Before the experiment, we should carefully preview, clarify the purpose and requirements of the experiment, master the performance and operation method of the instrument used, and make all preparations according to the requirements. The experiment can only be carried out after the teacher's inspection permission.

(3) Do not be late for the experimental lesson. Do not enter the laboratory in disordered clothes. Do not bring anything unrelated to the experimental class into the laboratory.

(4) Conduct experiments in strict accordance with operating procedures, record all kinds of experimental data truthfully, and do not leave the operating post without permission.

(5) After the experiment, students may only leave the laboratory after the teacher has checked the instruments, tools, utensils and experiment records.

(6) If any equipment is found to be damage, it should be reported immediately, and the cause should be investigated. Who violate the operation rules and cause accidents shall be dealt with according to relevant regulations.

(7) Pay attention to safety. In case of an accident, cut off the power and fire immediately, report to the teacher truthfully, and take emergency measures.

(8) Take good care of all facilities in the laboratory. Do not scribble. Do not use instruments, equipment and related facilities unrelated to the experiment.

(9) Be diligent and thrifty. Do not waste water, electricity or materials.

(10) The teachers have the right to stop or redo the experiment for those who do not obey the discipline and do not do the experiment seriously.

3. Laboratory safety management regulations

(1) Laboratory personnel must strictly adhere to the laboratory safety management provisions. According to the specific situation of the laboratory, the establishment of laboratory safety operation provisions, anti-theft safety system and fire prevention convention, with clear responsibilities and tasks assigned to people.

(2) Each laboratory and warehouse shall appoint a comrade with a strong sense of responsibility and familiar with the situation as a safety officer, who is fully responsible for the safety work of the laboratory. In case of an accident, emergency measures shall be taken immediately, and the accident shall be handled and reported in time. In the case of major accidents, the scene must be preserved for investigation.

(3) Any experiment with dangerous effects must be protected by safety measures and supervised by teachers or laboratory technicians, otherwise it may not be carried out.

(4) Inflammable, explosive, high pressure, high-temperature, toxic, harmful and other dangerous goods must be stored in designated storage rooms as per regulations, with a dedicated person reponsible for their proper management, and strict produres for their use must be alloved.

(5) Safe, electricity, gas, must be used safely, and electrical wires must not be carelessly pulled. Switches should be turned off immediately after use, and the debris should be cleaned in time.

(6) Gas furnaces, electric furnaces, electric heaters and high-power heaters are not allowed to be installed and used without authorization. If necessary for teaching and scientific research, approval shall be obtained from the supervisor and the power supply shall be cut off in time when the personnel leave.

(7) Fire equipment and facilities should be placed in an obvious position, frequent inspection, found fault in time to remove and repair, so that they are in good condition.

(8) The key of each laboratory should be strengthened management, not privately equipped or lent to others.

4. Rules for laboratory staff

(1) We will strengthen the construction of spiritual civilization and optimize the education system.

(2) Strictly abide by the school rules and regulations, and establish good professional ethics.

(3) Carefully prepare for eoperiments and strictly fulfill job responsibilities.

(4) Hold students to high standards, conducting experiments according to experimental procedures and operating rules.

(5) Do a good job of instruments, equipment, experimental equipment account, material management work, and clear monthly settlement. Keep accounts, material consistent.

(6) Take good care of instruments, equipment and experimental facilities. It shall not be freely borrowed or let alone private occupation.

(7) Master the operation skills and skills of instruments and equipment, diligently maintain and service them, improve the maintenance technology, and ensure the completeness rate of instruments and equipment and the opening rate of experimental projects.

(8) Conscientiously study experimental education theory, assiduously study business technology, and constantly improve the level of experimental teaching.

5. Rules for the use, storage and operation of drugs

(1) When handling dangerous chemicals, please be sure to abide by the operating rules, and follow the teacher's operating procedures. Do not change the experimental procedures by yourself.

(2) When receiving the drug, confirm that the Chinese name on the container corresponds to the required experimental drug.

(3) When receiving the medicine, please see clearly the hazard labels and symbols to confirm any potential dangers.

(4) When using volatile organic solvents, strong acids, strong alkalis, high corrosion and toxic drugs, must be sure to operate under special exhaust cabinets and table smoking pipes.

(5) Organic solvents, solid chemicals, acids and alkali compounds must be stored separately. Volatile chemicals must be stored in a medicine cabinet with an air extraction device.

(6) Highly volatile or easily oxidized chemicals must be stored in the refrigerator or freezer.

(7) Avoid conducting dangerous experiments alone in the lab.

(8) If unsupervised experiments are necessary, the experimental equipment should take into account fire, explosion and flood protection, and keep the laboratory lights on. Leave the emergency contact number and possible disaster on the door.

(9) Dangerous experiments must be approved by the laboratory director and can only be carried out with more than two people present. Dangerous experiments are strictly prohibited on holidays and at night.

(10) Experiments on dangerous gases must be carried out in fume hoods.

(11) For experiments involving radiation, laser and other serious harm to human body experiments, should establish strict safety measures, and do a good job of personal protection.

(12) The waste, expired liquid medicine or waste must be clearly marked according to the classification. The waste (liquid) from used chemicals should not be poured into the sink or ditch. It shall be poured into the special collection container for recycling.

II. Laboratory gas procurement and gas cylinder safety management regulations

1. Safety management

The school is the main responsible for the safety management of laboratory gas cylinders, and the responsible person of each laboratory has the responsibility for the safety management of the gas cylinders of its own laboratory. The responsible person must pay attention to check the safety situation of the gas cylinder at all times, master the use status of the gas cylinder, ensure that the relevant use conditions and emergency measures meet the standard requirements, discover the hidden danger of insecurity in time, and provide safety technical education to the gas cylinder users.

2. Archives management

Experimental Teaching Management Center is responsible for establishment the gas cylinder management files, which include:

(1) Laboratory gas cylinder safety management system and accident emergency treatment plan.

(2) Laboratory gas cylinder ledger, including cylinder quantity, storage location, filling gas composition, manager, *etc*.

(3) Qualification materials of laboratory gas suppliers, including copies of enterprise business license, hazardous chemical business license, gas cylinder filling license, *etc*., with official seal.

3. Purchase

In principle, the laboratory are not alloved to purchase gas cylinders. Article 26 of the Provisions on the Supervision of the Safety of Gas Cylinders stipulates that gas cylinder filling units shall have the obligation to provide gas cylinders to gas consumers and shall be fully responsible for the safety of gas cylinders. When the laboratory purchases experimental gas, the supplier shall provide the gas cylinder within the qualified testing period and issue the periodic testing certificate of the gas cylinder.

Users need to fill cut a purchase application form for experimental gas, which can be purchased only after approvl. The application form shall be archived by the Experimental Teaching Management Center.

4. Acceptance

Gas cylinders must be inspected and accepted before being put into normal use.

(1) Check whether the gas cylinder is correctly filled and the gas name label should be clearly marked.

(2) Check whether the cylinder is regularly inspected, whether there is steel seal, and whether the information is complete (manufacturer, manufacturing date, cylinder model, working pressure,

air pressure test pressure, air pressure test date and next inspection date, gas volume, cylinder weight).

(3) Check whether the gas cylinder has a factory certificate.

(4) Check whether the gas cylinder accessories (shock-proof rings, protective caps, and valves) are complete and intact, whether the pipes are made of appropriate materials, and whether they are damaged or aging.

(5) Check whether the appearance of the gas cylinder is normal, whether the gas nozzle is deformed. Whether the cylinder body is corroded, deformed, worn, cracked and other defects.

(6) After passing the gas cylinder inspection, the laboratory gas cylinder acceptance registration form shall be handled, and the registration form shall be stored in the laboratory for future reference.

5. Storage

(1) The number of gas cylinders stored in the laboratory should be reasonable. Neither oxygen nor combustible gas should be more than 1 bottle, and they should not be mixed.

(2) The storage point of the gas cylinder must be ventilated, away from heat source, avoid exposure to the sun and strong vibration, with a flat and dry ground.

(3) All gas cylinders should be passed through the cylinder cabinet, cylinder anti-falling-down chain, anti-falling-down fence, *etc*. Empty and full cylinders should be stared separately and clearly marked.

(4) It is strictly prohibited to mix with inflammable, explosive, toxic and dangerous chemicals, solid waste, and to avoid all kinds of radioactive sources.

(5) Laboratories that use highly toxic, inflammable and explosive gases shall be equipped with ventilation facilities, monitoring and alarm devices, and post necessary safety warning signs.

6. Use

Users shall strictly follow the operating procedures (instructions, precautions, *etc*.) of gas cylinders and instruments.

Before students use gas cylinders by themselves, instructors should inform them of potential risk factors, consequences and emergency measures. Students can only proceed with the relevant experiments after being approved and signed by instructors (application form for use of laboratory/instruments and equipment).

The person responsible for the laboratory must confirm the safety knowledge and use qualification of the student user.

(1) The gas cylinder must be equipped with the corresponding pressure reducer and set up correctly. Inflammable gas and may cause backflow occasions must be equipped with anti-backflow devices, such as check valve, check valve, buffer tank, and check its effectiveness at any time.

(2) The operator should stand at a position perpendicular to the cylinder interface. The operation of the decompression device and the on-off valve must be slow. When using the gas cylinder, open

the valve first, then open the pressure reducing valve. When closing the gas cylinder, close the valve first, and then close the decompression device after exhausting the remaining gas.

(3) The gas in the cylinder shall not be exhausted, and the residual pressure or weight must be left. The residual pressure of the permanent gas cylinder shall be no less than 0.05MPa (gauge pressure), and the residual gas of the liquefied gas cylinder shall be no less than 0.5%~1.0% of the prescribed filling amount.

(4) Gas cylinders must be regularly checked for leaks. Gas cylinders with gas leaks are strictly prohibited from use.

(5) The gas cylinder must be finished or submitted for inspection within the specified inspection period.

(6) The acetylene cylinder should not be placed on the insulator. Before use, it must stand upright for 20 minutes and then connect the reducing valve for use.

(7) Oxygen cylinders or hydrogen cylinders are strictly prohibited from contacting with oil, and the operator shall not wear grease or greasy work clothes and gloves to operate, so as not to cause combustion or explosion.

7. Handling

Before moving the gas cylinder, the should be tightly closed, the pressure reducer removed, and the labels of each component checked for integrity. Operators must be familtar with the name and properties, of the gas inside the cylinder, as well as the precautions for handing, and should have the appropriate toals and protective equipment ready.

(1) When moving the cylinder in a close distance (within 5 m), support the shoulder of the cylinder by hand, turn the bottom of the cylinder, and use gloves. When moving far away, special trolley or hanging basket should be used.

(2) When using special gas cylinder trucks for long-distance transportation, the protective cap should be securely tightened, and shock-proof washers should be installed. Forklift trucks and forklifts are strictly prohibited for transportation, and chemical loading and transportation are not allowed.

(3) When loading and unloading gas cylinders, they should be loaded and unloaded lightly. It is forbidden to lift the gas cylinders by holding the on-off valve. And it is forbidden to carry them by throwing, sliding, falling, rolling or touching. When unloading the vehicle, soft pads or rubber pads should be laid on the landing site of the gas cylinders, and it is forbidden to slip them away one by one to prevent accidents caused by illegal or improper operations.

III. Measures for the control of hazardous chemicals in laboratories

1. General rules

(1) For the purpose of further standardizing and strengthening the safety management of laboratory chemical dangerous goods, strictly preventing accidents, ensuring the safety of life and property of teachers, students and staff, and better serving teaching and scientific research, these measures are formulated in accordance with the relevant provisions of the Regulations on the Safety Management of Hazardous Chemicals and the Regulations on the Management of Precursor Chemicals.

(2) In these administrative measures, the dangerous chemicals referred to include the national standard GB 6944-2012 Classification and Name Number of Dangerous Goods and GB12268-2012 Name List of Dangerous Goods, which divides dangerous chemicals into nine categories:

(a) Explosives.
(b) Gas.
(c) Flammable liquids.
(d) Flammable solids, substances prone to spontaneous combustion, substances that release flammable gases in contact with water.
(e) Oxidizing substances and organic peroxides.
(f) Toxic substances and infectious substances.
(g) Radioactive material.
(h) Corrosive substances.
(i) Miscellaneous hazardous substances and articles, including substances harmful to the environment.

(3) All departments and individuals who purchase, transport, store, use and destroy chemical dangerous goods must comply with these measures.

2. Management system

(1) The safety management of dangerous chemical products in the school adopts a hierarchical responsibility system among the school, the laboratory and the user. Under the leadership of the leading group for the management of Dangerous Chemical Products and Radioactive Sources in the school, the Security Department and the experimental management Center are responsible for the centralized management of dangerous chemical products and perform the safety management function on behalf of the school.

(2) Division of management functions

(a) The Security Department is responsible for the safety supervision and management of chemical hazardous materials across the school, and signs the responsibility letter of chemical dangerous goods safety management with each colleges on behalf of the school.

(b) The Experimental Management Center is responsible for the approval procedures for the purchase of chemical dangerous goods and for the centralized procurement, transportation, and storage management of nationally controlled chemical hazardous meterials.

(c) The school shall be responsible for the safety management of chemical dangerous goods. Designate a leader as the responsible person and appoint a full-time (part-time) manager to assist the responsible person to carry out the work, formulate the safety management rules and regulations of the department for chemical dangerous goods, implement the safety responsibility system and responsible person, be responsible for the procurement of non-state controlled chemical dangerous goods, establish the corresponding storage warehouse for chemical dangerous goods, and take charge of the storage of chemical dangerous goods in the department. Urge all laboratories to strengthen the safety management of chemical dangerous goods to prevent the occurrence of safety accidents.

(d) Laboratories are responsible for the safe use and management of chemical dangerous goods within their departments. Establish and improve the responsibility system and safe operation procedures for the safety management of chemical dangerous goods in the laboratory, formulate corresponding emergency measures for the treatment of chemical dangerous goods accidents, and designate special personnel to manage chemical dangerous goods. Conduct safety education for teachers, students and staff in the laboratory, organize necessary safety learning and technical training, and improve the safety management awareness and safety use level of all staff.

3. Purchase and transport of chemical hazardous materials

(1) In order to ensure the safety and purchase of chemical hazardous materials, the procurement of chemical hazardous materials is subject to a tiered purchasing system for state-controlled and general chemical hazardous materials.

(a) The non-state-controlled chemical hazardous materials shall be purchased by the qualified chemical hazardous materials sales company designated by the school after the approval of the school leadership.

(b) For the procurement of chemical hazardous materials controlled by the State, the application and approval system shall be implemented in accordance with the requirements of public security organs. The school first submits the purchase application to the experimental management Center, which summarizes the application, and then reports it to the public security organ for approval after being examined and approved by the security Department of the school (the purchase of explosives, highly toxic chemicals and narcotics shall be submitted to the public security branch for approval, and the purchase of precursor chemicals shall be submitted to the public security anti-drug detachment for approval).

(c) For chemical hazardous materials controlled by the state, the experimental management Center shall assign special personnel to be responsible for centralized procurement, handle the relevant purchase procedures in accordance with the relevant regulations of the state, purchase them at the designated place, and be responsible for transportation.

(2) Transport of chemical hazardous materials must comply with the relevant state regulations on the transport of chemical hazardous materials

(a) Chemical hazardous materials must be transported with care to prevent shock, impact, friction, heavy pressure and tilt. Gas cylinders must be loaded with tight caps and unloaded lightly to prevent collision.

(b) Chemical hazardous materials with conflicting properties are strictly prohibited from mixed transportation.

(c) Inflammable goods, grease or articles with oil pollution are strictly prohibited from being loaded with oxygen cylinders and strong oxidants.

(d) Containers for storing hazardous materials must be strictly inspected in advance to ensure their safety before they can be used.

(e) Flammable, explosive and toxic chemical hazardous materials must be transported by vehicles with special permits.

(f) When transporting hazardous materials, corresponding warning signs shall be hung on the vehicle according to the regulations, and fireworks are strictly prohibited on the vehicle.

(g) When transporting hazardous materials, necessary fire-fighting and protective facilities must be equipped. When transporting hazardous materials in summer and autumn, safety measures such as shading or dampness must be taken.

(h) It is strictly prohibited to carry chemical dangerous goods on public transport.

4. Storage and custody of chemical hazardous materials

(1) The school sets up two levels of chemical hazardous materials storage warehouse, and carries out hierarchical management of chemical hazardous materials. The university-level warehouse is responsible for the storage of highly toxic chemicals and other state-controlled chemical hazardous materials, while the college warehouse is responsible for the storage of the department's purchase of chemical hazardous materials and the department's current (or recent) use of state-controlled chemical hazardous materials.

(2) Storage and safekeeping of chemical hazardous materials must comply with relevant state regulations, and designated special personnel for safekeeping.

(a) Establish the corresponding rules and regulations for the storage and management of chemical hazardous materials and implement the safety responsibility system.

(b) Special (part-time) warehouse keepers of chemical hazardous materials shall be set up. The warehouse keepers shall be trained by public security organs and have corresponding qualifications.

(c) The warehouse custodian must check and accept the newly purchased chemical hazardous materials strictly in accordance with the purchase plan and contract, and store them on time.

(d) When explosive products and highly toxic chemicals are taken out or into the warehouse, two persons must be present at the same time to carry out strict inspection and acceptance, and do a good job in the registration of receipt and delivery.

(e) Fireworks are strictly prohibited within 30 meters inside and outside the chemical hazardous materials warehouse. Enter the chemical hazardous materials warehouse, must hand over the fire to you, put an end to all the factors that may produce sparks.

(f) The warehouse of chemical hazardous materials must be equipped with necessary and suitable fire fighting equipment and alarm and protection facilities.

(g) It is strictly prohibited to mix chemical hazardous materials with conflicting properties or different fire extinguishing methods together.

(h) Steam toxic or steam mixed with air, easy to cause explosion of items, must be tightly sealed bottle stopper, and placed in the shade, at the same time pay attention to ventilation.

(i) It is strictly prohibited to put items (such as perchloric acid, *etc*.) that catch fire in wood directly on the wooden shelf.

(j) It is strictly prohibited to store chemical hazardous materials that are burning in contact with water, afraid of freezing and afraid of sun outdoors.

(k) Gas cylinders containing flammable or self-igniting gas, grease or articles with oil and cylinders containing compressed gas are strictly prohibited. Technical inspections must be carried out regularly according to regulations.

(l) For warehouses storing inflammable and explosive goods, heat prevention and cooling measures must be taken in summer.

(m) Explosives and highly toxic chemicals must be placed in the safe, and "double lock" is strictly implemented.

(n) The stored chemical hazardous materials shall be inspected regularly to prevent deterioration, spontaneous combustion or explosion accidents, and the spoiled and expired explosives, pyro chemicals and explosives to be destroyed shall be reported and sent to the units qualified for disposal for recovery.

(o) In case of loss or theft of chemical dangerous goods, report to the superior leadership and security department immediately.

5. Management of the receipt and use of chemical hazardous materials

(1) After purchasing state-controlled chemical hazardous materials according to the application plan, the experimental management Center shall timely notify the relevant college, and the college shall assign 2 people to the university-level chemical hazardous materials warehouse to handle relevant procedures and obtain them uniformly.

(2) The College should strengthen the daily use management of chemical hazardous materials, establish the corresponding collection and use system of chemical hazardous materials, and keep traceable records of the use process of chemical hazardous materials to ensure the safety of the use of chemical hazardous materials.

(a) The College must designate two persons with knowledge of hazardous chemicals business to be responsible for the use and distribution of explosives and highly toxic chemicals, and record them carefully.

(b) The laboratory needs to receive chemical hazardous materials, which must be responsible for by special personnel according to the demand of the experiment.

(c) Laboratory use of explosives and highly toxic chemicals, must be specified in detail the purpose, receive the minimum amount, signed by the person in charge of the laboratory, by two people at the same time to the hospital level chemical hazardous materials warehouse, go through the registration procedures before receiving.

(d) When students use dangerous chemicals, teachers must give detailed guidance, teach safe operation methods, and take necessary safety protection measures.

(e) For each bottle of highly toxic chemicals, the user must have a complete record of use, including time, place, person, quantity, *etc*.

(f) If the explosives and highly toxic chemicals used by the laboratory are in excess or cannot be used up by the same day, they must be returned to the warehouse of chemical dangerous goods at the hospital level immediately. Users are forbidden to keep them by themselves.

(g) It is strictly prohibited to transfer or borrow state-controlled chemical dangerous goods.

6. Waste chemical hazardous materials treatment

(1) The discharge and treatment of waste gas, liquid waste and slag generated by the experiment must meet the national environmental protection requirements.

(2) Waste chemical hazardous materials shall not be disposed of by the users themselves, but shall be collected by classification and stored properly. The containers shall be labeled with the contents and name of the waste and sent to the qualified units for recycling.

(3) The containers of chemical hazardous materials that have been used shall not be discarded or used for other purposes, but shall be sent to the qualified units for recycling.

Ⅳ. Experimental records and reports

Experiment is a scientific practice under the guidance of relevant theories. Its main purpose is to master the basic methods and skills of scientific observation through effective practice, so as to cultivate students' scientific thinking, analysis and judgment and the ability to solve practical problems. It is also an important link to cultivating the study style of seeking true knowledge, paying attention to scientific facts and truth, and cultivating a scientific attitude.

Therefore, students are required to preview and understand the basic principles, basic operation steps and precautions of the experiment before carrying out the experiment, list the required reagents and instruments, organize and arrange the time for the experiment, conduct the operation seriously, observe the changes carefully, truthfully make records, write the experiment report, *etc.*

1. Experimental records

(1) The experimental records are clean and legible.

(2) The phenomena, results and data observed in the experiment should be directly recorded in the record book in time, and the original records must be recorded precise, concise, detailed and clear.

(3) When recording, the experimental results should be recorded truthfully and correctly without subjective factors. Phenomena observed under experimental conditions should also be truthfully and carefully recorded. The data observed in the quantitative experiment, such as the weight of the weighing object, burette reading, spectrophotometer value, *etc.*, should be designed in a certain form for accurate record. Effective figures shall be recorded accurately according to the accuracy of the instrument. Each result of the experiment should be recorded correctly without omission.

(4) The type of instrument used in the experiment, the specification of reagents, chemical reaction reagents, molecular weight, concentration, *etc.*, should be recorded clearly. Complete lab records should include date, lab title, purpose, operation, results.

2. Experimental report

Experimental report is a written material where researchers truthfully record the whole process and results of the experiment in writing through observation, analysis, synthesis and judgment in order to test a certain scientific theory or hypothesis in scientific research activities. Experimental reports have the function of information exchange and data retention.

Science and technology experiment report is a kind of science and technology application style that describes and records the process and results of a scientific research subject. Writing experiment report is an indispensable part of science and technology experiment work. Although both experimental reports and scientific papers illustrate the results of scientific research in written form, they are different in content and expression. Scientific papers are usually based on the results of successful experiments. The experimental report objectively records the process and results of the experiment, focusing on a scientific fact, without the subjective views of the experimenter.

(1) After the experiment, the experimental results and records should be sorted out and summarized in time, and the experimental report should be written.

(2) The experimental report shall include the name of the experiment, the purpose and requirements of the experiment, the principle of the experiment, the preparation of reagents, the instrument and equipment, the operation method, the experimental results and the discussion, *etc*.

(3) When writing the experimental report, the purpose and requirements, principles and operation methods should be described briefly and to the point, but the key links of the experimental conditions and operation must be clearly written.

For the experimental results, according to the specific requirements of the experiment, the experimental results and data should be sorted, summarized, analyzed and compared, and summarized into various charts as far as possible, such as the standard curve chart and the experimental results of the experimental group and control group comparison table, *etc*. At the same time, the necessary explanation and analysis of the experimental results. The discussion part is not a restatement of the results, but a discussion and comment on the experimental methods, experimental results and abnormal phenomena, as well as the experimental design and understanding, experience and suggestions.

Generally speaking, the experiment should have a conclusion. The conclusion should be summarized, concise and comprehensive. And it can effectively explain the results of the experiment.

Common lab report format

Experimental title:_____

Experimental title:_____Experimental Date:_____Weather:_____Room temperature:_____

Ⅰ. Purpose of the experiment

Ⅱ. experimental principle

Ⅲ. Main instruments and reagents

Ⅳ. Experimental steps

Ⅴ. Precautions

Ⅵ. Think questions

第二部分 有机化学实验

Part Ⅱ　Organic chemistry experiment

实验一　1-溴丁烷的合成

一、实验目的

1. 掌握 1-溴丁烷合成反应机理及方法。
2. 熟悉蒸馏操作的方法。
3. 了解 1-溴丁烷的理化性质。
4. 了解尾气处理装置及操作过程。

二、实验原理

（一）化合物简介

1-溴丁烷又称为溴丁烷，是一种重要的有机化合物，其化学式为 C_4H_9Br，为无色透明液体，不溶于水，微溶于四氯化碳，溶于氯仿，易溶于乙醇、乙醚、丙酮，主要用作烷化剂、溶剂、稀有元素的萃取剂，在有机合成方面应用也非常广泛。

（二）合成工艺路线

1-溴丁烷的合成，主要通过正丁醇与氢溴酸作用制备，其合成工艺路线如下：
主反应

$$NaBr + H_2SO_4 \longrightarrow HBr + NaHSO_4$$

$$n\text{-}C_4H_9OH + HBr \longrightarrow n\text{-}C_4H_9Br + H_2O$$

副反应

$$n\text{-}C_4H_9OH \xrightarrow[\triangle]{H_2SO_4} C_4H_8 + H_2O$$

$$2\, n\text{-}C_4H_9OH \xrightarrow[\triangle]{H_2SO_4} (n\text{-}C_4H_9)_2O + H_2O$$

三、主要仪器和试剂

1. 主要仪器：回流装置、直形冷凝管、沸石、分液漏斗、锥形瓶、三颈烧瓶、200℃温度计、分析天平、磁力搅拌器、蒸馏烧瓶、烧杯、量筒、玻璃棒、气体吸收装置等。
2. 主要试剂：溴化钠、浓硫酸、正丁醇、10%碳酸钠、无水硫酸钠等。

四、实验步骤

在 250 mL 三颈烧瓶中，加入正丁醇 62 mL、研细的溴化钠 8.3 g 以及几粒沸石，随后安装好回流装置。在一小锥形瓶内加入水 10 mL，将锥形瓶放在冷水浴中冷却，一边摇荡，一边缓慢地加入浓硫酸 10 mL，配成稀硫酸。将稀释的硫酸分 4 次从冷凝管上端加入反应瓶内，快速搅拌，使反应物混合均匀。在冷凝管上口连接好气体吸收装置，将烧瓶放在加热套上，升温，保持回流反应 30 min。

回流反应结束后，待反应物稍微冷却，拆下回流冷凝管，再加入 1~2 粒沸石，改成蒸馏装置进行蒸馏。观察馏出液，直到无油滴蒸出时，停止反应。

将馏出液倒入分液漏斗中，静置后将油层即（下层）转入干燥的小锥形瓶中，然后将浓硫酸 4 mL 分两次加入锥形瓶内，同时不断摇动锥形瓶，若混合物发热，可用冰水浴进行冷却。将混合物慢慢倒入分液漏斗中，静置分层，放出下层的浓硫酸。油层依次用水 10 mL、10% 碳酸钠 5 mL 和水 10 mL 充分洗涤。将下层的粗 1-溴丁烷放入干燥的小锥形瓶中，加入适量的无水硫酸钠进行充分干燥，过夜。抽滤，滤液转至蒸馏烧瓶中，加入几粒沸石，安装好蒸馏装置，开始蒸馏，收集 99~103℃的馏分。

五、注意事项与思考题

（一）注意事项

1. 反应试剂若用含结晶水的溴化钠，要注意计算结晶水的量。
2. 冷凝管上口接的尾气吸收装置中，不要将漏斗全部置入水中，以免倒吸。
3. 要注意反应中的回流时间，时间太短，则反应不完全，回流时间太长，则会增加副产物。
4. 馏出液分为两层，要注意下层为粗 1-溴丁烷。

（二）思考题

1. 反应中使用的尾气吸收装置，主要吸收什么气体？
2. 由正丁醇制备正溴丁烷的反应，具体反应机理是什么？

Experiment 1   Synthesis of 1-bromobutane

Ⅰ. Purpose of the experiment

i. To master the mechanism and methods of synthesis 1-bromobutane.

ii. To familiar with the methods of distillation operation.

iii. To understand the physical and chemical properties of 1-bromobutane.

iv. To understand the exhaust gas treatment devices and operation process.

Ⅱ. Experimental principle

i. Compound introduction

1-bromobutane, also known as bromobutane, is an important organic compound with the chemical formula C_4H_9Br. It's colorless, transparent liquid that is insoluble in water, slightly soluble in carbon tetrachloride, soluble in chloroform, easily soluble in ethanol, ether. It is mainly used as alkylating agent, solvent, and rare element extraction agent. In organic synthesis, it is also very widely used.

ii. Synthetic process route

The synthesis of 1-bromobutane is mainly prepared by the action of n-butanol and hydrobromic acid. The synthesis process is as follows:

Main reaction:

$$NaBr + H_2SO_4 \longrightarrow HBr + NaHSO_4$$

$$n\text{-}C_4H_9OH + HBr \longrightarrow n\text{-}C_4H_9Br + H_2O$$

side reaction:

$$n\text{-}C_4H_9OH \xrightarrow[\triangle]{H_2SO_4} C_4H_8 + H_2O$$

$$2\ n\text{-}C_4H_9OH \xrightarrow[\triangle]{H_2SO_4} (n\text{-}C_4H_9)_2O + H_2O$$

Ⅲ. Main instruments and reagents

i. Main instruments: reflux device, straight condensing tube, zeolite, liquid separation funnel, conical flask, three-neck round-bottoming flask, 200℃ thermometer, analytical balance, magnetic stirrer, distillation flask, beaker, measuring cylinder, glass rod, gas absorption device, *etc*.

ii. Main reagents: sodium bromide, concentrated sulfuric acid, n-butanol, sodium carbonate, *etc*.

Ⅳ. Experimental steps

Add 62 mL of *n*-butanol, 8.3 g of finely ground sodium bromide and several grains of zeolite into a 250 mL three-necked round-bottomed. Then install the reflux device. Add 10 mL of water into a small conical bottle. Put the conical bottle in a cold bath to cool. While shaking, slowly add 10 mL of concentrated sulfuric acid. Make dilute sulfuric acid. Dilute sulfuric acid 4 times, from the top of the condensing tube into the bottle, stirring quickly, so that the reactants are mixed evenly. Connect the gas absorption device at the top of the condensate pipe, place the flask on the heating sleeve, and heat to maintain reflux reaction for 30 minutes.

After the reflux reaction, allow the reactants to cool down slightly, emove the reflux condensing tube. Add 1-2 zeolite pieces of boiling stones, then switch a distillation device for distillation. Observe the distillate and stop the reaction when no more oil droplets are being distilled.

Pour the distillate into the separator funnel, and put the oil layer into the small dry conical bottle from below. Then add 4 mL of concentrated sulfuric acid into the conical bottle in twice portions while constantly shaking the conical bottle. If the mixture is hot, cool it by an ice water bath. Slowly pour the mixture into a separator funnel. Let it stand to separate. And release the lower layer of concentrated sulfuric acid. Wash the reservoir in turn with 10 mL of water, 5 mL of 10% sodium carbonate, and 10 mL of water. Place the crude 1-bromobutane in the lower layer into a small conical bottle of dry. Add the appropriate amount of anhydrous sodium sulfate for thorough drying overnight. Filter under vacuum, transfer the filtrate to a distillation flask, Add a few pieces of boiling stones, and set up the distillation apparatus. Begin distillation collecting at 99-103 ℃.

V. Attention or thinking questions

i. Attention

(i) If using sodium bromide containing crystal water as the reaction reagent, pay attention to calculate the amount of crystal water.

(ii) In the exhaust gas absorption device connected to the upper end of the condensate pipe, do not put all the funnel into the water to avoid suction.

(iii) Pay attention to the reflux time in the reaction. If the time is too short, the reaction will not be complete. If the reflux time is too long, the by-product will be increased.

(iv) The distillate separates into two layers, noting that the lower layer is crude 1-bromobutane.

ii. Thinking questions

(i) What gas is mainly absorbed by the exhaust gas absorption device used in the reaction?

(ii) What is the specific reaction mechanism for the preparation of n-bromobutane from *n*-butanol?

实验二　对硝基苯乙酸的合成

一、实验目的

1. 掌握对硝基苯乙酸的合成反应机理。
2. 熟悉对硝基苯乙酸的合成实验操作过程。
3. 了解对硝基苯乙酸的理化性质。

二、实验原理

（一）化合物简介

对硝基苯乙酸，又名 4-硝基苯乙酸，分子式为 $C_8H_7NO_4$。其为淡黄色晶体，溶于热水、乙醇、氯仿、乙醚和苯，微溶于冷水。

对硝基苯乙酸是需求量极大的化工原料，也是一种重要的医药中间体。常用于合成对氨基苯乙酸、联苯乙酸、高效降压药氨酰新胺等，还可以与稀土金属形成具有良好性能的发光材料，因此在稀土功能材料方面也有广泛应用。

（二）合成工艺路线

对硝基苯乙酸的合成，主要通过对硝基苯乙腈在浓硫酸的作用下制备，其合成工艺路线如下：

$$NC-C_6H_4-NO_2 \xrightarrow[\triangle]{H^+} HOOC-CH_2-C_6H_4-NO_2 + NH_4HSO_4$$

三、主要仪器和试剂

1. 主要仪器：三颈烧瓶、分析天平、磁力搅拌器、200℃温度计、恒压滴液漏斗、冷凝管、烧杯、橡胶管、抽滤瓶、布氏漏斗、滤纸、真空泵、水浴装置等。
2. 主要试剂：对硝基苯乙腈、浓硫酸等。

四、实验步骤

在配有磁力搅拌器、温度计、冷凝管以及恒压滴液漏斗的 250 mL 三颈烧瓶中，加入对硝基苯乙腈 10 g，滴加稀硫酸（浓硫酸 30 mL＋水 28 mL），剧烈搅拌，升温至沸，溶液颜色缓慢变深，保温反应 30 min，反应结束后，将反应液全部倒入烧杯中，加入等体积（约 70 mL）的冷水，并置于冰水浴中，冷却至 0 ℃以下，抽滤，滤饼用冰水充分洗涤，得粗品。粗品用水进行重结晶，得对硝基苯乙酸纯品。

五、注意事项与思考题

（一）注意事项

1. 试写出本反应的反应机理。
2. 在配制稀酸的过程中，要注意浓硫酸向水中加入。

（二）思考题

本反应中用酸进行催化，还有哪些催化剂？

Experiment 2 Synthesis of *p*-nitrophenyl acetic acid

I. Purpose of the experiment

i. To master the synthetic reaction mechanism of *p*-nitrophenyl acetic acid.

ii. To familiar with the experimental operation process of *p*-nitrophenyl acetic acid synthesis.

iii. To understand the physical and chemical properties of *p*-nitrophenyl acetic acid.

II. Experimental principle

i. Compound introduction

p-nitrophenyl acetic acid, also known as 4-nitrophenylacetic acid, has the formula $C_8H_7NO_4$. It is a light yellow crystal. It is soluble in hot water, ethanol, chloroform, ether and benzene, and slightly soluble in cold water.

p-nitrophenyl acetic acid is a chemical raw material in great demand and an important medical intermediate. It is often used in the synthesis of *p*-amino phenyl acetic acid, biphenyl acetic acid, high efficiency antihypertensive drug aminoamide, *etc*. It can also form luminescent materials with good properties with rare earth metals, so it is also widely used in rare earth functional materials.

ii. Synthetic process route

The synthesis of *p*-nitrophenyl acetic acid is mainly through the preparation of *p*-nitrophenyl acetonitrile under the action of concentrated sulfuric acid. The synthesis process is as follows:

$$NC\text{-}C_6H_4\text{-}NO_2 \xrightarrow[\Delta]{H^+} HOOC\text{-}C_6H_4\text{-}NO_2 + NH_4HSO_4$$

III. Main instruments and reagents

i. Main instruments: three-necked flask, analysis balance, magnetic stirrer, 200 ℃ thermometer, constant pressure drip funnel, condensing tube, beaker, rubber tube, suction bottle, Brinell funnel, filter paper, vacuum pump, *etc*.

ii. Main reagents: *p*-nitro phenyl acetonitrile, concentrated sulfuric acid, *etc*.

IV. Experimental steps

Add 10 g of *p*-nitrophenyl acetonitrile into a 250 mL three-neck round-bottom flask equipped with a magnetic agitator, a thermometer, a condensing tube and a constant pressure drip funnel. Add diluted acid (30 mL of concentrated sulfuric acid + 28 mL of water) while stirring vigorously. Heat to boiling. And the color of the solution slowly darkens. Maintain the reaction for 30 min. After the reaction, pour all the reaction liquid into a beaker. Add the same volume (about 70 mL) of cold water, and place it in an ice-water bath to cool below 0 ℃, Perform suction filtration, thoroughly wash the filter cake with ice water to obtain the crude product. Recrystallise the crude product with water to obtain pure *p*-nitrophenyl acetic acid.

V. Attention or thinking questions

i. Attention

(i) Try to write the reaction mechanism of this reaction.

(ii) When preparing dilute acid, it is important to add concentrated sulfuric acid to water.

ii. Thinking questions

Acid is used to catalyze in this reaction. What other catalysts are there?

实验三　乙酸松油酯的合成

一、实验目的

1. 掌握乙酸松油酯的合成反应机理。
2. 熟悉乙酸松油酯的合成实验操作过程。
3. 了解乙酸松油酯的理化性质及应用。

二、实验原理

（一）化合物简介

乙酸松油酯为无色透明液体，主要为 α-异构体，分子式为 $C_{12}H_{20}O_2$，是异构的松油醇乙酸酯混合物。

乙酸松油酯是松油醇的酯化产物，具清香带甜，似香柠檬、薰衣草气息，留香时间较长。其广泛用于薰衣草、辛香、柑橘香型等日用香精，也用于白柠檬、樱桃、辛香、肉香等食用香精，在调味精油中起增强辛香的作用，是配制香柠檬油、薰衣草油、橙叶油等的重要原料。

（二）合成工艺路线

在工业生产中乙酸松油酯，主要是用松油醇和乙酸酐在硫酸存在下作用制备，其合成工艺路线如下：

$$\underset{\substack{HO \\ H_3C}}{\overset{H_3C}{\diagdown}}\!\!\!\!\diagup\!\!\!\!-\!\!\!\bigcirc\!\!\!-\!\!\!CH_3 \xrightarrow[H_2SO_4]{(CH_3CO)_2O} H_3CCOO\underset{H_3C}{\overset{H_3C}{\diagdown}}\!\!\!\!\diagup\!\!\!\!-\!\!\!\bigcirc\!\!\!-\!\!\!CH_3 + CH_3COOH$$

三、主要仪器和试剂

1. 主要仪器：三颈烧瓶、恒压滴液漏斗、分液漏斗、分析天平、磁力搅拌器、烧杯、锥形瓶、玻璃棒、蒸馏烧瓶、直形冷凝管、200℃温度计、蒸馏头、真空泵等。
2. 主要试剂：硫酸、乙酸酐、松油醇、15%氯化钠、碳酸氢钠等。

四、实验步骤

在配有磁力搅拌器、温度计、冷凝管、恒压滴液漏斗的 100 mL 三颈烧瓶中，加入乙酸酐 2.8 mL 和浓硫酸 2 滴，室温下搅拌下，加入松油醇 4.3 mL 进行乙酰化反应。反应 4h，把反应体系中的反应液转移到烧杯中，加入温热的 15%氯化钠溶液 10mL，进行充分洗涤，分解剩余的乙酸酐，分出水层，得到粗产品。把粗产品倒至 100 mL 烧瓶内，加入碳酸氢钠 0.05 g 进行减压蒸馏，收取 80～110℃的馏分。

五、注意事项与思考题

（一）注意事项

1. 由于本反应是放热的，同时反应对温度较为敏感，本反应温度要小于28℃。
2. 本反应的粗产品，无需碳酸氢钠溶液中和，否则易产生凝聚状沉淀，较难分离。

（二）思考题

1. 反应中的浓 H_2SO_4 主要作用是什么？
2. 为什么用热氯化钠溶液洗涤产物，而不是用热水洗？

Experiment 3 Synthesis of terpinyl acetate

I. Purpose of the experiment

    i. To master the synthetic reaction mechanism of terpineyl acetate.

    ii. To familiar with the experimental operation process of synthesis of terpineyl acetate.

    iii. To understand the physicochemical properties and applications of terpineyl acetate.

II. Experimental principle

i. Compound introduction

Terpineyl acetate is a colourless and transparent liquid, primarily consisting of α-isomer with molecular form $C_{12}H_{20}O_2$. It is a mixture of heterogeneous terpineyl acetate.

Terpineol acetate is an esterification product of terpineol. It has a sweet fragrance. It likes fragrant lemon and lavender, and lasts a long time. It is widely used in lavender, spice, citrus flavor and other daily flavor, but also used in white lemon, cherry, spice, meat flavor and other edible flavor. It plays a role in flavor essential oil to enhance the spice. It is an important raw material preparation of bergamot oil, lavender oil, orange leaf oil, *etc*.

ii. Synthetic process route

In industrial production, terpineol acetate is mainly prepared by the action of terpineol and acetic anhydride in the presence of sulfuric acid. The synthesis process is as follows:

$$\text{terpineol} \xrightarrow[H_2SO_4]{(CH_3CO)_2O} \text{terpinyl acetate} + CH_3COOH$$

III. Main instruments and reagents

    i. Main instruments: three-neck round bottom flask, constant pressure drip funnel, liquid separation funnel, analysis balance, magnetic stirrer, beaker, conical flask, glass rod, straight condensing tube, distillation flask, 200℃ thermometer, distillation head, vacuum pump, *etc*.

    ii. Main reagents: sulfuric acid, acetic anhydride, terpineol, sodium chloride, sodium bicarbonate, *etc*.

IV. Experimental steps

Add 2.8 mL of acetic anhydride and 2 drops of concentrated sulfuric acid into a 100 mL three-neck round-bottom flask equipped with a magnetic agitator, a thermometer, a condensing tube and a constant pressure drip funnel. After stirring at room temperature, add 4.3 mL of terpinol for acetylation. After 4h, transfer the reaction liquid in the reaction system to the beaker. Add 10mL of warm 15% sodium chloride solution for thorogh washing to decompose the remaining acetic anhydride. Separate the aquecus layer to obtain the crude product. Pour the crude product into a 100 mL flask. Add 0.05 g of sodium bicarbonate for vacuum distillation, and collect the fraction at 80~110℃.

V. Attention or thinking questions

i. Attention

(i) Because the reaction is exothermic and sensitive to temperature, the reaction temperature should be below 28℃.

(ii) The crude product of this reaction does not need to be neutralized by sodium bicarbonate solution, otherwise it is easy to produce condensed precipitation and difficult to separate.

ii. Thinking questions

(i) What is the main function of concentrated H_2SO_4 in the reaction?

(ii) Why wash the product with hot sodium chloride solution instead of hot water?

实验四　氯代环己烷的合成

一、实验目的

1. 掌握氯代环己烷的合成反应机理。
2. 熟悉氯代环己烷的合成实验操作过程。
3. 了解氯代环己烷的理化性质及用途。

二、实验原理

（一）化合物简介

氯代环己烷是一种重要的有机化合物，又名环己基氯，分子式为 $C_6H_{11}Cl$。其为无色液体，具有窒息性气味，不溶于水，溶于乙醇。

氯代环己烷可用于生产农药、橡胶防焦剂、药物等，在农药上用于合成杀螨剂三环锡和三唑锡的中间体三环己基氯化锡。此外，也可用于合成盐酸苯海索，还可用于制防焦剂等产品。

氯代环己烷一般储存于阴凉、干燥、通风良好的库房，远离火种、热源，防止阳光直射，包装密封。同时应与酸类、食用化学品分开存放，切忌混储。储存区域应备有合适的材料，用于收容泄漏物。

（二）合成工艺路线

氯代环己烷的合成，主要通过环己醇在浓盐酸催化的条件下制备，其合成工艺路线具体如下：

$$\text{环己醇} \xrightarrow{\text{浓盐酸}} \text{氯代环己烷}$$

三、主要仪器和试剂

1. 主要仪器：三颈烧瓶、磁力搅拌器、温度计、分析天平、恒压滴液漏斗、冷凝管、分液漏斗、烧杯、橡胶管、锥形瓶、抽滤瓶、分馏柱、布氏漏斗、滤纸、真空泵等。
2. 主要试剂：浓盐酸、环己醇、饱和氯化钠溶液、饱和碳酸氢钠溶液、无水硫酸镁等。

四、实验步骤

在配有磁力搅拌器、温度计、尾气回收装置、冷凝管、干燥管以及恒压滴液漏斗的 250 mL 三颈烧瓶中，加入环己醇 33 mL、浓盐酸 5 mL，快速搅拌，升温至回流，保温反应 3～4 h。

待反应结束后，将反应体系冷却至室温，静置分层，其中，有机相分别用饱和碳酸氢钠溶液 10 mL、饱和氯化钠溶液 10 mL，充分洗涤。用无水硫酸镁充分干燥，过夜，抽滤，用分馏柱分馏，收集 138℃以上的馏分，得到无色液体，即为氯代环己烷。

五、注意事项与思考题

（一）注意事项

1. 试写出本反应的反应机理。
2. 本反应回流温度不能太高，以防 HCl 逸出太多。开始回流温度在 85℃左右为宜，最后温度不超过 110℃。

（二）思考题

1. 本反应中除了用浓盐酸作为催化剂以及氯化剂外，还有哪些氯化剂可以使用？
2. 本反应可能会产生何种副反应？

Experiment 4 Synthesis of chloro-cyclohexane

I. Purpose of the experiment

i. To master the synthesis mechanism of chloro-cyclohexane.

ii. To familiar with the synthesis experiment of chloro-cyclohexane.

iii. To understand the physical and chemical properties and uses of chloro-cyclohexane.

II. Experimental principle

i. Compound introduction

Chloro-cyclohexane is an important organic compound, also known as cyclohexyl chloride, with the formula $C_6H_{11}Cl$. It is a colorless liquid with an asphyxiating odor. It is insoluble in water, and soluble in ethanol.

Chloro-cyclohexane can be used in the production of pesticides, rubber anti-coke agents, medicine, *etc.*. In pesticides, it is used in the synthesis of acaricide tricyclohexyl tin and triazole tin intermediate tin chloride. In addition, it can also be used in the synthesis of medicine trihexyphenidyl hydrochloride, and used in the production of anti-coke agents and other products.

Chloro-cyclohexane and is generally stored in a cool, dry, well-ventilated warehouse, away from fire, heat sources, direct sunlight, sealed packaging. At the same time, it should be stored separately from acids and edible chemicals, and mixing storage is strictly prohibited. Storage areas should be equipped with suitable materials for containing spills.

ii. Synthetic process route

The synthesis of chloro-cyclohexane is mainly prepared by cyclohexanol catalyzed by concentrated hydrochloric acid. The synthesis process is as follows:

$$\text{C}_6\text{H}_{11}\text{OH} \xrightarrow{\text{concentrated hydrochloric acid}} \text{C}_6\text{H}_{11}\text{Cl}$$

III. Main instruments and reagents

i. Main instruments: three-necked flask, magnetic stirrer, thermometer, analytical balance, constant pressure drip funnel, condensing tube, separation funnel, beaker, rubber tube, conical bottle, suction bottle, Brinell funnel, fractionation column filter paper, vacuum pump, *etc*.

ii. Main reagents: concentrated hydrochloric acid, cyclohexanol, saturated sodium chloride solution, saturated sodium bicarbonate solution, anhydrous magnesium sulfate, *etc*.

IV. Experimental steps

Add 33 mL of cyclohexanol and 5 mL of concentrated hydrochloric acid into a 250 mL three-neck round-bottom flask equipped with a magnetic stirrer, a thermometer, a tail gas recovery device, a condensing tube, a drying tube and a constant pressure drip funnel. The flask is rapidly mixed and heated to reflux. Maintain the reaction for 3-4 h.

After the reaction, cool the reaction system to room temperature and alllow it to stand for separation. Among them, the organic phase is fully washed with 10 mL of saturated sodium bicarbonate solution and 10 mL of saturated sodium chloride solution. Dry with anhydrous magnesium sulfate overnight. Filter by suction. tstil using a fractionating column. Collect the fractions above 138 ℃ to obtain a colorless liquid, which is cyclohexane chloride.

V. Attention or thinking questions

i. Attentions

(i) Try to write the reaction mechanism of this reaction.

(ii) The reflux temperature of this reaction should not be too high to prevent too much HCl escape. It is advisable to start reflux at about 85 ℃, and the final temperature should not exceed 110 ℃.

ii. Thinking questions

(i) In addition to using concentrated hydrochloric acid as catalyst and chlorination agent, what other chlorination agents can be used in this reaction?

(ii) What might be the side effects of this reaction?

实验五　苄基三乙基氯化铵的合成

一、实验目的

1. 掌握相转移催化反应的机理。
2. 熟悉相转移催化剂苄基三乙基氯化铵的合成操作过程。
3. 了解苄基三乙基氯化铵的理化性质及用途。

二、实验原理

（一）化合物简介

苄基三乙基氯化铵是一种重要的有机化合物，分子式是 $C_{13}H_{22}ClN$。纯品为白色结晶或粉末，有吸湿性，溶于水、乙醇、甲醇、异丙醇、二甲基甲酰胺、丙酮和二氯甲烷，常用作烷基化反应催化剂。

通过相转移催化迈克尔加成反应，合成多取代环丙烷，用作杀菌剂，亦可用于亲核取代、卡宾反应、C-烷化、N-烷化、O-烷化以及 S-烷化等烷基化反应中，同时不能与阴离子物共混。

常见的相转移催化包括季铵盐类，如苄基三乙基溴化铵(TEBA)、四丁基溴化铵(TBAB)等，冠醚类，如 18-冠-6，二环己基-18-冠-6 等，其他的相转移催化剂有开链聚醚，如聚乙二醇、聚乙醇醚等。

（二）合成工艺路线

苄基三乙基氯化铵的合成，主要通过氯化苄与三乙胺为原料合成，其合成工艺路线如下：

$$(C_2H_5)_3N + C_6H_5CH_2Cl \longrightarrow (C_2H_5)_3N^+CH_2C_6H_5CH_2Cl^-$$

三、主要仪器和试剂

1. 主要仪器：三颈烧瓶、回流冷凝器、磁力搅拌器、温度计、分析天平、锥形瓶、烧杯、恒压滴液漏斗、真空泵等。
2. 主要试剂：氯化苄、三乙胺、三乙基苄基氯化铵、1,2-二氯乙烷、乙酸乙酯、N,N-二甲基甲酰胺、苯等。

四、实验步骤

（一）第一种方法

在配有磁力搅拌器、温度计、冷凝管的 100 mL 三颈烧瓶中，加入氯化苄 6.3 g、三乙胺 5 g 以及 1,2-二氯乙烷 19 mL，快速搅拌下，升温至回流，保温反应 1 h。反应结束后，将反应体系置于室温，冷却，有固体晶体析出，随后，放置冰箱中冷却，过夜，待晶体全部析出后，抽滤，滤饼用少量的 1,2-二氯乙烷充分洗涤 2~3 次，干燥，得产物。

（二）第二种方法

在配有磁力搅拌器、温度计、冷凝管的 100 mL 三颈烧瓶中，加入三乙胺 12.6 mL、氯化苄 10 mL、N,N-二甲基甲酰胺 6.7 mL 以及乙酸乙酯 2 mL，升温，快速搅拌，控制温度在 105℃，保温反应 1 h，反应结束后，冷却至 80℃时，在搅拌下缓慢加入苯 8 g，进而使得铵盐沉淀，抽滤，滤饼用冷的苯试剂，充分洗涤 3 次，干燥，得产物。

五、注意事项与思考题

（一）注意事项

1. 试写出本反应的反应机理。
2. 本反应中，由于 1,2-二氯乙烷分解生成毒性更大的光气，所以在使用时需在通风橱中进行。
3. 苯试剂有剧毒，易致癌，需在通风橱中操作。

（二）思考题

如果在反应中将氯化苄改为氯苯，反应是否能进行？

Experiment 5 Synthesis of benzyl triethyl ammonium chloride

I. Purpose of the experiment

i. To master the basic principle of phase transfer catalytic reaction.

ii. To familiar with the synthesis process of benzyl triethyl ammonium chloride as a phase transfer catalyst.

iii. To understand the physicochemical properties of benzyl triethyl ammonium chloride.

II. Experimental principle

i. Compound introduction

Benzyl triethylammonium chloride is an important organic compound with the formula $C_{13}H_{22}ClN$. The pure product is white crystal or powder. It is hygrometric. It is soluble in water, ethanol, methanol, isopropyl alcohol, dimethylformamide, acetone and dichloromethane, which is commonly used as a catalyst for alkylation reaction.

Polysubstituted cyclopropane is synthesized by the Michael addition reaction catalyzed by phase transfer. It is used as a fungicide. It can be used in nucleophilic substitution, carbene reaction, c-alkylation, n-alkylation, o-alkylation and s-alkylation reactions. It cann't be mixed with anions at the same time.

Common phase transfer catalysts include quaternary ferric salts, such as benzyl triethylammonium bromide (TEBA) and tetrabutylammonium bromide (TBAB), crown ethers, such as 18-crown-6 and dicyclohexyl-18-crown-6, *etc*. Other phase transfer catalysts include open-chain polyethers, such as polyethylene glycol and polyethyl alcohol ether.

ii. Synthetic process route

The synthesis of benzyl triethyl ammonium chloride is mainly prepared by the synthesis of benzyl chloride and triethylamine. The synthesis process is as follows:

$$(C_2H_5)_3N + C_6H_5CH_2Cl \longrightarrow (C_2H_5)_3N^+CH_2C_6H_5CH_2Cl^-$$

III. Main instruments and reagents

i. Main instruments: three-necked flask, reflux condenser, agitator, thermometer, analytical balance, conical flask, beaker, constant pressure drip funnel, vacuum pump, *etc*.

ii. Main reagents: benzyl chloride, triethylamine, triethyl benzyl ammonium chloride, 1, 2-dichloroethane, ethyl acetate, *N, N*-dimethylformamide, benzene, *etc*.

IV. Experimental steps

i. The first method

Add 6.3 g of benzyl chloride, 5 g of triethylamine and 19 mL of 1, 2-dichloroethane into a 100 mL three-neck round-bottom flask equipped with a magnetic stirrer, a thermometer and a condensate tube. Under rapid stirring, heat to reflux, and the maintaining reaction for 1 h. At the end of the reaction, allow the reaction system to cool to room temperature, Where solid crystals will precipitate, Then place it in the refrigerator to cool overnight until all crystals precipitation. Filter by suction, and wash the filter cake thoroughly 2-3 times with a small amount of 1, 2-dichloroethane, then dry to obtain the product.

ii. The second method

Add 12.6 mL of triethylamine, 10 mL of benzyl chloride, 6.7 mL of *N*, *N*-dimethyl formamide and 2 mL of ethyl acetate into a 100mL three-neck round-bottomed flask equipped with magnetic stirrers, thermometers and condensation tubes. Heat and stir quickly. Control the temperature at 105 ℃, and hold the flask for 1h. Cool down to 80 ℃ after the reaction. Under stirring, is slowly add 8 g of benzene to precipitate the ammonium salt. Filter by suction .Wash the filter cake thoroughly 3 times with cold benzene. Dry to obtain the product.

V. Attention or thinking questions

i. Attention

(i) Try to write the reaction mechanism of this reaction.

(ii) In this reaction, due to the decomposition of 1, 2-dichloroethane into phosgene, which is more toxic, the experiment should be carried out in the fume hood when used.

(iii) Benzene reagent is highly toxic, easy to cause cancer, needs to be operated in the fume hood.

ii. Thinking questions

Can the reaction proceed if benzyl chloride is changed to chlorobenzene in the reaction?

实验六 乳酸丁酯的合成

一、实验目的

1. 掌握乳酸丁酯的合成反应机理。
2. 熟悉乳酸丁酯的实验合成操作过程。
3. 了解乳酸丁酯的理化性质及用途。

二、实验原理

（一）化合物简介

乳酸丁酯是一种重要的有机化合物，分子式为 $C_7H_{14}O_3$，微溶于水，能与烃类、油脂混溶。能溶解硝酸纤维素、醋酸纤维素、天然树脂与合成树脂等。20℃时在水中溶解 4.0%，25℃时溶解 3.4%。

乳酸丁酯与氧化剂可发生反应，其能扩散到相当远的地方，遇火源会着火回燃。若遇高热，容器内压增大，有开裂和爆炸的危险。用作生产胺基、硝基涂料的溶剂，也用于生产香料、合成树脂、黏合剂等。

乳酸丁酯用作硝酸纤维素漆、印刷油墨、天然及合成树脂等的溶剂。也用于干洗液、黏结剂、防结皮剂和香料等。为高沸点溶剂，用于天然树脂、合成树脂、油漆、印刷油墨。

乳酸丁酯用作化妆品溶剂，主要用作指甲油等化妆品的主溶剂，对硝化纤维素等皮膜形成剂具有优良的溶解性，通常需加入助溶剂，以调节黏度和挥发速度。

（二）合成工艺路线

乳酸丁酯的合成，主要是通过硫酸氢钠催化乳酸与正丁醇进行制备，其合成工艺路线如下：

$$CH_3CH(OH)COOH + CH_3(CH_2)_2CH_2OH \xrightarrow{NaHSO_4} CH_3CH(OH)COOC_4H_9 + H_2O$$

三、主要仪器和试剂

1. 主要仪器：三颈烧瓶、磁力搅拌器、温度计、恒压滴液漏斗、冷凝管、分水器、干燥管、分液漏斗、烧杯、橡胶管、锥形瓶、抽滤瓶、布氏漏斗、分析天平、滤纸、真空泵等。
2. 主要试剂：乳酸、正丁醇、硫酸氢钠、无水硫酸镁等。

四、实验步骤

在配有磁力搅拌器、温度计、分水器、冷凝管、干燥管的 250 mL 三颈烧瓶中，加入乳酸 10.5 mL、正丁醇 27 mL 以及硫酸氢钠 0.5 g，快速搅拌，升温至回流，温度约为 120℃，发现分水器中无水珠，持续回流 30 min。停止加热，冷却至室温，静置，待硫酸氢钠完全沉淀后，随后将反应体系中的液体，全部倾倒入分液漏斗中，进行分液，分出水层，有机相用水充分洗涤至中性，随后用无水硫酸镁干燥，过夜，抽滤。

将有机相转入圆底烧瓶中，进行常压蒸馏，先蒸出未反应的正丁醇，然后换用空气冷凝管收集170～190℃馏分，得粗产物，即为乳酸正丁酯。粗品经过简单分馏，收集185～187℃馏分。

五、注意事项与思考题

（一）注意事项

1. 试写出本反应的反应机理。
2. 注意反应中正丁醇既是反应物又是带水剂。

（二）思考题

1. 在常压蒸馏收集馏分时，为什么换用空气冷凝管？
2. 本反应中，采用了硫酸氢钠作为催化剂，还有哪些催化剂可以使用？

Experiment 6　Synthesis of butyl lactate

I. Purpose of the experiment

i. To master the synthetic reaction mechanism of butyl lactate.

ii. To familiar with the synthetic process of butyl lactate.

iii. To understand the physicochemical properties and applications of butyl lactate.

II. Experimental principle

i. Compound introduction

Butyl lactate is an important organic compound with the molecular formula $C_7H_{14}O_3$. It is slightly soluble in water and can be miscible with hydrocarbons and oils. It can dissolve cellulose nitrate, cellulose acetate, natural resin and synthetic resin. Dissolve 4.0% in water at 20℃ and 3.4% at 25℃.

Butyl lactate can react with an oxidizing agent, which can spread over a considerable distance. It will ignite when it encounters a fire source. In case of high heat, the pressure inside the container increases, Posing a risk of cracking and explosion. It is used for the production of amine, nitro coating solvent, also used in spices, synthetic resins, adhesives and other aspects.

Butyl lactate is used as a solvent for cellulose nitrate paint, printing ink, natural and synthetic resins, *etc*. It is also used in dry lotion, adhesive, anti-crusting agents and spice. As a high boiling solvent, it is used for natural resin, synthetic resin, paint, printing ink.

Butyl lactate is used as the solvent for cosmetics, mainly used as the main solvent for nail polish and other cosmetics. It has excellent solubility to nitrocellulose and other skin film forming agents. It is usually necessary to add cosolvent to adjust the viscosity and volatilization rate.

ii. Synthetic process route

The synthesis of butyl lactate mainly involves the preparation of lactic acid and *n*-butanol catalyzed by sodium bisulfate. The synthesis process is as follows:

$$CH_3CH(OH)COOH + CH_3(CH_2)_2CH_2OH \xrightarrow{NaHSO_4} CH_3CH(OH)COOC_4H_9 + H_2O$$

III. Main instruments and reagents

i. Main instruments: three-necked flask, magnetic stirrer, thermometer, constant pressure drip funnel, condensing tube, separation funnel, beaker, rubber tube, conical flask, suction bottle, Brinell funnel, analytical balance, filter paper, vacuum pump, *etc*.

ii. Main reagents: lactic acid, n-butanol, sodium bisulfate, anhydrous magnesium sulfate, *etc*.

Ⅳ. Experimental steps

Add 10.5 mL of lactic acid, 27 mL of *n*-butanol and 0.5 g of sodium bisulfate into a 250 mL three-neck round-bottom flask equipped with a magnetic stirrer, a thermometer, a water separator, a condensing tube and a drying tube. Stir quickly and heat to reflux at about 120℃, noting that no water droplets apper in the separator funnel. Continus refluxing for 30 min. Stop heating and cool to room temperature. Let it stand until the sodium bisulfate has completely precipitated. Then, pour all the liquid in the reaction system into the liquid separation funnel for liquid separation, separating the aqueous layer. Wash the organic phase with water until neutral, then dry with anhydrous magnesium sulfate overnight and filter.

The organic phase to a round-bottom flask for atmospheric distillation. First distilling off the unreacted *n*-butanol, then switch to an air condensing tube to obtain the crude product at 170-190℃, which is *n*-butyl lactate. After simple fractionation, collect the fraction at 185-187℃.

Ⅴ. Attention or thinking questions

i. Attention

(i) Try to write the reaction mechanism of this reaction.

(ii) Note that *n*-butanol is both a reactant and a water-carrying agent in the reaction.

ii. Thinking questions

(i) Why do you use air condensing tubes when collecting fractions in atmospheric distillation?

(ii) In this reaction, sodium bisulfate is used as the catalyst. What other catalysts can be used?

实验七　桂皮酰哌啶的合成

一、实验目的

1. 掌握桂皮酰哌啶反应中的氯化、酰化的基本原理。
2. 熟悉桂皮酰哌啶的无水操作及产品精制的实验过程。
3. 熟悉桂皮酰哌啶的合成工艺路线。
4. 了解桂皮酰哌啶的理化性质及用途。

二、实验原理

（一）化合物简介

桂皮酰哌啶是丙戊酸钠（抗癫灵）的简化物，分子式为 $C_{14}H_{17}NO$。现代药理实验证明，其抗惊作用不低于丙戊酸钠，并具有广谱抗惊作用。研究发现，肉桂酰胺类药物是蛋白络氨酸激酶的抑制剂，并证明蛋白络氨酸激酶抑制剂有抗癌活性，能够诱导白血病细胞的分化。

（二）合成工艺路线

桂皮酰哌啶的合成，主要通过羟醛缩合反应制得。芳香醛和酸酐在酸酐相应的碱金属盐存在下，发生类似醇醛缩合反应得到 α, β–不饱和芳香酸。这个反应用于合成桂皮酸，称为 Perkin 反应。生成的桂皮酸与二氯亚砜进行酰氯化反应，得到酰氯化物，最后与哌啶缩合，得到产物桂皮酰哌啶，其合成工艺路线如下：

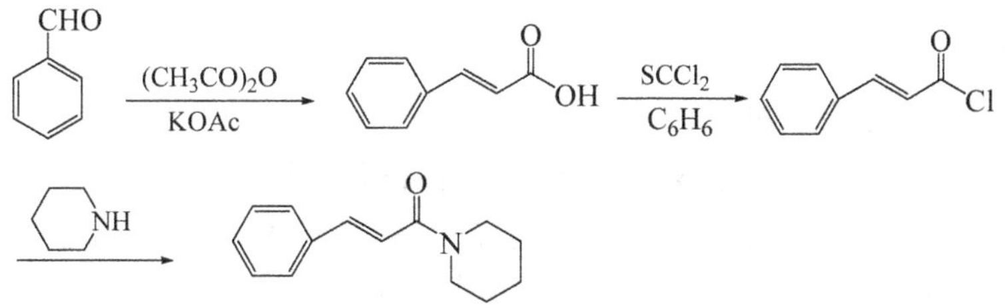

三、主要仪器和试剂

1. 主要仪器：三颈烧瓶、200℃温度计、分析天平、空气冷凝管、油浴锅、蒸馏烧瓶、搅拌子、磁力搅拌器、玻璃棒、布氏漏斗、抽滤瓶、滤纸、干燥管、加热套、烧杯、橡胶管、锥形瓶、真空泵等。

2. 主要试剂：苯甲醛、乙酸酐、乙酸钾、碳酸钠、活性炭、乙醇、浓盐酸、氯化亚砜、苯、哌啶、氯化钙、无水硫酸钠等。

四、实验步骤

（一）桂皮酸的合成

在配有磁力搅拌器、温度计、冷凝管、干燥管以及恒压滴液漏斗的 250 mL 三颈烧瓶中，加入苯甲醛 20 g、乙酸酐 20 mL 以及乙酸钾 12 g，快速搅拌，升温至回流，保温反应 4 h。反应结束后，将反应液倒入热水 125 mL 中，加入适量碳酸钠，调节 pH 至 8，倒入蒸馏烧瓶中，进行水蒸气蒸馏，除去未反应、过量的苯甲醛后，加入适量活性炭，煮沸回流 30 min，趁热抽滤，冷却后，缓慢滴加浓盐酸酸化，边加边搅拌，有桂皮酸固体析出，抽滤，滤饼用水充分洗涤，干燥，得桂皮酸粗品。可以用乙醇重结晶，得桂皮酸纯品。

（二）桂皮酰氯的合成

在配有磁力搅拌器、温度计、冷凝管、干燥管、气体吸收装置以及恒压滴液漏斗的 250 mL 三颈烧瓶中，加入桂皮酸 7.4 g、苯 60 mL 以及氯化亚砜 4 mL，快速搅拌，升温至回流，直至无氯化氢气体生成为止，反应约 3 h，反应结束后，将反应装置换成蒸馏装置，减压蒸去苯溶剂，得到桂皮酰氯的固体结晶产物。

（三）桂皮酰胺的合成

将上步制得的桂皮酰氯加入圆底烧瓶中，加无水苯 100 mL，升温溶解，分批次加入哌啶 10 mL，快速搅拌，室温反应 2 h。密闭，室温放置 2 h，完成胺解反应。有固体产生，抽滤，除去固体，滤液用水充分洗涤，有机层再用 10%盐酸洗涤，有机层再用饱和碳酸钠洗涤，再用水洗涤至中性，分出有机层，用无水硫酸钠充分干燥，过夜，抽滤，减压浓缩，除去有机溶剂，得固体，即为桂皮酰哌啶产物。

五、注意事项与思考题

（一）注意事项

1. 试写出本反应的反应机理。
2. 苯甲醛容易被空气氧化生成苯甲酸，使用前需要重新蒸。
3. 本反应中桂皮酸的合成，对无水要求非常严格，涉及的试剂、仪器等均需做无水处理。
4. 乙酸酐、氯化亚砜均容易吸水，在使用时，均需盖紧瓶塞，在通风柜中量取。

（二）思考题

1. 桂皮酸合成，为什么对无水的要求特别严格？
2. 本反应在使用氯化亚砜作为酰化剂时，有哪些注意事项？还有哪些酰化剂？

Experiment 7 Synthesis of cinnamyl piperidine

Ⅰ. Purpose of the experiment

i. To master the basic principle of chlorination and acylation in cinnamyl piperidine reaction.

ii. To familiar with the anhydrous operation of cinnamyl piperidine and the experimental process of product refining.

iii. To familiar with the synthesis process of cinnamyl piperidine.

iv. To understand the physicochemical properties and applications of cinnamyl piperidine.

Ⅱ. Experimental principle

i. Compound introduction

Cinnamyl piperidine is a simplified anti-purpurin with the molecular formula $C_{14}H_{17}NO$. Modern pharmacological experiments have proved that its anti-shock effect is no inferior than antiepileptica. It has a broad spectrum of anti-shock effect. It is found that cinnamamide drugs are inhibitors of protein tyrosine kinase. And demonstrated that protein tyrosine kinase inhibitors have anticancer activity and can induce differentiation of leukemia cells.

ii. Synthetic process route

The synthesis of cinnamyl piperidine is mainly achieved through a hydroxy aldehyde condensation reaction. Aromatic aldehydes and anhydrides undergo a similar aldehyde-like condensation reaction in the presence of the corresponding alkali metal salts of the anhydride to obtain α, β-unsaturated aromatic acids. This reaction is used to synthesize cinnamic acid and called the Perkin reaction. Cinnamic acid is chlorinated with thionyl chloride to obtain acyl chloride. Finally, it is condensed with piperidine to obtain cinnamyl piperidine. The synthesis process is as follows:

Ⅲ. Main instruments and reagents

i. Main instruments: three necked flask, distillation flask thermometer, analytical balance, condensing tube, oil bath, stirrer, magnetic stirrer, glass rod, Brinell funnel, filter bottle, filter paper, drying tube, heating sleeve, beaker, rubber tube, conical flask, vacuum pump, *etc*.

ii. Main reagents: benzaldehyde, acetic anhydride, potassium acetate, sodium carbonate, activated carbon, concentrated hydrochloric acid, sulfoxide chloride, benzene, piperidine, calcium chloride, anhydrous sodium sulfate, *etc.*

IV. Experimental steps

i. Synthesis of cinnamic acid

Add 20 g benzaldehyde, 20 mL of acetic anhydride and 12 g potassium acetate into a 250 mL three-necked round-bottom flask equipped with a magnetic agitator, a thermometer, a condensing tube, a drying tube and a constant pressure drip funnel. Stir quickly and heat to reflux. Maintain the reaction for 4 hours. After the reaction, pour the reaction liquid into 125 mL of hot water. Add an appropriate amount of sodium carbonate. Adjust the pH to 8. Pour it into a round-bottomed flask. Steam distillation. Remove the unreacted. Excessive benzaldehyde. Add an appropriate amount of activated carbon. Boil reflux 30 min. Filter while hot, cool and slowly add concentrated hydrochloric acid acidification while stirring. Solid cinnamic acid will precipitate. Filter. Wash filter the cake with water. Dry to obtain crude cinnamic acid. It can be recrystallized with ethanol to produce pure cinnamic acid.

ii. Synthesis of cinnamyl chloride

Add 7.4 g of cinnamic acid, 60 mL of benzene and 4 mL of sulfoxide chloride into a 250 mL three-necked round-bottom flask equipped with a magnetic agitator, a thermometer, a condensing tube, a drying tube, a gas absorption device and a constant pressure drip funnel. Stir quickly. Heat to reflux until no hydrogen chloride gas is generated, for about 3 h. After the reaction is complete, replace the reaction with a distillation unit. Remove the benzene solvent under reduced pressure to obtain the solid crystalline product of cinnamyl chloride.

iii. Synthesis of cinnamamide

Add the cinnamyl chloride prepared in the previous step into a round-bottomed flask. Add 100 mL of anhydrous benzene. Dissolve at high temperature. Add 10 mL of piperidine in batches. Stir quickly, and react at room temperature for 2 h. Seal and let it sit at room temperature for 2 h to complete the aminolysis reaction. Wash the organic layer with 10% hydrochloric acid. Wash the organic layer with saturated sodium carbonate. Wash the organic layer with water until it is neutral. Separate the organic layer. Dry with anhydrous sodium sulfate overnight. Filter and concentrate under reduced pressure to remove the organic solvent. And a solid is obtained, which is cinnamyl piperidine product.

V. Attention or thinking questions

i. Attention

(i) Try to write the reaction mechanism of this reaction.

(ii) Benzaldehyde is easily oxidized by air to produce benzoic acid, which needs to be re-steamed before use.

(iii) The synthesis of cinnamic acid in this reaction requires very strict anhydrous treatment, and the reagents and instruments involved need to be treated anhydrous.

(iv) Acetic anhydride and sulfoxide chloride are easy to absorb water. When in use, it is necessary to cover the bottle plug tightly and measure it in the fume hood.

ii. Thinking questions

(i) Cinnamic acid synthesis, why is the requirement of anhydrous especially strict?

(ii) In this reaction, when using sulfoxide chloride as an acylating agent, what are the precautions? What other acylates are there?

实验八 3-苯甲酰基丙烯酸的合成

一、实验目的

1. 掌握 3-苯甲酰基丙烯酸的合成反应机理。
2. 熟悉 3-苯甲酰基丙烯酸的实验操作过程。
3. 了解 3-苯甲酰基丙烯酸的理化性质及用途。

二、实验原理

（一）化合物简介

3-苯甲酰基丙烯酸又称为 γ-苯基-γ-氧代-α-丁烯酸，分子式为 $C_{10}H_8O_3$，是食品防腐剂、农药的重要中间体，以及是血管紧张素转化酶抑制剂依那普利药物合成的中间体。

（二）合成工艺路线

3-苯甲酰基丙烯酸的合成，主要是通过苯与顺丁烯二酸酐为原料，在路易斯酸条件下制得，其合成工艺路线如下：

$$\text{C}_6\text{H}_6 + \text{顺丁烯二酸酐} \xrightarrow[\text{C}_6\text{H}_6]{\text{AlCl}_3} \text{C}_6\text{H}_5\text{-CO-CH=CH-COOH}$$

三、主要仪器和试剂

1. 主要仪器：三颈烧瓶、磁力搅拌器、温度计、恒压滴液漏斗、冷凝管、干燥管、水浴装置、分液漏斗、烧杯、橡胶管、分析天平、锥形瓶、抽滤瓶、布氏漏斗、滤纸、真空泵等。
2. 主要试剂：苯、无水 $AlCl_3$、顺丁烯二酸酐、2mol/L 盐酸等。

四、实验步骤

在配有磁力搅拌器、温度计、尾气吸收装置、冷凝管、干燥管以及恒压滴液漏斗的 500 mL 三颈烧瓶中，加入苯 50 mL，置于冰盐浴中冷却，分批加入无水 $AlCl_3$ 10.5 g，控制反应体系温度，使其小于 0℃，缓慢滴加顺丁烯二酸酐 5 g，滴加的过程中，温度小于 5℃，滴加结束后，撤去冰盐浴，使得反应自然升至室温后，升温至 100℃，保温搅拌反应 1 h，反应结束后，冷却至室温，随后，用冰浴冷却，向反应体系中缓慢滴加水 15 mL，注意滴加的速度，避免反应过于剧烈。滴加结束后，剧烈搅拌反应 30 min，向反应体系中加入 2 mol/L 盐酸 50 mL，搅拌反应 30 min，冷却，抽滤，用 2 mol/L 盐酸充分洗涤滤饼，随后，用水充分洗涤滤饼至中性，抽干，真空干燥，得到粗产物。同时，粗产物可以使用苯试剂进行重结晶。

五、注意事项与思考题

（一）注意事项

1. 试写出本反应的反应机理。
2. 本反应中，需使用苯试剂，有毒，应在通风橱中进行试验。
3. 本反应属于傅克酰基化反应，滴加原料时，速度必须慢，否则放热太多，易使苯或其他物料溢出。

（二）思考题

本反应用无水氯化铝作为催化剂，还有哪些催化剂在本反应适用？

Experiment 8 Synthesis of 3-benzoyl acrylic acid

I. Purpose of the experiment

i. To master the synthesis reaction mechanism of 3-benzoyl acrylic acid.

ii. To familiar with the experimental operation process of 3-benzoyl acrylic acid.

iii. To understand the physicochemical properties and applications of 3-benzoyl acrylic acid.

II. Experimental principle

i. Compound introduction

3-benzoyl acrylic acid, also known as γ-phenyl-γ-oxy-α-butenoic acid, with molecular formula $C_{10}H_8O_3$, is an important intermediate as a food preservative, pesticide, as well as the angiotensin converting enzyme inhibitor enalapril synthesis intermediate.

ii. Synthetic process route

The synthesis of 3-benzoyl acrylic acid is mainly prepared by benzene and maleic anhydride under the condition of Lewis acid. The synthesis process is as follows:

$$\text{C}_6\text{H}_6 + \text{maleic anhydride} \xrightarrow[C_6H_6]{AlCl_3} \text{C}_6\text{H}_5\text{-CO-CH=CH-COOH}$$

III. Main instruments and reagents

i. Main instruments: three-necked flask, magnetic stirrer, thermometer, constant pressure drip funnel, condensing tube, drying tube, water bath equipment, separation funnel, beaker, rubber tube, analytical balance, conical flask, suction bottle, Brinell funnel, filter paper, vacuum pump, *etc.*

ii. Main reagents: benzene, anhydrous AlCl$_3$, maleic anhydride, 2mol/L hydrochloric acid, *etc.*

IV. Experimental steps

Add 50 mL of benzene into a 500 mL three-neck round-bottom flask equipped with a magnetic agitators, a thermometers, a tail gas absorbent devices, a condensing tubes, a drying tubes and a constant pressure drip funnel. Cool it in an ice salt bath. Add anhydrous 10.5 g of AlCl$_3$ in batches to control the reaction system temperature to below 0°C. Slowly add 5 g of maleic anhydride, ensuring the temperature remains below 5°C during the addition. After the additon, the ice salt bath is removed, so that the reaction naturally rises to room temperature. Then heat to 100°C. and maintain with strring for 1 h. After the reaction, the water is cooled to room temperature, and then the water is cooled with the ice bath. Slowly add 15 mL of water. Take care to control the addition rate to avoid excessive reaction. After dripping, the mixture is stirred vigorously for 30 min. Add

50 mL of 2 mol/L hydrochloric acid into the reaction system for 30 min. Then, the mixture is cooled and filtered. Wash the filter cake with 2 mol/L hydrochloric acid. Follow by washing with water until neutral. Then drain. Vacuum dry to obtain crude products. Meanwhile, the crude product can be recrystallized using a benzene reagent.

V. Attention or thinking questions

i. Attention

(i) Try to write the reaction mechanism of this reaction.

(ii) In this reaction, benzene reagent is used, which is toxic and should be tested in the fume hood.

(iii) This reaction belongs to the Friedel-Crafts acylation, drop raw materials. The speed must be slow, otherwise too much heat release, easy make benzene or other materials overflow.

ii. Thinking questions

This reaction uses anhydrous aluminum chloride as the catalyst. What other catalysts are suitable for this reaction?

实验九　D-葡萄糖酸-δ-内脂的合成

一、实验目的

1. 掌握葡萄糖酸内酯的合成反应机理。
2. 熟悉减压浓缩的实验操作过程。
3. 了解葡萄糖酸内酯的理化性质以及用途。

二、实验原理

（一）化合物简介

葡萄糖酸-δ-内酯，简称内酯或 GDL，分子式为 $C_6H_{10}O_6$，白色结晶或白色结晶性粉末，几乎无臭，味先甜后酸。易溶于水，稍溶于乙醇，几乎不溶于乙醚，在水中水解为葡萄糖酸及其 δ-内酯的混合物。1%水溶液 pH 3.5，在 2 h 后变为 pH 2.5。

葡萄糖酸-δ-内酯用作凝固剂，主要用于豆腐的生产，也可作为奶类制品蛋白质凝固剂。葡萄糖酸内酯是以葡萄糖酸为原料合成的多功能食品添加剂，无毒，使用安全，主要用作牛奶蛋白和大豆蛋白的凝固剂。例如，用它制作的豆腐保水性好，细腻滑嫩、可口。加入鱼、禽畜的肉中作为保鲜剂，可使其外观保持光泽和肉质保持弹性。

同时，葡萄糖酸-δ-内酯为色素稳定剂，使午餐肉和香肠等肉制品色泽鲜艳。其还可以作为疏松剂用于糕点、面包，改善质感和风味，还可作为酸味剂。

（二）合成工艺路线

本反应通过葡萄糖酸钙为原料，用草酸脱钙生成葡萄糖酸，随后，葡萄糖酸在加热浓缩时，发生分子内酯化得到葡萄糖酸内酯。

三、主要仪器和试剂

1. 主要仪器：三颈烧瓶、温度计、冷凝管、烧杯、分析天平、干燥管、抽滤瓶、布氏漏斗、滤纸、真空泵等。
2. 主要试剂：葡萄糖酸钙、二水合草酸、硅藻土、95%乙醇等。

四、实验步骤

在配有磁力搅拌器、温度计、冷凝管的 100 mL 三颈烧瓶中，加入葡萄糖酸钙 15 g、二水合草酸 4.5 g，快速搅拌，当反应体系温度至 60℃时，向其中加入水 18 mL，于 60℃下，快速搅拌，保温反应 2 h。反应结束后，向三颈烧瓶中加入 1.5 g 硅藻土，剧烈搅拌，随后趁热抽滤，滤饼用 60℃热水 10 mL 洗涤 3 次，抽滤，合并滤液和洗涤液。

将滤液转入减压蒸馏装置的烧瓶中，在小于 45℃的条件下减压浓缩，浓缩至 8 mL 左右时，停止浓缩。向反应体系中加入约葡萄糖酸内酯晶种 1 g，继续减压浓缩，至瓶内出现大量细小晶粒时停止，在 20℃以下，静置结晶。随后抽滤，同时用 95%乙醇洗涤晶体，抽滤，干燥。

五、注意事项与思考题

1. 试写出本实验的反应机理。

2. 本反应中，由于草酸钙的结晶颗粒非常细小，较难过滤，所以在过滤时，需加入硅藻土，有助于过滤。

3. 本反应在后处理的过程中，浓缩的温度不宜过高，过高的温度，会导致产物颜色变深。

4. 葡萄糖酸内酯不容易结晶，加入晶种有助于结晶，同时，结晶时最好静置过夜，有助于晶型的生长。

Experiment 9 Synthesis of D-gluconic acid -δ-endolipin

I. Purpose of the experiment

i. To master the synthesis and reaction mechanism of gluconolactone.

ii. To familiar with the experimental operation process of decompression concentration.

iii. To understand the physicochemical properties and uses of gluconolactone.

II. Experimental principle

i. Compound introduction

Gluconate -δ-lactone, abbreviates as lactone or GDL with the molecular formula $C_6H_{10}O_6$. It is white crystal or white crystalline powder. It is almost odorless, sweet after sour taste. It is easily soluble in water, slightly soluble in ethanol, almost insoluble in ether, hydrolyzed in water to a mixture of gluconic acid and its δ-lactone. A 1% aqueous solution has a pH of 3.5, Which changes to pH 2.5 after 2 h.

Gluconate-δ-lactone is used as a coagulant, mainly used in the production of tofu, but also as a protein coagulant for milk products. Gluconolactone is a multifunctional food additive synthesized from gluconic acid. It is non-toxic and safe to use. It is mainly used as the coagulant of milk protein and soybean protein. For example, tofu made with it holds water well, delicate, tender and delicious. Adding to fish, livestock meat as preservative, it can make its appearance to maintain luster and meat quality to maintain elasticity.

At the same time, gluconate-δ-lactone is the pigment stabilizer, which makes luncheon meat and sausage and other meat products bright color. It can also be used as a loose agent for pastry, bread, improve texture and flavor, also can be used as a sour flavor agent.

ii. Synthetic process route

In this reaction, gluconolactone is obtained by calcium gluconate as raw material, which is decalcified by oxalate. When gluconic acid is heated and concentrated, gluconolactone is obtained by intramolecular esterification.

III. Main instruments and reagents

i. Main instruments: three-necked flask, thermometer, condensing tube, beaker, analytical balance, drying tube, suction bottle, Brinell funnel, filter paper, vacuum pump, *etc*.

ii. Main reagents: calcium gluconate, oxalic acid, diatomite, 95% ethanol, *etc*.

IV. Experimental steps

Add 15 g of calcium gluconate and 4.5 g of oxalic acid dihydrate into a 100 mL round-bottomed flask equipped with a magnetic stirrer, a thermometer and a condensing tube. Stir quickly. When the reaction temperature of the flask reaches 60 ℃, add 18 mL of water into it. Stir quickly at 60 ℃. Hold the flask for 2 h. After the reaction, add 1.5 g of diatomaceous earth into the three-necked flask. Stir violently. Then pump and filter while hot. Wash the filter cake 3 times with 10 mL of 60 ℃ hot water. Pump and filter, and combine the filtrate and washing liquid.

Transfer the filtrate to a vacuum distillation unit, and concentrate under pressure at below 45 ℃. Stop the concentration when the concentration reaches about 8 mL. Add about 1 g gluconolactone seed into the reaction system. Continue the concentration under reduced pressure until a large number of fine grains appeared in the bottle. Stop and allow crystallisation to occur below 20 ℃. Perform suction filtration washing the crystals with 95% ethanol and dry.

V. Attention or thinking questions

i. Try to write the reaction mechanism of this experiment.

ii. In this reaction, because the crystalline particles of calcium oxalate are very small, it is difficult to filter, so diatomaceous earth should be added during filtration to facilitate filtration.

iii. In the post-processing process of this reaction, the concentration temperature should not be too high, excessive temperature will lead to dark product color.

iv. Gluconolactone is not easy to crystallize. Adding seed is conducive to crystallization. At the same time, it is best to stand overnight during crystallization, conducive to the growth of crystal type.

实验十 N-苄基乙酰苯胺的合成

一、实验目的

1. 掌握 N-苄基乙酰苯胺的合成反应机理。
2. 熟悉 N-苄基乙酰苯胺的实验操作过程。
3. 了解 N-苄基乙酰苯胺的理化性质及用途。

二、实验原理

（一）化合物简介

N-苄基乙酰苯胺，其分子式为 $C_{15}H_{15}NO$，是一种非常重要的医药、化工中间体。

（二）合成工艺路线

N-苄基乙酰苯胺的合成，是通过苯胺、乙酸酐等为原料制备，其合成工艺路线如下：

$$\text{PhNH}_2 \xrightarrow{Ac_2O} \text{PhNHCOCH}_3 \xrightarrow[\text{Bu}_4\text{NCl, NaHCO}_3]{C_6H_5CH_2Cl} \text{PhN(CH}_2\text{C}_6\text{H}_5)\text{COCH}_3$$

三、主要仪器和试剂

1. 主要仪器：三颈烧瓶、磁力搅拌器、干燥管、温度计、分析天平、恒压滴液漏斗、冷凝管、分液漏斗、烧杯、橡胶管、锥形瓶、抽滤瓶、布氏漏斗、滤纸、真空泵等。
2. 主要试剂：苯胺、乙酸酐、四丁基氯化铵、无水碳酸氢钠、苄氯、丙酮、乙酸乙酯、饱和氯化钠溶液、无水硫酸钠等。

四、实验步骤

（一）乙酰苯胺的合成

在配有磁力搅拌器、温度计、干燥管以及恒压滴液漏斗的 250 mL 三颈烧瓶中，加入苯胺 8 mL 以及水 120 mL，快速搅拌下，缓慢滴加乙酸酐 12 mL，随后放置冰水中，搅拌 5 min，有固体析出，抽滤，用少量冷水充分洗涤，抽干，干燥，得乙酰苯胺粗品。

（二）N-苄基乙酰苯胺的合成

在配有磁力搅拌器、温度计、干燥管以及恒压滴液漏斗的 250 mL 三颈烧瓶中，加入上步制备的乙酰苯胺 5 g、四丁基氯化铵 0.1 g 以及无水碳酸氢钠 4 g，然后加入无水丙酮 100 mL，快速搅拌，升温至 60～70℃，缓慢滴加溶液（苄氯 4.68 g + 丙酮 50 mL），滴加结束后，保温搅拌反应 6 h，反应结束后，冷却，抽滤，回收溶剂。反应瓶中剩余物溶解于 100 mL 乙酸乙酯中，用饱和氯化钠溶液充分洗涤 3 次，随后有机相用无水硫酸钠充分干燥，过夜，抽滤，减压回收溶剂，得到浅黄色油状产物。

五、注意事项与思考题

1. 试写出本反应的反应机理。
2. 本反应中的苯胺、乙酸酐等具有刺激气味，需要在通风橱中操作。

Experiment 10 Synthesis of *N*-benzyl acetanilide

I. Purpose of the experiment

i. To master the synthesis mechanism of *N*-benzyl acetanilide.
ii. To familiar with the experimental operation process of *N*-benzyl acetanilide.
iii. To understand the physicochemical properties and applications of *N*-benzyl acetanilide.

II. Experimental principle

i. Compound introduction

N-benzyl acetanilide, With the molecular formula $C_{15}H_{15}NO$, is a very important intermediate in medicine and chemical industry.

ii. Synthetic process route

The synthesis of *N*-benzyl acetanilide is made from aniline and acetic anhydride. The synthesis process is as follows:

$$\text{C}_6\text{H}_5\text{NH}_2 \xrightarrow{Ac_2O} \text{C}_6\text{H}_5\text{NHCOCH}_3 \xrightarrow[\text{Bu}_4\text{NCl, NaHCO}_3]{\text{C}_6\text{H}_5\text{CH}_2\text{Cl}} \text{C}_6\text{H}_5\text{N(CH}_2\text{C}_6\text{H}_5)\text{COCH}_3$$

III. Main instruments and reagents

i. Main instruments: three-necked flask, magnetic stirrer, thermometer, drying tube, analytical balance, constant pressure drip funnel, condensing tube, separation funnel, beaker, rubber tube, conical bottle, suction bottle, Brinell funnel, filter paper, vacuum pump, *etc.*

ii. Main reagents: aniline, acetic anhydride, tetrabutylammonium chloride, anhydrous sodium bicarbonate, benzyl chloride, acetone, ethyl acetate, saturated sodium chloride solution, anhydrous sodium sulfate, *etc.*

IV. Experimental steps

i. Synthesis of acetanilide

Add 8 mL of aniline and 120 mL of water into a 250 mL three-necked flask equipped with a magnetic stirrer, a thermometer, a drying tube, and a constant pressure dropping funnel. Under rapid stirring, slowly add 12 mL of acetic anhydride, then place in an ice-water bath and stir for 5 minutes. A solid will precipitate, which is then filtered, thoroughly washed with a small amount of cold water, dried, and the crude acetanilide is obtained.

ii. Synthesis of *N*-benzyl acetanilide

Add 5 g of acetanilide prepared in the previous step, 0.1 g of tetrabutylammonium chloride and 4 g of anhydrous sodium bicarbonate into a 250 mL three-necked round-bottom flask equipped with a magnetic stirrer, a thermometer, a drying tube and a constant pressure drip funder. Add 100 mL of anhydrous acetone. Stir quickly and heat to 60~70℃. Slowly add the solution (4.68 g of benzyl chloride +50 mL of acetone). After the addition, heat preservation and stir reaction for 6 h. After the reaction, cool, filter, and recover the solvent. The remaining residue in the reaction bottle is dissolved in 100 mL of ethyl acetate, and washed fully with saturated sodium chloride solution 3 times. Then the organic phase is fully dried with anhydrous sodium sulfate overnight, pumped and decompressed to recover the solvent and obtain the yellowish oily product.

Ⅴ. Attention or thinking questions

i. Try to write the reaction mechanism of this reaction.

ii. Aniline and acetic anhydride involved in this reaction have an irritating odor and need to be operated on the fume hood.

实验十一　对硝基乙酰苯胺的合成

一、实验目的

1. 掌握对硝基乙酰苯胺的硝化反应机理。
2. 熟悉硝化反应的实验操作过程。
3. 了解对硝基乙酰苯的理化性质及用途。

二、实验原理

（一）化合物简介

对硝基乙酰苯胺，又称为 N-乙酰基对苯二胺，分子式为 $C_8H_8N_2O_3$，无色晶体，熔点 215.6℃，沸点 100℃。溶于热水、醇、醚，溶于氢氧化钾溶液呈橙色，水解生成对硝基苯胺，还原生成对氨基乙酰苯胺。几乎不溶于冷水。

对硝基乙酰苯胺是重要的染料中间体，主要用于制备染料等，如分散黄 G、直接耐酸朱红 4BS、酸性品红 6B、活性蓝 AG 等。

（二）合成工艺路线

对硝基乙酰苯胺的合成，主要是通过乙酰苯胺的硝化反应制备，其合成工艺路线如下：

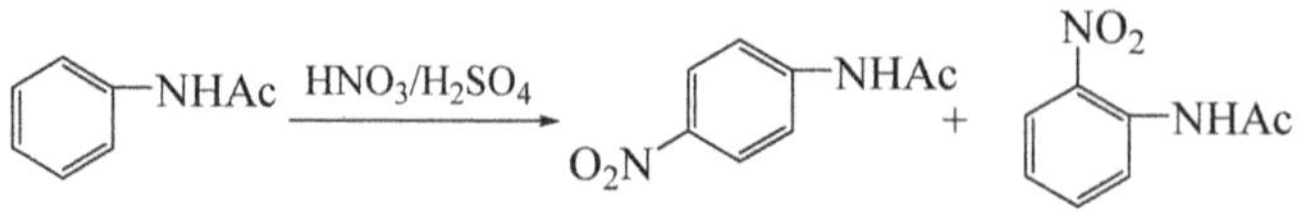

三、主要仪器和试剂

1. 主要仪器：三颈烧瓶、磁力搅拌器、温度计、分析天平、回流冷凝管、干燥管、恒压滴液漏斗、烧杯、锥形瓶、真空泵等。
2. 主要试剂：乙酰苯胺、冰乙酸、浓硫酸、浓硝酸、乙醇等。

四、实验步骤

在配有磁力搅拌器、温度计、冷凝管、干燥管以及恒压滴液漏斗的 250 mL 三颈烧瓶中，加入乙酰苯胺 13.5 g、冰乙酸 13.5 mL，快速搅拌，在冰水浴冷却下，缓慢滴加浓硫酸 27 mL，在滴加浓硫酸的过程中，保证反应体系的温度小于 30℃。随后，用冰盐浴将反应体系降至 0℃，然后，通过恒压滴液漏斗，缓慢滴加混酸（浓硫酸 6 mL 和浓硝酸 6.9 mL 混合），滴加的过程中，一定要缓慢滴加，使得反应体系的温度不能超过 10℃，滴加结束后，于室温下反应 1 h。将反应混合物，在快速搅拌下，倒入装有碎冰 130 g 的烧杯中，有固体沉淀析出，抽滤，滤饼用冰水充分洗涤至中性，抽干得到粗产物。粗产物可以使用乙醇重结晶，得对硝基乙酰苯胺纯品。

五、注意事项与思考题

（一）注意事项

1. 本反应在滴加浓硫酸时，剧烈放热，因此要控制滴加速度，防止反应体系的温度剧烈上升，应缓慢滴加。
2. 本反应所需的混酸溶液，在配制时，注意配制时的温度，要将浓硫酸滴加至硝酸中，且要缓慢滴加。

（二）思考题

1. 在配制混酸时，可能得到的溶液带有一定的浅棕色，为什么？
2. 试写出硝化反应的机理？

Experiment 11 Synthesis of *p*-nitro acetanilide

I. Purpose of the experiment

i. To master the nitration reaction mechanism of *p*-nitro acetanilide.
ii. To familiar with the experimental operation process of nitration reaction.
iii. To understand the physicochemical properties and uses of *p*-nitroacetyl benzene.

II. Experimental principle

i. Compound introduction

p-nitro acetanilide, also known as *n*-acetyl-*p*-phenylenediamine, has the molecular formula $C_8H_8N_2O_3$. It is a colorless crystal with a melting point of 215.6℃ and a boiling point of 100℃. It is soluble in hot water, alcohol, ether, It is soluble in potassium hydroxide solution orange. It hydrolysis to *p*-nitroaniline and reduces to *p*-amino-acetanilide. It is almost insoluble in cold water.

p-nitro acetanilide is an important dye intermediate, mainly used in the preparation of dyes, such as disperse yellow G, direct acid-resistant vermilion 4BS, acid fuchsin 6B, active blue AG, *etc*.

ii. Synthetic process route

The synthesis of *p*-nitro acetanilide is mainly prepared by nitrification of acetanilide. The specific synthesis process is as follows:

III. Main instruments and reagents

i. Main instruments: three-necked flask, magnetic stirrer, thermometer, drying tube, analytical balance, reflux condensing tube, drip funnel, beaker, conical flask, vacuum pump, *etc*.
ii. Main reagents: acetanilide, glacial acetic acid, concentrated sulfuric acid, concentrated nitric acid, ethanol, *etc*.

IV. Experimental steps

Add 13.5 g of acetanilide and 13.5 mL of glacial acetic acid into a 250 mL three-necked round-bottom flask equipped with a magnetic agitator, a thermometer, a condensing tube, a drying tube and a constant pressure drip funnel. Stir rapidly. Under the cooling of the ice water bath, slowly add 27 mL of concentrated sulfuric acid. In the process of drip, add concentrated sulfuric acid, and ensure that the temperature of the reaction system is below 30℃. Then, the reaction system is lowered to 0℃ with an ice salt bath. Then, through a constant pressure drip funnel, slowly add the mixed acid (6 mL of concentrated sulfuric acid and 6.9 mL of concentrated nitric acid). During the addition, it must be slowly added so that the temperature of the reaction system could not exceed 10℃. Pour the reaction mixture into a beaker with 130 g of crushed ice while stirring rapidly, resulting in the precipitation of solid. Filter the mixture, thoroughly wash the filter cake with ice water to neutral. Dry to obtain the crude products. The crude product can be recrystallized by ethanol to produce pure *p*-nitro acetanilide.

V. Attention or thinking questions

i. Attention

(i) In this reaction, when concentrated sulfuric acid is added by drops, heat release is severe. Therefore, the drop acceleration should be controlled to prevent the temperature of the reaction system from rising sharply.

(ii) When preparing the mixed acid solution required by this reaction, pay attention to the temperature at the time of preparation. Concentrated sulfuric acid should be added to nitric acid by drops, and slowly.

ii. Thinking questions

(i) When preparing mixed acids, why might the solution have a certain light brown color?

(ii) Write down the mechanism of nitration reaction.

实验十二　4-氨基-1,2,4-三唑-5-酮的合成

一、实验目的

1. 掌握 4-氨基-1,2,4-三唑-5-酮的合成反应机理。
2. 熟悉 4-氨基-1,2,4-三唑-5-酮的实验操作流程。
3. 了解 4-氨基-1,2,4-三唑-5-酮的理化性质及用途。

二、实验原理

(一) 化合物简介

4-氨基-1,2,4-三唑-5-酮为含氮杂环化合物，分子式为 $C_2H_4N_4O$，具有含氮高、结构致密等特点，与相应的少氮和纯碳环状化合物相比，摩尔体积小、密度高，自身可作为高能钝感炸药以及传爆药的组分，是一类具有广泛应用前景的含能材料。

(二) 合成工艺路线

4-氨基-1,2,4-三唑-5-酮的合成，是通过原甲酸三乙酯与碳酰肼为原料制得，其合成工艺路线如下：

$$C_2H_5O-CH(OC_2H_5)_2 \xrightarrow{NH_2NHCONHNH_2} \underset{\underset{H}{N}-NH}{\overset{NH_2}{\underset{|}{N}}}\!\!\!\!\!\!\!\!\!\!\!\!=\!\!O \;+\; 3\,C_2H_5OH$$

三、主要仪器和试剂

1. 主要仪器：三颈烧瓶、磁力搅拌器、温度计、恒压滴液漏斗、分析天平、冷凝管、分液漏斗、烧杯、橡胶管、锥形瓶、抽滤瓶、布氏漏斗、滤纸、真空泵等。
2. 主要试剂：碳酰肼、原甲酸三乙酯、无水乙醇、蒸馏水等。

四、实验步骤

在配有磁力搅拌器、温度计、冷凝管、干燥管以及恒压滴液漏斗的 250 mL 三颈烧瓶中，加入碳酰肼 5 g、原甲酸三乙酯 20 mL 以及水 2 mL。缓慢升温至 65~85℃，保温回流反应 2 h，反应结束后，随后蒸馏，蒸出反应体系中的副产物乙醇以及水。反应结束后，快速搅拌，冷却，有粉红色固体生成，抽滤，干燥，待用。产物用水溶解后，同时加入 2 倍量的无水乙醇，升温至回流，趁热抽滤，冷却，得到白色固体，即为 4-氨基-1,2,4-三唑-5-酮。

五、注意事项与思考题

1. 试写出本反应的反应机理。
2. 原甲酸三乙酯具有刺激性气味，需在通风橱中进行操作。

Experiment 12   Synthesis of 4-amino-1, 2, 4-triazole-5-ketone

Ⅰ. Purpose of the experiment

i. To master the synthetic reaction mechanism of 4-amino-1, 2, 4-triazol-5-ketone.

ii. To familiar with the experimental operation process of 4-amino-1, 2, 4-triazol-5-ketone.

iii. To understand the physicochemical properties and uses of 4-amino-1, 2, 4-triazo-5-ketone.

Ⅱ. Experimental principle

i. Compound introduction

4-amino-1, 2, 4-triazol-5-ketone is a heterocyclic compound containing nitrogen with the molecular formula $C_2H_4N_4O$. It is characterised of a high nitrogen content, a dense structure. Compared with the corresponding less nitrogen and pure carbon ring compounds, it has a smaller molar and higher density. It can be used as a high energy explosive and explosive delivery components. It's a kind of energetic materials with wide application prospects.

ii. Synthetic process route

The synthesis of 4-amino-1, 2, 4-triazole-5-ketone is obtained by triethyl orthoformate and carbamide hydrazine. The synthesis process is as follows:

Ⅲ. Main instruments and reagents

i. Main instruments: three-necked flask, magnetic stirrer, thermometer, constant pressure drip funnel, analytical balance, condensing tube, separation funnel, beaker, rubber tube, conical bottle, filter bottle, Brinell funnel, filter paper, vacuum pump, *etc.*

ii. Main reagents: carbonyl hydrazine, triethyl orthoformate, anhydrous ethanol, distilled water, *etc.*

Ⅳ. Experimental steps

Add 5 g of carbohydrazide, 20 mL of triethyl orthoformate and 2 mL of water into a 250 mL three-necked round-bottoming flask equipped with a magnetic agitator, a thermometer, a condensing tube, a drying tube and a constant pressure drip funnel. Gradually heat to 65-85°C and maintain reflus for 2 h. After the reaction, distil to remove the by-products of ethanol and water from the reaction system. At the end of the reaction, stir quickly and cool, resulting in the formation of a pink solid. Filter and dry. Then set aside. Dissolve the product in water, add twice the amount of anhydrous ethanol. Heat to reflux. Pump while hot. Cool to obtained a white solid, which is 4-amino-1, 2, 4-triazol-5-ketone.

V. Attention or thinking questions

i. Try to write the reaction mechanism of this reaction.

ii. Triethyl orthoformate has a pungent smell and operates in a fume hood.

实验十三 丙酰氯的合成

一、实验目的

1. 掌握丙酰氯的合成反应机理。
2. 熟悉丙酰氯合成操作过程。
3. 了解氯化剂的种类以及特点。
4. 了解丙酰氯的理化性质及用途。

二、实验原理

（一）化合物简介

丙酰氯是一种重要的有机化合物，化学式为 C_3H_5ClO，其为无色透明液体，有刺激性气味。用作有机合成的丙酰化剂和丙酰的引入剂、香料和医药的原料、农药除草剂原料、聚合引发剂的原料。

丙酰氯化合物操作注意事项：密闭操作，提供充分的局部排风；操作人员必须经过专门培训，严格遵守操作规程；建议操作人员佩戴自吸过滤式防毒面具，穿胶布防毒衣，戴橡胶耐油手套。远离火种、热源，工作场所严禁吸烟；使用防爆型的通风系统和设备；避免产生烟雾。防止烟雾和蒸气释放到工作场所空气中；避免与氧化剂、醇类、碱类接触，尤其要注意避免与水接触，搬运时要轻装轻卸，防止包装及容器损坏；配备相应品种和数量的消防器材、泄漏应急处理设备。

丙酰氯化合物储存注意事项：储存于阴凉、干燥、通风良好的库房；远离火种、热源；库温不宜超过 30℃；保持容器密封；应与氧化剂、醇类、碱类、食用化学品分开存放，切忌混储；不宜久存，以免变质；采用防爆型照明、通风设施；禁止使用易产生火花的机械设备和工具；储存区域应备有泄漏应急处理设备和合适的收容材料。

（二）合成工艺路线

丙酰氯的合成，主要是通过丙酸为原料与三氯化磷反应制备，具体合成工艺路线如下：

$$3\ CH_3CH_2COOH + PCl_3 \longrightarrow 3\ CH_3CH_2COCl + H_3PO_3$$

三、主要仪器和试剂

1. 主要仪器：磁力搅拌器、分析天平、三颈烧瓶、温度计、橡胶管、玻璃棒、回流冷凝器、干燥管、油浴锅、烧杯等。
2. 主要试剂：丙酸、三氯化磷等。

四、实验步骤

在配有磁力搅拌器、温度计、冷凝管、尾气吸收装置以及干燥管的 250 mL 三颈烧瓶中，加入丙酸 37.3 mL、三氯化磷 17.5 mL，快速搅拌，升温至 50℃，保温反应 4 h。反应结束后，冷却至室温后，常压蒸馏，收集 76~80℃的馏分，得产物。

五、注意事项与思考题

（一）注意事项

1. 试写出本实验的反应机理。
2. 本反应是放热反应，特别是在反应开始阶段，反应剧烈，需要控制反应温度。
3. 由于在反应中产生 HCl 气体，反应装置要接尾气吸收装置。

（二）思考题

1. 本反应除了用三氯化磷作为氯化剂，还可以用哪些氯化剂？
2. 常压蒸馏过程有哪些注意事项？

Experiment 13 Synthesis of propionyl chloride

I. Purpose of the experiment

　　i. To master the synthetic reaction mechanism of propionyl chloride.

　　ii. To familiar with the propionyl chloride synthesis process.

　　iii. To understand the types and characteristics of chlorination agents.

　　iv. To understand the physicochemical properties and uses of propionyl chloride.

II. Experimental principle

i. Compound introduction

　　Propionyl chloride is an important organic compound with the chemical formula C_3H_5ClO. It is a clear and colorless liquid with a pungent odor. It is used as a propionylation agent and a propionic anhydride introducer for organic synthesis, a raw material for spices and medicines, a raw material for pesticide herbicides, and a raw material for polymerization initiator.

　　Precautions for operation of propionyl chloride compounds: Closed operation to provide adequate local exhaust air. Operators must be specially trained and strictly abide by operating procedures. It is recommended that operators wear self-priming filter gas masks, adhesive cloth anti-poison jackets, and rubber oil-resistant gloves. Keep away from fire, heat source, no smoking in the workplace. Use explosion-proof ventilation systems and equipment. Avoid creating smoke. Prevent the release of fumes and vapors into the workplace air. Avoid contact with oxidant, alcohol, alkali, especially to avoid contact with water, handling light loading and unloading, to prevent damage to packaging and containers. Equipped with the corresponding variety and quantity of fire equipment, leakage emergency treatment equipment.

　　Precautions for storage of propionyl chloride compounds: Store in a cool, dry and well-ventilated warehouse. Keep away from fire and heat sources. Warehouse temperature should not exceed 30℃. Keep the container sealed. They should be stored separately from oxidants, alcohols, alkalis and edible chemicals, and should not be mixed. Do not be long, so as not to deteriorate. Explosion-proof lighting and ventilation facilities are adopted. It is forbidden to use mechanical equipment and tools that are easy to produce sparks. The storage area should be equipped with emergency spill treatment equipment and suitable holding materials.

ii. Synthetic process route

　　The synthesis of propionyl chloride is mainly prepared by reacting propionic acid with phosphorus trichloride. The specific synthesis process is as follows:

$$3\ CH_3CH_2COOH + PCl_3 \longrightarrow 3\ CH_3CH_2COCl + H_3PO_3$$

III. Main instruments and reagents

i. Main instruments: magnetic stirrer, analytical balance, three necked flask, thermometer, rubber tube, glass rod, reflux condenser, drying tube, oil bath, beaker, *etc.*

ii. Main reagents: propionic acid, phosphorus trichloride, *etc.*

IV. Experimental steps

Add 37.3 mL of propionic acid and 17.5 mL of phosphorus trichloride into a 250 mL three-neck round-bottom flask equipped with a magnetic stirrer, a thermometer, a condensate tube, a tail gas absorption device and a drying tube. Stir quickly. Heat to 50 ℃, and hold for 4 h. After the reaction, cool to room temperature and atmospheric distillation. Collect the fraction at 76-80 ℃ to obtain the product.

V. Attention or thinking questions

i. Attention

(i) Try to write the reaction mechanism of this experiment.

(ii) This reaction is an exothermic reaction, especially at the beginning of the reaction. The reaction is violent. It needs to control the reaction temperature.

(iii) Because hydrogen chloride gas is produced in the reaction, the reaction device should be connected with the tail gas absorption device.

ii. Thinking questions

(i) In addition to using phosphorus trichloride as a chlorination agent, which chlorination agents can be used in this reaction?

(ii) What are the precautions of the atmospheric distillation process?

实验十四　紫罗兰酮的合成

一、实验目的

1. 掌握紫罗兰酮的合成反应机理。
2. 熟悉紫罗兰酮的实验操作过程。
3. 了解紫罗兰酮的理化性质及用途。

二、实验原理

（一）化合物简介

紫罗兰酮是一种有机化合物，分子式为 $C_{13}H_{20}O$，无色至微黄色液体。沸点 237℃，闪点 115℃。不溶于水和甘油，溶于乙醇、丙二醇、大多数非挥发性油和矿物油。具有较强的紫罗兰香气，稀释后呈鸢尾根香气，再与乙醇混合，则呈紫罗兰香气，香味比紫罗兰酮好。

天然紫罗兰酮存在于金合欢油、桂花浸膏等中。主要用以配制龙眼、树莓、黑莓、樱桃、柑橘等型香精。

（二）合成工艺路线

紫罗兰酮的合成，主要通过柠檬醛与丙酮，在稀氢氧化钠溶液中缩合，得到中间体假紫罗兰酮，中间体不需要完全纯化，即可在硫酸作用下，环化成为紫罗兰酮。其合成工艺路线如下：

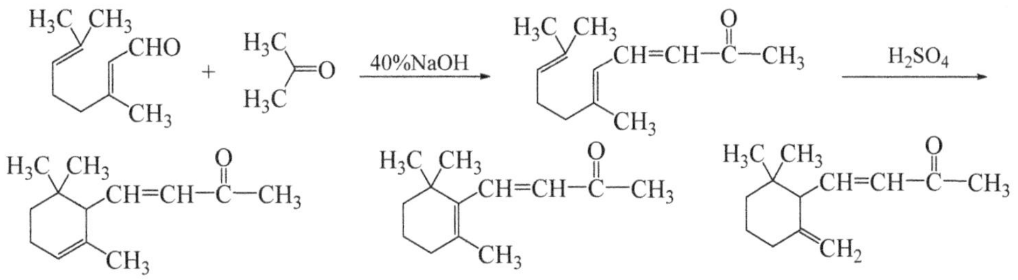

三、主要仪器和试剂

1. 主要仪器：三颈烧瓶、磁力搅拌器、温度计、分析天平、恒压滴液漏斗、冷凝管、分液漏斗、烧杯、橡胶管、锥形瓶、抽滤瓶、布氏漏斗、滤纸、真空泵等。
2. 主要试剂：浓硫酸、甲苯、15%碳酸钠、50%乙酸、氯化钠、柠檬醛、丙酮、45%氢氧化钠等。

四、实验步骤

在配有磁力搅拌器、温度计、冷凝管、干燥管以及恒压滴液漏斗的 100 mL 三颈烧瓶中，加入丙酮 30 g、45%氢氧化钠 2 mL，快速搅拌，升温至 50℃，随后，缓慢滴加柠檬醛 10 g，滴加结束后，保温反应 2 h，用 50%乙酸溶液调节 pH 至 5，常压蒸馏，除去过量的丙酮。剩余物用 10%氯化钠 20 mL 充分洗涤后，蒸馏，得到粗产物，待用。

将上一步制得粗产物，加至 55%硫酸溶液 8 mL、甲苯 14 mL，在室温下，快速搅拌 10 min，加水适量，用分液漏斗分出有机层，随后用 15%碳酸钠溶液 5 mL，进行洗涤，蒸去甲苯，剩余反应物，减压蒸馏，收集 120~130℃的馏分。

五、注意事项与思考题

1. 试写出本反应的反应机理。
2. 复习减压蒸馏的操作过程。

Experiment 14 Synthesis of ionone

I. Purpose of the experiment

i. To master the synthetic reaction mechanism of ionone.

ii. To familiar with the experimental operation of ionone.

iii. To understand the physicochemical properties and uses of ionone.

II. Experimental principle

i. Compound introduction

Ionone is an organic compound with the formula $C_{13}H_{20}O$. It is a colorless to yellowish liquid. It has a boiling point of 237℃ and a flash point of 115℃. It is insoluble in water and glycerol, soluble in ethanol, propylene glycol, most non-volatile and mineral oils. It has a strong violet aroma, diluted with iris root aroma, and mixed with ethanol. It has a violet aroma, which is better than ionone.

Natural ionone is found in acacia oil, osmanthus extract, *etc*. It is mainly used to prepare longan, raspberry, blackberry, cherry, citrus and other flavors.

ii. Synthetic process route

The synthesis of ionone is mainly through the condensation of citral and acetone in a dilute sodium hydroxide solution to obtain the intermediate pseudo ionone. The intermediate does not need to be completely purified, but can be cycled into ionone under sulfuric acid. The synthesis process is as follows:

III. Main instruments and reagents

i. Main instruments: three-necked round-bottom flask, magnetic stirrer, thermometer, analytical balance, constant pressure drip funnel, condensing tube, separation funnel, beaker, rubber tube, conical bottle, suction bottle, Brinell funnel, filter paper, vacuum pump, *etc*.

ii. Main reagents: concentrated sulfuric acid, toluene, 15% sodium carbonate, 50% acetic acid, sodium chloride, citral, acetone, 45% sodium hydroxide, *etc*.

IV. Experimental steps

Add 30 g of acetone and 2 mL of 45% sodium hydroxide into a 100 mL three-necked round-bottom flask equipped with a magnetic stirrer, a thermometer, a condensing tube, a drying tube and a constant pressure drip funnel. Stir quickly. Heat to 50 ℃. Then slowly add 10 g of citric aldehyde. After the addition, maintain the reaction for 2 h. Adjust the pH to 5 with 50% acetic acid solution. Remove the excess acetone by atmospheric distillation. Wash the residue thoroughly with 20 mL of 10% sodium chloride, then distill to obtain the crude product for use.

Add the crude product prepared in the previous step to 8 mL of 55% sulfuric acid solution and 14 mL of toluene. Stir rapidly at room temperature for 10 minutes, then add appropriate amount of water and separate the organic layer by a liquid separation funnel. Wash with 5 mL of 15% sodium carbonate solution, evaporate the toluene, and perform a reduced pressure distillation to collect the fraction at 120-130 ℃.

V. Attention or thinking questions

i. Try to write the reaction mechanism of this reaction.
ii. Review the operation process of vacuum distillation.

实验十五 α-呋喃丙烯酸的合成

一、实验目的

1. 掌握α-呋喃丙烯酸的合成反应机理。
2. 熟悉反应中的调 pH 值、水洗滤饼的基本操作过程。
3. 了解α-呋喃丙烯酸的理化性质及用途。

二、实验原理

（一）化合物简介

α-呋喃丙烯酸，化学品名为 3-(2-呋喃基)-2-丙烯酸，分子式为 $C_7H_6O_3$，分子量 138.12，白色粉末或针状结晶，溶于乙醇、乙醚、苯和乙酸。

α-呋喃丙烯酸是合成呋喃类香料，α-呋喃丙烯酸和其酯化产物——呋喃丙烯酸酯类。在食用和日化产品中，有广泛用途的呋喃杂环类合成香料，香气特征强，阈值低，用量少，增香效果明显。以焦糖甜香和水果香气为其香气特征。呋喃类香料可以作为多种食用类香精的调香原料，广泛用于糖果、软饮料、冰制食品、烘焙食品中，是国内香精香料及日化产业，研究开发的重要品种。

合成抗血吸虫病口服药——呋喃苯胺，对于治疗血吸虫病有较好效果。合成其他化工原料呋喃丙烯酸，还可用于合成庚酮二酸、庚二酸、乙烯呋喃及其酯类等重要化工原料。该物质对环境可能有危害，对水体应给予特别注意。

（二）合成工艺路线

α-呋喃丙烯酸的合成，主要通过呋喃甲醛与丙二酸在吡啶存在的情况下反应制备，其合成工艺路线如下：

$$\text{furan-CHO} + \begin{matrix}\text{COOH}\\ \text{CH}_2\\ \text{COOH}\end{matrix} \xrightarrow{\text{吡啶}} \text{furan-CH=CHOOH}$$

三、主要仪器和试剂

1. 主要仪器：三颈烧瓶、回流冷凝器、温度计、抽滤瓶、布氏漏斗、分析天平、烧杯、真空泵等。
2. 主要试剂：呋喃甲醛、丙二酸、吡啶、浓盐酸、18%盐酸、浓氨水、乙醇等。

四、实验步骤

在配有磁力搅拌器、温度计、冷凝管、干燥管的 100 mL 三颈烧瓶中，加入呋喃甲醛 4.9 g、丙二酸 5.3 g、吡啶 2.5 mL，快速搅拌，升温至 100℃，反应 2 h。反应结束后，冷却至室温，将反应液倒入烧杯中，同时，向烧杯中加水 50 mL，加浓氨水，最终使得反应物溶解，抽滤，水洗滤纸，合并滤液。随后，在快速搅拌下，加入约 10 mL 的稀盐酸（浓度 18%），调 pH 至 3，冷却后，抽滤，将滤饼用水洗 3 次，干燥，得到 α-呋喃丙烯酸粗产品，同时，可以用乙醇进行重结晶。

五、注意事项与思考题

（一）注意事项

1. 试写出本实验的反应机理。
2. 本反应对无水要求特别严格，对所用到的试剂、反应器皿等需要做无水处理。

（二）思考题

本实验中用到的吡啶试剂作用是什么？

Experiment 15 Synthesis of α-furanacrylic acid

I. Purpose of the experiment

i. To master the synthetic reaction mechanism of α-furanacrylic acid.

ii. To familiar with the basic operation process of pH adjustment and washing filter cake in reaction.

iii. To understand the physicochemical properties and applications of α-furanacrylic acid.

II. Experimental principle

i. Compound introduction

α-furanacrylic acid, chemically named 3-(2-furanyl) -2-acrylic acid, has the formula $C_7H_6O_3$ and molecular weight of 138.12, as a white powder or acicular crystal. It can be dissolved in ethanol, ether, benzene, and acetic acid.

α-furan acrylic acid is a synthetic furan flavor. α-furan acrylic acid and its esterification product-furan acrylic acid esters are widely used in food and daily chemical products. Furan heterocyclic synthetic flavor, strong aroma characteristics, low threshold, less dosage, increasing fragrance effect is obvious. It is characterized by caramel sweetness and fruit aroma. Furan flavor can be used as a variety of edible flavor flavoring raw materials, widely used in candy, soft drinks, ice food, baked food, the domestic flavor and chemical industry, research and development of important varieties.

The synthetic oral drug against schistosomiasis, furan aniline, has a good effect on the treatment of schistosomiasis. Synthesis of other chemical raw materials of furan acrylic acid, but also can be used in the synthesis of hydrochelidonic acid, pimelic acid, vinyl furan and its esters and other important chemical raw materials. The substance may be harmful to the environment, and special attention should be paid to the water body.

ii. Synthetic process route

The synthesis of α-furan acrylic acid is mainly prepared by the reaction of furan formaldehyde and malonic acid in the presence of pyridine. The specific synthesis process is as follows:

$$\text{furan-CHO} + \underset{\text{COOH}}{\overset{\text{COOH}}{\text{CH}_2}} \xrightarrow{\text{pyridine}} \text{furan-CH=CHOOH}$$

III. Main instruments and reagents

i. Main instruments: round-bottoming flask, reflux condenser, thermometer, filter flask, Brinell funnel, analytical balance, beaker, vacuum pump *etc*.

ii. Main reagents: furan formaldehyde, malonic acid, pyridine, concentrated hydrochloric acid, 18% hydrochloric acid, ammonia, ethanol, *etc*.

IV. Experimental steps

Add 4.9 g of furan formaldehyde, 5.3 g of malonic acid and 2.5 mL of pyridine into a 100 mL three-necked round-bottom flask equipped with a magnetic agitator, thermometer, condensing tube and drying tube. Stir quickly and heated to 100℃ for 2 h. After the reaction, cool to room temperature, and pour the reaction liquid into the beaker. At the same time, add 50 mL of water into the beaker. Add concentrated ammonia water, and finally make the reactants dissolve. Filter, wash the filter paper with water, and combine the filtrate. Then, under rapid stirring, add about 10 mL of dilute hydrochloric acid (concentration 18%) to adjust the pH to 3. After cooling, filter, wash the filter cake with water 3 times. Dry to obtain crude α-furanacrylic acid. At the same time, it can be recrystallized with ethanol.

V. Attention or thinking questions

i. Attention

(i) Try to write the reaction mechanism of this experiment.

(ii) This reaction is very strict on anhydrous requirements, reagents used, reaction vessels, *etc*. It needs to do anhydrous treatment.

ii. Thinking questions

What is the action of the pyridine reagent used in this experiment?

实验十六　4-溴-2-萘酚的合成

一、实验目的

1. 掌握 4-溴-2-萘酚的合成反应机理。
2. 熟悉 4-溴-2-萘酚的实验操作流程。
3. 了解 4-溴-2-萘酚的理化性质及用途。

二、实验原理

（一）化合物简介

4-溴-2-萘酚又名 1-溴-3-羟基萘，化学式为 $C_{10}H_7BrO$，是一种非常重要的医药、化工中间体。

（二）合成工艺路线

4-溴-2-萘酚的合成，主要通过 α-萘胺为原料制得，其合成工艺路线如下：

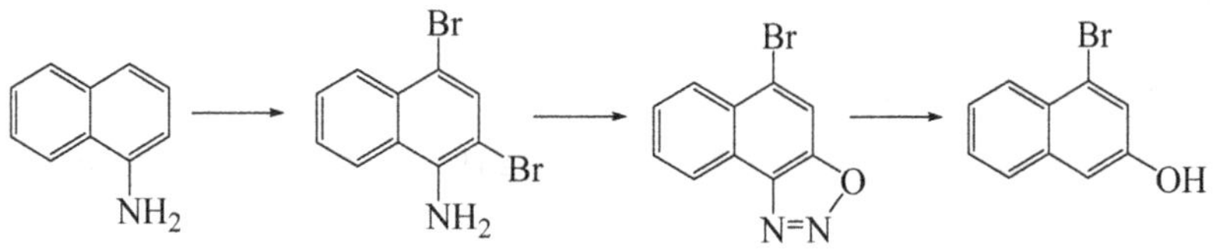

三、主要仪器和试剂

1. 主要仪器：三颈烧瓶、磁力搅拌器、温度计、恒压滴液漏斗、冷凝管、干燥管、分液漏斗、烧杯、橡胶管、锥形瓶、抽滤瓶、布氏漏斗、分析天平、滤纸、真空泵等。
2. 主要试剂：溴、羧甲基纤维素、石油醚、乙醇、α-萘胺、二氯甲烷、乙酸乙酯、乙酸、NaOH、$NaNO_2$、$NaBH_4$、浓盐酸、丙酸、Br_2、硅胶等。

四、实验步骤

（一）2,4-二溴萘胺的合成

在配有磁力搅拌器、温度计、冷凝管、干燥管以及恒压滴液漏斗的 100 mL 三颈烧瓶中，加入 α-萘胺 5.6 g、乙酸 40 mL，快速搅拌，用冰水浴冷却至 0~5℃，再滴入 Br_2 与乙酸的混合溶液（Br_2 9 mL +乙酸 80 mL），随后，再向反应体系中加入乙酸 40 mL，升温至 60℃，保温反应 30 min，反应结束后，用冰水浴冷却，过滤，滤饼用冰乙酸充分洗涤后，滤饼加入过量稀 NaOH 溶液，剧烈搅拌 30 min，抽滤，用水充分洗涤滤饼，得到固体，可以用乙醇进行重结晶，即为 2,4-二溴萘胺纯品。

（二）4-溴-2-萘酚的合成

在配有磁力搅拌器、温度计、冷凝管、干燥管的 250 mL 三颈烧瓶中，加入上步制得的 2,4-二溴萘胺 10 g、乙酸 150 mL 以及丙酸 30 mL，快速搅拌，冰浴降温至 8～10℃，随后加入 2.7 g $NaNO_2$，随后搅拌反应 30 min 后，反应液呈黄褐色，将其全部倒入 200 mL 冰水中，搅拌过滤，将滤液加至 3000 mL 水中，有黄色固体析出，过滤，得粗产物，供下步反应使用。

在配有磁力搅拌器、温度计、冷凝管、干燥管的 250 mL 三颈烧瓶中，加入上步制得的待用产物 5.1 g、乙醇 90 mL，将反应体系冷却至 10℃ 以下，随后加入与上步制得的待用产物相同物质量的 $NaBH_4$，快速搅拌反应 3 h，直至无气体产生。将反应液倒入盐酸溶液中（浓盐酸 5 mL+水 500 mL），加入 10% NaOH 至溶液呈强碱性，再用二氯甲烷萃取其杂质（每次 50 mL，共 3 次），水层酸化至酸性，有固体析出，抽滤，干燥，得粗产物。若无固体析出，则用二氯甲烷进行萃取，减压蒸干溶剂，即得到粗产物。

五、注意事项与思考题

（一）注意事项

试写出本反应的反应机理。

（二）思考题

1. 加入丙酸的主要作用是什么？
2. 本反应中使用 $NaBH_4$ 作为还原剂，还有哪些还原剂在本反应中适用？

Experiment 16　Synthesis of 4-bromo-2-naphthol

I. Purpose of the experiment

　　i. To master the synthesis mechanism of 4-bromo-2-naphthol.
　　ii. To familiar with the experimental operation process of 4-bromo-2-naphthol.
　　iii. To understand the physicochemical properties and applications of 4-bromo-2-naphthol.

II. Experimental principle

i. Compound introduction

　　4-bromo-2-naphthol, also known as 1-bromo-3-hydroxyl naphthol, has the chemical formula $C_{10}H_7BrO$. It is a very important intermediate in medicine and chemical industry.

ii. Synthetic process route

　　4-bromo-2-naphthol is synthesized mainly by α-naphthol, and the synthesis process is as follows:

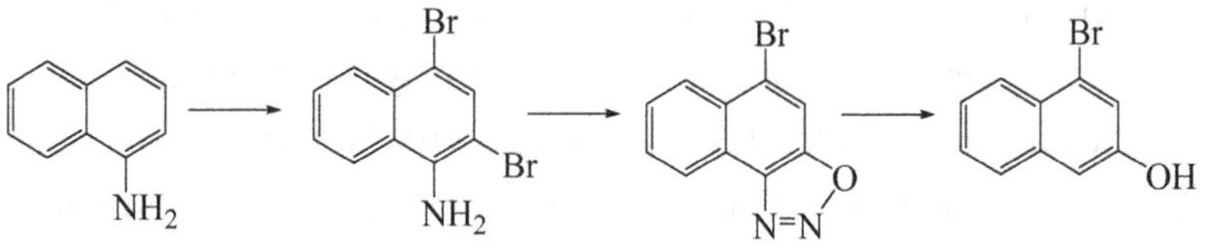

III. Main instruments and reagents

　　i. Main instruments: three-necked round-bottom flask, magnetic stirrer, thermometer, constant pressure drip funnel, condensing tube, drying tube, separation funnel, beaker, rubber tube, conical flask, suction bottle, Brinell funnel, analytical balance, filter paper, vacuum pump, *etc.*
　　ii. Main reagents: bromine, carboxymethyl cellulose, petroleum ether, ethanol, α-naphthalamine, dichloromethane, ethyl acetate, acetic acid, NaOH, $NaNO_2$, $NaBH_4$, concentrated hydrochloric acid, propionic acid, Br_2, silica gel, *etc.*

IV. Experimental steps

i. Synthesis of 2, 4-dibromonaphthylamine

　　Add 5.6 g of α-naphthylamine and 40 mL of acetic acid into a 100 mL three-necked round-bottomed flask equipped with a magnetic agitator, a thermometer, a condensing tube, a drying tube and a constant-pressure drip funnel. Stir rapidly and cool to 0-5℃ using an ice water bath. Drop the mixture of Br_2 and acetic acid (9 mL of Br_2 +80 mL of acetic acid). Then add 40 mL acetic acid to the reaction system, heat to 60 ℃, and maintain the reaction for 30 min. After the reaction, cool with an ice water bath and filter. After thoroughly wash the cake with ice acetic acid, add an excess of dilute NaOH solution to the cake, and stir vigorously for 30 min. Wash the filter cake thoroughly with water to obtain a solid, which can be recrystallised using ethanol to pure 2, 4-dibromonaphthylamine.

ii. Synthesis of 4-bromo-2-naphthol

Add 10 g of 2, 4-dibromonaphthylamine, 150 mL of acetic acid and 30 mL of propionic acid prepared in the previous step into a 250 mL three-neck round-bottom flask equipped with a magnetic stirrer, a thermometer, a condensation tube and a drying tube. Stir quickly and cool to 8-10 ℃ with an ice water bath. Then add 2.65 g of $NaNO_2$. After stirring and reaction for 30 min, the reaction liquid is yellowish brown. Pour the entire mixture into 200 mL of ice water. Stir and filter. Add the filtrate to 3000 mL of water, where a yellow solid precipitates. Filter to obtain the crude product for the next reaction.

Add 5.1g of the suspended product prepared in the previous step and 90 mL of ethanol into a 250 mL three-neck round-bottom flask equipped with a magnetic agitator, a thermometer, a condensing tube and a drying tube. Cool the reaction system to below 10 ℃. Add $NaBH_4$ of the same weight as the suspended product prepared in the previous step for rapid stirring and reaction for 3 h until no gas produced. Pour the reaction solution into hydrochloric acid solution (5 mL of concentrated hydrochloric acid +500 mL of water). Add 10% NaOH until the solution is strongly alkaline, and then extract impurities (50 mL × 3). Acidify the aqueous layer to acidic, where a solid precipitates. Perform suction filtration, and obtain the crude product. If no solid precipitates, extract with dichloromethane and evaporate the solvent under reduced pressure to obtain the crude product.

V. Attention or thinking questions

i. Attention

Try to write the reaction mechanism of this reaction.

ii. Thinking questions

(i) What is the main function of adding propionic acid?
(ii) Bestdes $NaBH_4$, what other reducing agents are suitable for this reaction?

实验十七　对硝基苯甲醛的合成

一、实验目的

1. 掌握对硝基苯甲醛的合成反应机理。
2. 熟悉对硝基苯甲醛的实验操作过程。
3. 了解对硝基苯甲醛的理化性质及用途。

二、实验原理

（一）化合物简介

对硝基苯甲醛是一种重要的有机化合物，化学式为 $C_7H_5NO_3$，为白色或淡黄色棱状结晶。熔点 105～108℃，相对密度 1.496。溶于乙醇、苯和乙酸，微溶于水、乙醚。能升华，能随水蒸气挥发。

对硝基苯甲醛为染料、医药等有机合成的中间体，在医药工业中用来生产硝基苯丁烯酮，还常用于合成氨苯硫脲、甲氧苄胺嘧啶和乙酰胺等。

（二）合成工艺路线

对硝基苯甲醛的合成，是通过对硝基甲苯、乙酸酐作为原料，经过氧化反应、水解反应制备，其合成工艺路线如下：

对硝基甲苯 $\xrightarrow{CrO_3/Ac_2O}$ 对硝基苄叉二乙酸酯 $\xrightarrow{H_2O/H_2SO_4}$ 对硝基苯甲醛

三、主要仪器和试剂

1. 主要仪器：三颈烧瓶、回流冷凝器、分析天平、滴液漏斗、温度计、水浴锅、抽滤瓶、布氏漏斗、烧杯、真空泵等。
2. 主要试剂：对硝基甲苯、乙酸酐、铬酸酐、浓硫酸、2%碳酸钠、乙醇等。

四、实验步骤

在配有磁力搅拌器、温度计、冷凝管、干燥管以及恒压滴液漏斗的 100 mL 三颈烧瓶中，加入对硝基甲苯 6.3 g、乙酸酐 50 mL，用冰盐浴冷却后，加入浓硫酸 10 mL，随后冷却至 0℃，快速搅拌，加入由铬酸酐与乙酸酐配制的溶液（铬酸酐 12.5 g +乙酸酐 57 mL），缓慢滴加，保证反应温度在 10℃以下，滴加结束后，温度在 5～10℃，保温反应 2 h。反应结束后，将反应液倒入 250 g 碎冰中，剧烈搅拌后，加冰水稀释至 750 mL，有固体析出，抽滤，将滤饼置于烧瓶中，向其中加入 40 mL 2%碳酸钠溶液，剧烈搅拌，抽滤，滤饼用水、乙醇分别充分洗涤，抽滤，干燥，得产物。

将上步得到的产物放入三颈烧瓶中，加入乙醇、水分别 20 mL，浓硫酸 2 mL，装上冷凝管，回流反应 1 h，回流结束后，将反应液趁热抽滤，滤液，冷却，析晶，随后抽滤，滤饼干燥，得产品。

五、注意事项与思考题

（一）注意事项

1. 试写出本实验的反应机理。
2. 注意铬酸酐溶液配制时的搅拌速度，同时，注意加入顺序，应是铬酸酐加至乙酸酐中。

（二）思考题

本反应中使用铬酸酐作为氧化剂，还可以用哪些氧化剂能够氧化本反应？

Experiment 17 Synthesis of *p*-nitro benzaldehyde

Ⅰ. Purpose of the experiment

i. To master the synthetic reaction mechanism of *p*-nitro benzaldehyde.

ii. To familiar with the experimental operation process of *p*-nitro benzaldehyde.

iii. To understand the physicochemical properties and applications of *p*-nitro benzaldehyde.

Ⅱ. Experimental principle

i. Compound introduction

p-nitro benzaldehyde is an important organic compound with the chemical formula $C_7H_5NO_3$. It is a white or pale yellow crystalline solids. Its melting point is 105-108 ℃ and relative density is 1.496. It is soluble in ethanol, benzene and acetic acid, slightly soluble in water, ether. It can sublimate, and can evaporate with water vapor.

p-nitro benzaldehyde is used for dye, medicine and other organic synthesis intermediates, used in the pharmaceutical industry to produce nitrophenyl butene ketone, in the pharmaceutical industry. And it is often used in the synthesis of phenylthiocarbamide, trimethoprim and acetamide.

ii. Synthetic process route

The synthesis of *p*-nitro benzaldehyde is prepared by using *p*-nitrotoluene and acetic anhydride as raw materials through oxidation reaction and hydrolysis reaction. The specific reaction process is as follows:

$$\underset{NO_2}{\underset{|}{C_6H_4}}-CH_3 \xrightarrow{CrO_3/Ac_2O} \underset{NO_2}{\underset{|}{C_6H_4}}-CH(OAc)_2 \xrightarrow{H_2O/H_2SO_4} \underset{NO_2}{\underset{|}{C_6H_4}}-CHO$$

Ⅲ. Main instruments and reagents

i. Main instruments: three-necked flask, reflux condenser, analytical balance, drip funnel, thermometer, water bath, filter flask, Brinell funnel, beaker, vacuum pump, *etc*.

ii. Main reagents: *p*-nitrotoluene, acetic anhydride, chromic anhydride, concentrated sulfuric acid, 2% sodium carbonate, ethanol *etc*.

Ⅳ. Experimental steps

Add 6.3 g of *p*-nitrotoluene and 50 mL of acetic anhydride into a 100 mL round-bottom flask equipped with a magnetic stirrer, a thermometer, a condensing tube, a drying tube and a constant pressure drip funnel. After cooling in an ice salt bath, add 10 mL of concentrated sulfuric acid, then cool to 0 ℃ and stir quickly. Add the solution prepared by chromic anhydride and acetic anhydride (12.5 g of chromic anhydride +57 mL of acetic anhydride). Slowly add to ensure that the reaction temperature is below 10 ℃. After the addition, maintain the temperature between 5-10 ℃, holding the reaction for 2 h. After the reaction, pour the reaction liquid into 250 g of crushed ice. Stir vigorously. Dilute with ice water to 750 mL. A solid will precipitate. Filter it cut and place the filter cake in the flask. Add 40 mL of 2% sodium carbonate solution. Stir vigorously. Filter and wash filter the cake thoroughly with water and ethanol. Dry to obtain the product.

Put the product obtained in the above step into a three-neck reaction bottle, add 20 mL of ethanol, 20mL of water and 2 mL of concentrated sulfuric acid. Set up a condenser and reflux for 1 h. After refluxing, filter the hot reaction mixture. cool the filtrate to crystallise. Then filter again and dry the filter cake to obtain the product.

Ⅴ. Attention or thinking questions

i. Attention

(i) Try to write the reaction mechanism of this experiment.

(ii) Pay attention to the stirring speed when the chromic anhydride solution is prepared, and at the same time, pay attention to the adding sequence, which should be chromic anhydride added to acetic anhydride.

ii. Thinking questions

Chromic anhydride is used as an oxidant in this reaction. What oxidants can be used to oxidize this reaction?

实验十八　2-庚酮的合成

一、实验目的

1. 掌握 2-庚酮的合成反应机理。
2. 熟悉 2-庚酮的实验操作过程。
3. 了解 2-庚酮的理化性质及用途。

二、实验原理

（一）化合物简介

2-庚酮又名甲基戊基酮,是一种重要的有机化合物,化学式为 $C_7H_{14}O$,为无色透明液体,有类似梨的水果香味,不溶于水,可混溶于多种有机溶剂,主要用作硝化纤维素的溶剂和涂料、惰性反应介质,也用作香料原料。

2-庚酮储存于阴凉、通风的库房,库温不宜超过 37℃,远离火种、热源,保持容器密封,应与氧化剂、还原剂、碱类分开存放,切忌混储。采用防爆型照明、通风设施,禁止使用易产生火花的机械设备和工具,储存区域应备有泄漏应急处理设备和合适的收容材料。

（二）合成工艺路线

2-庚酮的合成,主要通过正丁基乙酰乙酸乙酯为原料,进行皂化、酸化、脱羧形成 2-庚酮,其合成工艺路线如下：

$$CH_3COCHCO_2C_2H_5 + NaOH \longrightarrow CH_3COCHCO_2Na + C_2H_5OH$$
$$||$$
$$CH_2(CH_2)_2CH_3 CH_2(CH_2)_2CH_3$$

$$CH_3COCHCO_2Na + H_2SO_4 \longrightarrow 2\ CH_3CO(CH_2)_4CH_3 + 2\ CO_2 + Na_2SO_4$$
$$|$$
$$CH_2(CH_2)_2CH_3$$

三、主要仪器和试剂

1. 主要仪器：磁力搅拌器、温度计、三颈烧瓶、恒压滴液漏斗、冷凝管、干燥管、分析天平、分液漏斗、烧杯、橡胶管、锥形瓶、真空泵等。
2. 主要试剂：5% NaOH、浓硫酸、正丁基乙酰乙酸乙酯、氢氧化钠、饱和氯化钠、无水硫酸镁等。

四、实验步骤

在配有磁力搅拌器、温度计、冷凝管、干燥管的 250 mL 三颈烧瓶中,加入 5% NaOH 溶液 100 mL、正丁基乙酰乙酸乙酯 14 g,快速搅拌,升温至回流,保温反应 2 h。随后缓慢滴加 50%硫酸,直至反应体系为弱酸性。将反应液倒入烧瓶中,进行蒸馏,直至馏出液无油状物时，停止蒸馏。馏出液在搅拌下加适量氢氧化钠，使呈碱性，蒸馏至馏出液无油状物时，停止蒸馏。

馏出液用分液漏斗分离出有机层，随后用饱和氯化钠洗涤，除去多余的乙醇，有机层用无水硫酸镁进行充分干燥，过夜，抽滤，转至蒸馏装置中，进行蒸馏，收集 146~153℃馏分，即为 2-庚酮。

五、注意事项与思考题

1. 试写出本反应的反应机理。
2. 硫酸加入时会产生大量二氧化碳气体，一定要缓慢加入。

Experiment 18 Synthesis of 2-heptanone

I. Purpose of the experiment

i. To master the synthetic reaction mechanism of 2-heptanone.

ii. To familiar with the experimental operation process of 2-heptanone.

iii. To understand the physicochemical properties and uses of 2-heptanone.

II. Experimental principle

i. Compound introduction

2-heptanone, also known as methyl amyl ketone, is an important organic compound with the chemical formula $C_7H_{14}O$. It is a colorless, transparent liquid with a pear fruit flavor to pears. It is insoluble in water and miscible in a variety of organic solvents.

2-heptanone should be stored in a cool and ventilated warehouse. The temperature should not exceed 37℃. It should keep away from tinder and heat source. Keep the container sealed. It should be separated from oxidant, reducing agent, alkali storage. Do not mix storage. Explosion-proof lighting and ventilation facilities shall be adopted. Mechanical equipment and tools that are easy to produce sparks shall be prohibited. The storage area should be equipped with leakage emergency treatment equipment and appropriate holding materials.

ii. Synthetic process route

The synthesis of 2-heptanone mainly forms 2-heptanone through saponification, acidification and decarboxylation by ethyl n-butyl acetoacetate. The synthesis process is as follows:

$$CH_3COCH(CH_2(CH_2)_2CH_3)CO_2C_2H_5 + NaOH \longrightarrow CH_3COCH(CH_2(CH_2)_2CH_3)CO_2Na + C_2H_5OH$$

$$CH_3COCH(CH_2(CH_2)_2CH_3)CO_2Na + H_2SO_4 \longrightarrow 2\ CH_3CO(CH_2)_4CH_3 + 2\ CO_2 + Na_2SO_4$$

III. Main instruments and reagents

i. Main instruments: magnetic stirrer, thermometer, three-necked flask, constant pressure drip funnel, condensing tube, drying tube, analytical balance, separation funnel, beaker, rubber tube, conical flask, vacuum pump, *etc*.

ii. Main reagents: 5%NaOH, concentrated sulfuric acid, saturated sodium chloride solution, anhydrous magnesium sulfate, *n*-butyl acetoacetate, sodium hydroxide, *etc*.

IV. Experimental steps

Add 100 mL of 5% NaOH aqueous solution and 14 g of *n*-butyl ethyl acetoacetate into a 250 mL three-necked round-bottomed flask equipped with magnetic agitators, thermometers, condensing tubes and drying tubes. Stir quickly and heat to reflux, maintaining the reaction for 2 h. Then slowly add 50% sulfuric acid until the reaction system is weakly acidic. Pour the reaction liquid into a flask and distil until no aily is present in the distillate. Stop the distillation. Under stirring, add an appropriate amount of sodium hydroxide to make it alkaline, and distil until no aily is present in the distillate. Stop the distillation.

Separate the organic layer from the distillate with a liquid separation funnel, and then wash with saturated sodium chloride to remove excess ethanol. The organic layer is thoroughly dried with anhydrous magnesium sulfate overnight, filtered, and transferred to the distillation unit for distillation. Collect the 146-153℃ distillate.

V. Attention or thinking questions

i. Try to write the reaction mechanism of this reaction.
ii. Sulfuric acid will produce a lot of carbon dioxide gas. It must be added slowly.

实验十九　查尔酮的合成

一、实验目的

1. 掌握查尔酮缩合反应的机理。
2. 熟悉查尔酮合成中的实验操作过程。
3. 了解查尔酮的理化性质及用途。

二、实验原理

（一）化合物简介

查尔酮是一种重要的有机化合物，分子式为 $C_{15}H_{12}O$，分子量为 208.26。外观呈微淡黄色斜方或菱形结晶。易溶于醚、氯仿、二硫化碳和苯，微溶于醇，难溶于冷石油醚。

查尔酮可以在碱性条件下由苯乙酮与苯甲醛缩合而成。主要用作有机合成试剂和指示剂，同时有药物活性。

（二）合成工艺路线

查尔酮的合成，以苯甲醛与苯乙酮为原料，在碱性条件下制备，其合成工艺路线如下：

$$\text{PhCHO} + \text{PhCOCH}_3 \xrightarrow[-H_2O]{NaOH} \text{Ph-CH=CH-CO-Ph}$$

三、主要仪器和试剂

1. 主要仪器：三颈烧瓶、回流冷凝器、恒压滴液漏斗、温度计、分析天平、水浴锅、抽滤瓶、布氏漏斗、烧杯、玻璃棒、真空泵等。
2. 主要试剂：苯甲醛、苯乙酮、95%乙醇、氢氧化钠、查尔酮、乙酸乙酯等。

四、实验步骤

在配有磁力搅拌器、温度计、冷凝管以及恒压滴液漏斗的 100 mL 三颈烧瓶中，加入苯乙酮 10.4 g、95%乙醇 25 mL 以及氢氧化钠 4.4 g（溶于 40 mL 水），快速搅拌，缓慢滴加苯甲醛 9.2 g。在缓慢滴加中，保证反应的温度小于 25℃。滴加结束后，继续反应 1 h，加入查尔酮晶种少许，继续反应 1.5 h，有固体析出，减压抽滤，滤饼用水充分洗涤至中性，干燥，得粗产物。粗产物可以用乙酸乙酯进行重结晶，得到浅黄色固体。

五、注意事项与思考题

1. 本缩合反应的机理是什么？
2. 本反应析晶过程中，加入晶种的主要作用什么？

Experiment 19   Synthesis of chalcone

I. Purpose of the experiment

    i. To master the mechanism of Chalcone condensation reaction.
    ii. To familiar with the experimental operation process of chalcone synthesis.
    iii. To understand the physicochemical properties and uses of chalcone.

II. Experimental principle

i. Compound introduction

    Chalcone is an important organic compound with the molecular formula $C_{15}H_{12}O$ and a molecular weight of 208.26. It appears as pale yellow rhombic or prismatic crystal. It is easily soluble in ether, chloroform, carbon disulfide and benzene, slightly soluble in alcohol, and poorly soluble in cold petroleum ether.
    Chalcone can be formed by condensation of acetophenone with benzaldehyde under alkaline conditions. It is mainly used as an organic synthesis reagent and indicator, and has drug activity.

ii. Synthetic process route

    The synthesis of Chalcone is mainly through the preparation of benzaldehyde and acetophenone under alkaline conditions. The synthesis process is as follows:

$$\text{Ph-CHO} + \text{Ph-COCH}_3 \xrightarrow[-H_2O]{NaOH} \text{Ph-CH=CH-CO-Ph}$$

III. Main instruments and reagents

    i. Main instruments: three-necked flask, reflux condenser, drip funnel, thermometer, analytical balance, water bath, filter flask, Brinell funnel, beaker, glass rod, vacuum pump, *etc*.
    ii. Main reagents: benzaldehyde, acetophenone, 95% ethanol, sodium hydroxide, chalcone, ethyl acetate *etc*.

IV. Experimental steps

    Add 10.4 g of acetone, 25 mL of 95% ethanol and 4.4 g of sodium hydroxide (dissolved in 40 mL of water) into a 100 mL three-necked round bottling flask equipped with magnetic stirrers, thermometers, condensing tubes, drying tubes and constant pressure drip funnel. Stir rapidly. Slowly add 9.2 g of benzaldehyde. In the addition, the reaction temperature is ensured to be kept below 25℃. After the addition, the reaction continues for 1 h. Add a little chalcone seed. And the reaction continues for 1.5 h. Some solids are precipitate. Thoroughly wash the filter cake with water until neutral. Dry to obtain the crude product. The crude product can be recrystallized with ethyl acetate to produce a yellowish solid.

V. Attention or thinking questions

i. What is the mechanism of this condensation reaction?

ii. What is the main function of adding seeds in the process of crystallization in this reaction?

实验二十　己酸异戊酯的合成

一、实验目的

1. 掌握己酸异戊酯的合成反应机理。
2. 熟悉己酸异戊酯的实验操作过程。
3. 了解己酸异戊酯的理化性质及用途。

二、实验原理

（一）化合物简介

己酸异戊酯是一种重要的有机化合物，化学式为 $C_{11}H_{22}O_2$，无色液体，呈苹果和菠萝似香味。沸点 222℃，闪点 88℃。溶于乙醇、非挥发性油和矿物油，不溶于丙二醇、水和甘油。

己酸异戊酯有水果香气，天然存在于香蕉、苹果、草莓等水果中，还存在于甜橙油和多种酒类中。

（二）合成工艺路线

己酸异戊酯的合成，以己酸与异戊醇为原料，在对甲苯磺酸的催化条件下制得，其合成工艺路线如下：

三、主要仪器和试剂

1. 主要仪器：三颈烧瓶、加热套、分析天平、分液漏斗、回流冷凝管、分水器、温度计、冷凝管、搅拌子、磁力搅拌器、烧杯、橡胶管、锥形瓶、沸石、真空泵等。
2. 主要试剂：己酸、对甲苯磺酸、异戊醇、10%碳酸钠、饱和氯化钠、无水硫酸镁等。

四、实验步骤

在配有磁力搅拌器、温度计、冷凝管以及恒压滴液漏斗的 100 mL 三颈烧瓶中，分别加入己酸 11.6 g、异戊醇 26.4 g 以及对甲苯磺酸 0.7 g，快速搅拌，升温至 130～150℃，回流反应至无水蒸馏。

反应结束后，将反应体系冷却至室温，用 10%碳酸钠进行中和，然后用饱和氯化钠溶液充分洗涤至中性。用无水硫酸镁充分干燥，过夜，抽滤，常压蒸馏，得无色透明液体。

五、注意事项与思考题

（一）注意事项

1. 试写出本反应的反应机理。
2. 注意己酸异戊酯的合成实验操作过程

（二）思考题

本反应使用对甲苯磺酸作为催化剂，还有哪些催化剂适合本反应？

Experiment 20　Synthesis of isoamyl caproate

Ⅰ. Purpose of the experiment

i. To master the synthetic reaction mechanism of isoamyl caproate.

ii. To familiar with the experimental operation process of isoamyl caproate.

iii. To understand the physicochemical properties and applications of isoamyl caproate.

Ⅱ. Experimental principle

i. Compound introduction

Isoamyl caproate is an important organic compound with the chemical formula $C_{11}H_{22}O_2$, It is colorless liquid with an apple-like and pineapple-like fragrance. Its boiling point is 222℃ and its flash point is 88℃. It is soluble in ethanol, non-volatile oil and mineral oil, insoluble in propylene glycol, water and glycerol.

Isoamyl caproate has a fruity aroma. It is found naturally in bananas, apples, strawberries and other fruits, as well as in sweet orange oil and many alcoholic drinks.

ii. Synthetic process route

The synthesis of isoamyl caproate is mainly prepared by caproic acid and isoamyl alcohol catalyzed by *p*-toluene sulfonic acid. The synthesis process is as follows:

$$\text{CH}_3(\text{CH}_2)_4\text{COOH} + (\text{CH}_3)_2\text{CHCH}_2\text{CH}_2\text{OH} \xrightarrow{p\text{-toluene sulfonic acid}} \text{CH}_3(\text{CH}_2)_4\text{COOCH}_2\text{CH}_2\text{CH}(\text{CH}_3)_2$$

Ⅲ. Main instruments and reagents

i. Main instruments: three-necked flask, heating sleeve, analysis balance, liquid separation funnel, reflux condensing tube, water dispenser, thermometer, condensing tube, stirrer, magnetic stirrer, beaker, rubber tube, conical bottle, zeolite, vacuum pump, *etc*.

ii. Main reagents: caproic acid, *p*-toluene sulfonic acid, isoamyl alcohol, saturated sodium carbonate, anhydrous magnesium sulfate, *etc*.

Ⅳ. Experimental steps

Add 11.6 g of caproic acid, 26.4 g of isoamyl alcohol and 0.7 g of *p*-toluene sulfonic acid into a 100 mL three-neck round-bottom flask equipped with a magnetic agitator, a thermometer, a condensing tube and a constant-pressure drip funnel. Stir quickly and heated to 130-150℃, refluxing until no more water is distilled.

After the reaction, cool the reaction system to room temperature, neutralized with 10% sodium carbonate, and then wash thoroughly with saturated sodium chloride until neutral. Dry with anhydrous magnesium sulfate overnight. Filter under vacuum, and perform distillation atmospheric pressure to obtain a colorless transparent liquid.

V. Attention or thinking questions

i. Attention

(i) Try to write the reaction mechanism of this reaction.

(ii) Pay attention to the experimental operation process of isoamyl caproate synthesis.

ii. Thinking questions

This reaction uses *p*-toluene sulfonic acid as the catalyst. Which other catalysts are suitable for this reaction?

实验二十一　二苯甲醇的合成（1）

一、实验目的

1. 掌握二苯甲醇的合成反应机理。
2. 熟悉二苯甲醇的实验操作过程。
3. 了解二苯甲醇的理化性质及用途。

二、实验原理

（一）化合物简介

二苯甲醇又名二苯基甲醇、双苯甲醇、α-苯基苯甲醇、羟基-二苯基甲烷，分子式为 $C_{13}H_{12}O$。常温下为白色至浅米色结晶固体，易溶于乙醇、醚、氯仿和二硫化碳。20℃时水中溶解度仅为 0.5 g/L。低毒，避免与皮肤和眼睛接触，缺乏有关毒性数据，可参照甲醇毒性。遇明火、高温、强氧化剂可燃烧，释放出有毒气体。

二苯甲醇主要用于有机合成原料，在医药工业中作为苯甲托品、苯海拉明及乙酰唑胺的合成中间体。

（二）合成工艺路线

二苯甲醇的合成，主要通过硼氢化钠还原二苯甲酮制得，其主要合成工艺路线如下：

$$\text{Ph}_2\text{C=O} \xrightarrow{\text{NaBH}_4} \text{Ph}_2\text{CHOH}$$

三、主要仪器和试剂

1. 主要仪器：三颈烧瓶、回流冷凝器、滴液漏斗、温度计、水浴锅、抽滤瓶、布氏漏斗、烧杯、玻璃棒、分析天平、真空泵等。
2. 主要试剂：二苯酮、硼氢化钠、乙醇、稀盐酸等。

四、实验步骤

在配有磁力搅拌器、温度计、冷凝管、干燥管的 100 mL 三颈烧瓶中，加入二苯酮 18.2 g、95%乙醇 100 mL，升高温度，使得所加的反应物溶解，快速搅拌下，缓慢分批加入硼氢化钠 1.9 g，在加入的过程中，要保证反应体系的温度小于 50℃。随后升温至回流，保温反应 1 h。反应结束后，将反应体系降温至室温，然后向反应瓶中加入 100 mL 水，搅拌，再加入 10%稀盐酸（用以分解未反应的硼氢化钠），冷却至室温，有固体析出，减压抽滤，用水充分洗涤滤饼，干燥，得粗产物。可以用石油醚对粗产物进行重结晶。

五、注意事项与思考题

（一）注意事项

试写出本反应的反应机理。

（二）思考题

本反应中除了用硼氢化钠作为还原剂，还可以用哪些还原剂？

Experiment 21　Synthesis of diphenyl carbinol (1)

I. Purpose of the experiment

i. To master the synthetic reaction mechanism of diphenyl carbinol.

ii. To familiar with the experimental operation process of diphenyl carbinol.

iii. To understand the physicochemical properties and uses of diphenyl carbinol.

II. Experimental principle

i. Compound introduction

Diphenyl carbinol, also known as diphenyl carbinol, α-phenyl carbinol, hydroxy-diphenylmethane, has the molecular formula $C_{13}H_{12}O$. At room temperature, it is a white to light beige crystalline solid. It is easily soluble in ethanol, ether, chloroform and carbon disulfide, with a solubility of only 0.5 g/L in water at 20℃. It is low toxicity, avoiding contact with skin and eyes. There is a lack of toxicity data, but methanol toxicity can be reference. It is flammable when exposed open flame, high temperature, strong oxidant, release toxic gas.

Diphenyl carbinal is primarily used as a raw material for organic synthesis and as a synthetic intermediate in the pharmaceutical industry for the synthesis of benzotropine, Diphenhydramine and acetazolamide.

ii. Synthetic process route

The synthesis of dibenzyl alcohol is mainly produced by reducing dibenzophenone with sodium borohydride. The main synthesis process is as follows:

III. Main instruments and reagents

i. Main instruments: three-necked flask, reflux condenser, drip funnel, thermometer, water bath, filter flask, Brinell funnel, beaker, glass rod, analytical balance, vacuum pump, *etc*.

ii. Main reagents: diphenyl ketone, sodium borohydride, ethanol, dilute hydrochloric acid, *etc*.

IV. Experimental steps

Add 18.2 g of diphenyl ketone and 100 mL of 95% ethanol into a 100 mL three-necked round-bottomed flask equipped with a magnetic stirrer, a thermometer, a condensing tube and a drying tube. Increase the temperature to dissolve the added reactants. Under rapid stirring, slowly add 1.9 g of sodium borohydride in batches. During the addition, ensure the temperature of the reaction system below 50 ℃. Then raise the temperature to reflux and maintain the reaction for 1 h. After the reaction, cool the reaction system to room temperature. Add 100 mL of water into the reaction bottle, and stir. Add 10% dilute hydrochloric acid (used to decompose the unreacted sodium borohydride). Cool to room temperature. A solid will precipitated. Pressure vacuum filtration, wash the filter cake with water and dry to obtain the crude products. The crude product can be recrystallized with petroleum ether.

V. Attention or thinking questions

i. Attention

Try to write the reaction mechanism of this reaction.

ii. Thinking questions

In addition to sodium borohydride as a reducing agent in this reaction, what other reducing agents can be used?

实验二十二　正丁醛的合成

一、实验目的

1. 掌握正丁醛的合成反应机理。
2. 熟悉正丁醛合成的实验操作流程。
3. 了解正丁醛的理化性质及用途。

二、实验原理

（一）化合物简介

正丁醛是一种重要的有机化合物，化学式为 C_4H_8O，主要用作树脂、塑料增塑剂、硫化促进剂、杀虫剂等合成的中间体。

（二）合成工艺路线

正丁醛的合成，主要是通过正丁醇在重铬酸钠的氧化下制备。其合成工艺路线如下：

主反应

$$CH_3(CH_2)_2CH_2OH \xrightarrow[H_2SO_4]{Na_2Cr_2O_7} CH_3(CH_2)_2CHO + H_2O$$

副反应

$$CH_3(CH_2)_2CHO \xrightarrow[H_2SO_4]{Na_2Cr_2O_7} CH_3(CH_2)_2COOH$$

三、主要仪器和试剂

1. 主要仪器：三颈烧瓶、加热套、分析天平、温度计、恒压滴液漏斗、分液漏斗、搅拌子、磁力搅拌器、烧杯、橡胶管、锥形瓶、干燥管、沸石、真空泵等。
2. 主要试剂：重铬酸钠、浓硫酸、正丁醇、无水硫酸镁等。

四、实验步骤

在配有磁力搅拌器、温度计、冷凝管、分馏柱以及恒压滴液漏斗的 100 mL 三颈烧瓶中，加入正丁醇 11.1 g，沸石 3 粒。然后配制重铬酸钠硫酸溶液，在烧杯中加入重铬酸钠 15 g、水 85 mL，快速搅拌，使其溶解，缓慢加入浓硫酸 11 mL，冷却。将配制好的重铬酸钠硫酸溶液，加至恒压滴液漏斗中，反应体系开始升温至正丁醇沸腾，开始滴加重铬酸钠硫酸溶液。本反应为放热反应，注意控制反应温度，以分馏柱顶部温度小于 80℃为宜。在接收器中有正丁醛生成。将其转入分液漏斗中，分出有机层，用无水硫酸镁充分干燥，过夜，过滤。将反应装置转换成蒸馏装置，收集 70~80℃的馏分。

五、注意事项与思考题

1. 铬酸氧化法可以用于制备分子量低的醛,是将铬酸滴加到热的酸性醇溶液中,以防止反应混合物中有过量的氧化剂存在,使生成的醛被进一步氧化成酸。

2. 在操作上,应及时把较低沸点的醛,从反应混合物中蒸出,是避免醛被进一步氧化的重要措施。

3. 试写出本反应的反应机理。

Experiment 22 Synthesis of *n*-butyl aldehyde

Ⅰ. Purpose of the experiment

i. To master the synthesis mechanism of *n*-butyl aldehyde.

ii. To familiar with the experimental operation process of *n*-butyl aldehyde synthesis.

iii. To understand the physicochemical properties and uses of *n*-butyl aldehyde.

Ⅱ. Experimental principle

i. Compound introduction

n-butyl aldehyde is an important organic compound with the chemical formula C_4H_8O. It is mainly used as resin, plastic plasticizer, vulcanization accelerator, insecticide synthesis intermediates.

ii. Synthetic process route

The synthesis of *n*-butyl aldehyde is mainly prepared by the oxidation of *n*-butanol in sodium dichromate. The synthesis process is as follows:

Main reaction

$$CH_3(CH_2)_2CH_2OH \xrightarrow[H_2SO_4]{Na_2Cr_2O_7} CH_3(CH_2)_2CHO + H_2O$$

Side effects

$$CH_3(CH_2)_2CHO \xrightarrow[H_2SO_4]{Na_2Cr_2O_7} CH_3(CH_2)_2COOH$$

Ⅲ. Main instruments and reagents

i. Main instruments: three-necked flask, constant pressure drip funnel, heating sleeve, analysis balance, thermometer, separation funnel, stirrer, magnetic stirrer, beaker, rubber tube, conical bottle, drying tube, zeolite, vacuum pump, *etc.*

ii. Main reagents: sodium dichromate, concentrated sulfuric acid, *n*-butanol, anhydrous magnesium sulfate, *etc.*

Ⅳ. Experimental steps

Add 11.1g of *n*-butanol and 3 zeolite into a 100 mL triple-necked round-bottom flask equipped with a magnetic agitator, a thermometer, a condensing tube, a fractionation column, and a constant pressure drip funnel. Then prepare sodium dichromate sulfuric acid solution by adding 15 g of sodium dichromate and 85 mL of water into the beaker. Stir quickly to dissolve. Slowly add 11 mL of concentrated sulfuric acid while cooling. Add the prepared sodium dichromate sulfuric acid solution to the constant pressure drip funnel. The reaction system begins to heat up to *n*-butanol boiling, and begins to drop the aggravating sodium chromate sulfuric acid solution. This reaction is an exothermic reaction, so it at should be paid to controlling the reaction temperature. The temperature at the top of the fractionation column should be less than 80℃. *n*-Butyl aldehyde is generated in the receiver. Transfer it to the liquid separation funnel, and separate the organic layer. Dry with anhydrous magnesium sulfate overnight and filter. Convert the reaction unit into a distillation unit to collect the fraction at 70-80℃.

V. Attention or thinking questions

i. Chromic acid oxidation method can be used to prepare low molecular weight aldehydes by adding chromic acid drops to hot acidic alcohol solution to prevent the presence of excessive oxidant in the reaction mixture, so that the generated aldehyde is further oxidized into acid.

ii. In operation, the aldehyde with a lower boiling point should be steamed from the reaction mixture in time, which is an important measure to avoid the aldehyde being further oxidized.

iii. Try to write the reaction mechanism of this reaction.

实验二十三 2-氨基丙醇的合成

一、实验目的

1. 掌握 2-氨基丙醇的合成反应机理。
2. 熟悉 2-氨基丙醇的实验操作过程。
3. 了解 2-氨基丙醇的理化性质及用途。

二、实验原理

（一）化合物简介

2-氨基丙醇是一种重要的医药、化工中间体，分子式为 C_3H_9NO，分子量为 75.11，易溶于水、醇和醚，对湿度和空气较为敏感。

（二）合成工艺路线

2-氨基丙醇的合成，以丙氨酸为原料，经过氯化反应、还原反应制备，其合成工艺路线如下：

$$\underset{\text{(alanine)}}{\text{H}_3\text{C}-\underset{\underset{\text{O}}{\|}}{\text{C}}\text{H}(\text{NH}_2)-\text{COOH}} \xrightarrow[\text{C}_2\text{H}_5\text{OH}]{\text{SOCl}_2} \text{H}_3\text{C}-\underset{\underset{\text{O}}{\|}}{\text{C}}\text{H}(\text{NH}_2)-\text{COOC}_2\text{H}_5 \xrightarrow{\text{KBH}_4} \text{H}_3\text{C}-\text{CH}(\text{NH}_2)-\text{CH}_2\text{OH}$$

三、主要仪器和试剂

1. 主要仪器：三颈烧瓶、回流冷凝器、恒压滴液漏斗、分析天平、温度计、水浴锅、抽滤瓶、布氏漏斗、烧杯、玻璃棒、真空泵等。
2. 主要试剂：L-氨基丙酸、乙醇、氯化亚砜、硼氢化钠、浓盐酸等。

四、实验步骤

在配有磁力搅拌器、温度计、冷凝管、干燥管以及恒压滴液漏斗的 250 mL 三颈烧瓶中，加入乙醇 150 mL、L-氨基丙酸 8.9 g，快速搅拌，采用冰盐浴进行冷却，缓慢滴加 10 mL 氯化亚砜，保证滴加时的反应体系温度小于 10℃，滴加结束后，将反应体系置于室温下反应 5 h。反应结束后，减压浓缩后，向反应烧瓶中加入乙醇 50 mL，剧烈搅拌，得 L-丙氨酸乙酯盐酸盐的乙醇溶液。

接着在相同的反应装置中，加入硼氢化钠 14.2 g、水 55 mL，快速搅拌，冰浴降温至小于 10℃，滴加上步得到的产物的乙醇溶液，要缓慢滴加，保证反应温度小于 10℃，滴加完毕后，保证室温反应 4 h。反应结束后，将反应体系抽滤，滤饼用大约 30 mL 乙醇洗涤，合并滤液及洗液，改反应装置为蒸馏装置，随后减压蒸馏，收集 72~75℃的馏分，得无色黏稠状产物。

五、注意事项与思考题

（一）注意事项
试写出本反应的反应机理。

（二）思考题
本反应中除了用硼氢化钠进行还原，还有哪些还原剂？

Experiment 23 Synthesis of 2-amino-propyl alcohol

I. Purpose of the experiment

i. To master the synthesis mechanism of 2-amino-propyl alcohol.

ii. To familiar with the experimental operation process of 2-amino-propyl alcohol.

iii. To understand the physicochemical properties and uses of 2-amino-propyl alcohol.

II. Experimental principle

i. Compound introduction

2-aminopropanol is an important pharmaceutical and chemical intermediate with the molecular formula C_3H_9NO and a molecular weight of 75.11. It is easily soluble in water, alcohol and ether. It is sensitive to humidity and air.

ii. Synthetic process route

The synthesis of 2-aminopropanol is prepared by chlorination and reduction reaction of alanine. The synthesis process is as follows:

$$\underset{}{H_3C\text{-CH(NH}_2\text{)-COOH}} \xrightarrow[C_2H_5OH]{SOCl_2} \underset{}{H_3C\text{-CH(NH}_2\text{)-COOC}_2H_5} \xrightarrow{KBH_4} \underset{}{H_3C\text{-CH(NH}_2\text{)-CH}_2\text{OH}}$$

III. Main instruments and reagents

i. Main instruments: three-necked flask, reflux condenser, Constant pressure drip funnel, analytical balance, thermometer, water bath, filter flask, Brinell funnel, beaker, glass rod, vacuum pump, *etc*.

ii. Main reagents: L-amino, concentrated hydrochloric acid, ethanol, sulfoxide chloride, sodium borohydride, *etc*.

IV. Experimental steps

Add 150 mL of ethanol and 8.9 g of L-amino propanoic acid is a 250 mL three-neck flask with a magnetic agitator, a thermometer, a condensing tube, a drying tube and a constant pressure drip funnel. Stiry rapidly and cool by an ice salt bath. Slowly add 10 mL of sulfoxide chloride to ensure that the temperature of the reaction system remains below 10 ℃ during the addition. After the addition, allow the reaction system to react at room temperature for 5 h. After the reaction, add 50 mL of ethanol to the reaction flask after decompression and concentration. Stir vigcrously to obtain the ethanol solution of L-alanine ethyl ester hydrochloride.

Then, in the same reaction device, add 14.2 g of sodium borohydride and 55 mL of water. Stir quickly and cool in the ice bath to below 10 ℃. Slowly add the ethanol solution of the product made by step to ensure that the reaction temperature remains below 10 ℃. After the addition, allow the reaction to proceed at room temperature for 4 h. At the end of the reaction, filter the reaction system. Wash the filter cake with about 30 mL of ethanol. Combin the filtrate and lotion. And charge the reaction device into a distillation device. Then perform the decompressed distillation. to collect the fraction at 72-75 ℃, yielding the colorless and viscous product.

V. Attention or thinking questions

i. Attention

Try to write the reaction mechanism of this reaction.

ii. Thinking questions

In addition to sodium borohydride for reduction, what other reducing agents are used in this reaction?

实验二十四　3,4-二甲氧基硝基苯的合成

一、实验目的

1. 掌握 3,4-二甲氧基硝基苯硝化反应的反应机理。
2. 熟悉 3,4-二甲氧基硝基苯的实验操作流程。
3. 了解 3,4-二甲氧基硝基苯的理化性质及用途。
4. 了解硝化反应的注意事项。

二、实验原理

（一）化合物简介

3,4-二甲氧基硝基苯，化学式为 $C_8H_9NO_4$，是非常重要的医药化工中间体，其对眼睛、呼吸系统以及皮肤具有较强的刺激性，操作时应在通风橱中进行。

（二）合成工艺路线

3,4-二甲氧基硝基苯的合成，主要是通过藜芦醚经过硝化反应制备，由于苯环上相邻位置连有两个甲氧基，活性较强，在低温时，即可发生硝化反应。其合成工艺路线如下：

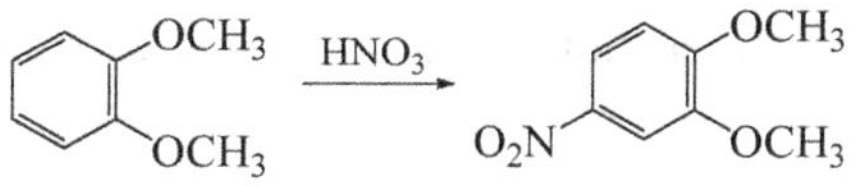

三、主要仪器和试剂

1. 主要仪器：分析天平、温度计、三颈烧瓶、恒压滴液漏斗、三颈烧瓶、搅拌子、磁力搅拌器、烧杯、橡胶管、锥形瓶、抽滤瓶、滤纸、布氏漏斗、沸石、真空泵等。
2. 主要试剂：浓硝酸、藜芦醚等。

四、实验步骤

在配有磁力搅拌器、温度计、冷凝管、干燥管以及恒压滴液漏斗的 100 mL 三颈烧瓶中，加入 65%硝酸 2.9 mL，冰浴冷却至 5℃以下，缓慢滴加 1.1 g 藜芦醚(1,2-二甲氧基苯)，反应液渐变成深红色。由于本反应是放热反应，注意滴加试剂的速度。滴加结束后，有黄色固体生成，保温反应 30 min，反应结束后，减压抽滤，加水 20 mL 充分洗涤，干燥，得到浅黄色 3,4-二甲氧基硝基苯产物。

五、注意事项与思考题

1. 试写出本反应的反应机理。
2. 注意硝化反应过程中，滴加速度、硝酸浓度以及反应温度。

Experiment 24　Synthesis of 3, 4-dimethoxy-nitrobenzene

Ⅰ. Purpose of the experiment

　　i. To master the reaction mechanism of 3, 4-dimethoxy-nitrobenzene nitrification.
　　ii. To familiar with the experimental operation process of 3, 4-dimethoxy-nitrobenzene.
　　iii. To understand the physicochemical properties and applications of 3, 4-dimethoxy-nitrobenzene.
　　iv. To understand the precautions for nitrification.

Ⅱ. Experimental principle

i. Compound introduction

　　3, 4-dimethoxy-nitrobenzene, with the chemical formula $C_8H_9NO_4$, is a very important chemical intermediate. It has strong irritant on the eyes, respiratory system and skin, and should be handle in the fume hood.

ii. Synthetic process route

　　The synthesis of 3, 4-dimethoxy-nitrobenzene is mainly produced by the digestive reaction of veratrol. Since there are two methoxyl groups in adjacent positions on the benzene ring, the activity is relatively large, and the nitration reaction can occur at low temperature. The synthesis process is as follows:

Ⅲ. Main instruments and reagents

　　i. Main instruments: analytical balance, thermometer, round-bottoming flask, constant pressure dripping funnel, three-necked flask, stirrer, magnetic stirrer, beaker, rubber tube, conical flask, filter bottle, filter paper, Brinell funnel, zeolite, vacuum pump, *etc*.
　　ii. Main reagents: concentrated nitric acid, veratrol, *etc*.

Ⅳ. Experimental steps

　　Add 2.9 mL of 65% nitric acid into a 100 mL three-necked flask equipped with a magnetic agitator, a thermometer, a condensing tube, a drying tube and a constant pressure drip funnel. Cool in an ice bath to below 5℃. Slowly add 1.1 g of veratrol (1, 2-dimethoxy-benzene), and the reaction liquid will gradually turn dark red. Pay attention to the rate of reagent addition. After the addition, a yellow solid will form, and main tain the reaction for 30 min. After the reaction, perform vacuum filtration, wash thoroughly with 20 mL of water, and dry to obtain a light yellow product of 3, 4-dimethoxy-nitrobenzene.

V. Attention or thinking questions

i. Try to write the reaction mechanism of this reaction.

ii. Pay attention to droplet acceleration, nitric acid concentration and reaction temperature in the process of nitrification.

实验二十五　对氨基苯甲酰-β-丙氨酸的合成

一、实验目的

1. 掌握对氨基苯甲酰-β-丙氨酸氢化还原反应的机理。
2. 熟悉对氨基苯甲酰-β-丙氨酸的实验操作过程。
3. 了解对氨基苯甲酰-β-丙氨酸的理化性质及用途。
4. 了解本反应涉及的还原剂的种类。

二、实验原理

（一）化合物简介

对氨基苯甲酰-β-丙氨酸，分子式为 $C_{10}H_{12}N_2O_3$，是一种重要的医药、化工中间体，可由 β-丙氨酸为原料通过两步反应制备得到，可用于制备巴柳氮钠。

（二）合成工艺路线

对氨基苯甲酰-β-丙氨酸的合成，主要是通过对硝基苯甲酰-β-丙氨酸加氢还原制得，其合成工艺路线如下：

$$\underset{NO_2}{\underset{|}{C_6H_4}}\text{—}CONHCH_2CH_2COOH \xrightarrow[C_2H_5OH]{H_2/PdCl} \underset{NH_2}{\underset{|}{C_6H_4}}\text{—}CONHCH_2CH_2COOH$$

三、主要仪器和试剂

1. 主要仪器：氢化反应装置、温度计、抽滤瓶、分析天平、布氏漏斗、烧杯、玻璃棒、真空泵等。
2. 主要试剂：对硝基苯甲酰-β-丙氨酸、氯化钯、乙醇等。

四、实验步骤

在常压氢化反应的实验装置中，加入对硝基苯甲酰-β-丙氨酸 6 g、催化剂氯化钯 0.1 g 以及乙醇 80 mL，剧烈搅拌，向反应装置中通入氢气，注意通入氢气的速度，直至充分反应为止。抽滤，将滤液进行减压浓缩，得到油状黏稠物，加入无水乙醚后，有固体析出，抽滤，得到白色固体，即为对氨基苯甲酰-β-丙氨酸粗品。

五、注意事项与思考题

（一）注意事项

试写出本反应的反应机理。

（二）思考题

本反应除了用氯化钯作为催化剂，还有哪些催化剂？

Experiment 25 Synthesis of *p*-amino benzoyl-*β*-alanine

I. Purpose of the experiment

i. To master the mechanism of hydrogenation reduction reaction of *p*-amino benzoyl-*β*-alanine.

ii. To familiar with the experimental operation process of *p*-amino benzoyl-*β*-alanine.

iii. To understand the physicochemical properties and applications of *p*-amino benzoyl-*β*-alanine.

iv. To understand the types of reducing agents involved in this reaction.

II. Experimental principle

i. Compound introduction

p-amino benzoyl-*β*-alanine, with the molecular formula $C_{10}H_{12}N_2O_3$, is an important pharmaceutical and chemical intermediate, which can be prepared from *β*-alanine as raw material by a two-step reaction. It can be used to prepare balsalazide sodium.

ii. Synthetic process route

The synthesis of *p*-amino benzoyl-*β*-alanine is mainly produced by hydrogenation reduction of *p*-amino benzoyl-*β*-alanine. The specific synthesis process is as follows:

$$\underset{NO_2}{\underset{|}{C_6H_4}}-CONHCH_2CH_2COOH \xrightarrow[C_2H_5OH]{H_2/PdCl} \underset{NH_2}{\underset{|}{C_6H_4}}-CONHCH_2CH_2COOH$$

III. Main instruments and reagents

i. Main instruments: hydrogenation reaction device, thermometer, filter bottle, analytical balance, Brinell funnel, beaker, glass rod, vacuum pump, *etc*.

ii. Main reagents: *p*-amino benzoyl-*β*-alanine, palladium chloride, ethanol, *etc*.

IV. Experimental steps

Add 6 g of *p*-amino benzoyl-*β*-alanine, 0.1 g of catalyst palladium chloride and 80 mL of ethanol into the experimental device of atmospheric hydrogenation reaction. And stir vigorously. Add hydrogen into the reaction device. Pay attention to the rate of hydrogen injection until the reaction complete. The filtrate is pumped and concentrated under pressure to get an oil-like gunk. Add anhydrous ether, and some solids will precipitate. Filter by suction to obtain a white solid, which is the crude product *p*-amino benzoyl-*β*-alanine.

V. Attention or thinking questions

i. Attention

Try to write the reaction mechanism of this reaction.

ii. Thinking questions

In addition to palladium chloride as catalyst, what other catalysts are used in this reaction?

实验二十六 $KMnO_4$ 氧化法对硝基苯甲酸的合成

一、实验目的

1. 掌握对硝基苯甲酸的合成反应机理。
2. 熟悉对硝基苯甲酸的实验操作过程。
3. 了解对硝基苯甲酸的理化性质及用途。

二、实验原理

（一）化合物简介

对硝基苯甲酸，又称为 4-硝基苯甲酸，是一种羧酸类有机化合物，外观为淡黄色无味结晶，显酸性。分子式为 $C_7H_5NO_4$。它是苯甲酸的对位被硝基取代而成的化合物，是硝基苯甲酸的一种，同时也是邻硝基苯甲酸和间硝基苯甲酸的同分异构体。

由于其苯环上有硝基，硝基是吸电子基团，通过共轭效应（具有单键-双键交替结构的体系，其中双键的 p 轨道，通过电子离域相互连接，通常会降低分子的总能量并增加其稳定性），使得其比苯甲酸的酸性更强。

本品可用于制备普鲁卡因胺酸盐酸盐、对氨甲基苯甲酸、叶酸、苯佐卡因、头孢菌素 V、对氨基苯甲酰谷氨酸以及滤光剂、彩色胶片成色剂、金属表面除锈剂、防晒剂等。

（二）合成工艺路线

对硝基苯甲酸的合成，主要是对硝基甲苯在高锰酸钾强氧化剂的作用下制备，其合成工艺路线如下：

$$\text{对硝基甲苯} \xrightarrow[H_2O]{KMnO_4} \text{对硝基苯甲酸钾} \xrightarrow{HCl} \text{对硝基苯甲酸}$$

三、主要仪器和试剂

1. 主要仪器：三颈烧瓶、回流冷凝器、恒压滴液漏斗、温度计、水浴锅、抽滤瓶、布氏漏斗、烧杯、分析天平、玻璃棒、真空泵等。
2. 主要试剂：对硝基甲苯、高锰酸钾、浓盐酸等。

四、实验步骤

在配有磁力搅拌器、温度计、回流冷凝管以及恒压滴液漏斗的 250 mL 三颈烧瓶中，加入高锰酸钾 10 g、对硝基甲苯 7 g 以及水 100 mL，快速搅拌，升温至 80℃，保温反应 1 h，再加入高锰酸钾 5 g，反应 1 h，随后再加入高锰酸钾 5 g，反应 0.5 h，然后将反应体系的温度升温至 100℃，持续反应，直至高锰酸钾的颜色褪去。反应结束后，将反应瓶冷却至室温，抽滤，滤饼用水 20 mL 充分洗涤，随后，滤液在快速搅拌下，加浓盐酸 10 mL 进行酸化，有固体析出，抽滤，水洗滤饼，得产物，可以用乙醇作为溶剂，进行重结晶，得到纯品。

五、注意事项与思考题

（一）注意事项

试写出本反应的反应机理。

（二）思考题

1. 本反应中高锰酸钾的颜色褪去后，说明什么？
2. 本反应中的高锰酸钾，为什么要分批加入？

Experiment 26 Synthesis of *p*-nitrobenzoic acid by KMnO₄ oxidation

I. Purpose of the experiment

    i. To master the synthetic reaction mechanism of *p*-nitrobenzoic acid.

    ii. To familiar with the experimental operation process of *p*-nitrobenzoic acid.

    iii. To understand the physicochemical properties and applications of *p*-nitrobenzoic acid.

II. Experimental principle

i. Compound introduction

    p-nitrobenzoic acid, also known as 4-nitrobenzoic acid, is an organic compound of the carboxylic acid class, appearing as pale yellow, odorless crystals with acidic properties. Its molecular formula is $C_7H_5NO_4$. It is a compound by the the para-position of a nitro group an benzaic. It is a kind of nitro benzoic acid. And it is also an isomer of *o*-nitrobenzoic acid and *m*-nitrobenzoic acid.

    Because of the nitro group on its benzene ring, which is an electron-absorbing group, it is more acidic than benzoic acid through conjugation (a system with an alternating single-double bond structure, in which the *p* orbitals of the double bond, connect to each other by electron delocalization, usually reduce the total energy of the molecule and increase its stability).

    It is used in the production of procaine hydrochloride, procaine amine hydrochloride, *p*-amino methyl benzoic acid, folic acid, benzocaine, cephalosporin V, *p*-amino benzoyl glutamic acid and filter, color film preparation agent, metal surface rust remover, sunscreen, *etc*.

ii. Synthetic process route

    The synthesis of *p*-nitrobenzoic acid is mainly prepared by *p*-nitrotoluene under the action of potassium permanganate strong oxidant. The specific synthesis process is as follows:

$$\underset{NO_2}{\underset{|}{C_6H_4}}-CH_3 \xrightarrow{\text{KMnO}_4 / H_2O} \underset{NO_2}{\underset{|}{C_6H_4}}-COOK \xrightarrow{\text{HCl}} \underset{NO_2}{\underset{|}{C_6H_4}}-COOH$$

III. Main instruments and reagents

    i. Main instruments: three-necked flask, reflux condenser, drip funnel, thermometer, water bath, filter flask, Brinell funnel, beaker, analytical balance, glass rod, vacuum pump, *etc*.

    ii. Main reagents: *p*-nitrotoluene, potassium permanganate, concentrated hydrochloric acid, *etc*.

IV. Experimental steps

Add 10 g of potassium permanganate, 7 g of *p*-nitrotoluene and 100 mL of water into a 250 mL three-necked round-bottom flask equipped with a magnetic agitator, a thermometer, a condensing tube and constant pressure drip funnel. Stir quickly. Heat to 80 ℃, and maintain the reaction for 1 h. Then add 5 g of potassium permanganate and react for 1 h. Then add 5 g of potassium permanganate and react for 0.5 h. Raise the temperature of the reaction system to 100 ℃ and continue the reaction until the color of potassium permanganate disappears. After the reaction, cool the reaction bottle to room temperature, perform suction filtration, and wash the filter cake thoroughly with 20 mL of water. Acidify the filtrate with 10 mL of concentrated hydrochloric acid under rapid stirting acidification, resulting in solid precipitation. Filter and wash the filter cake with water, and the product can be recrystaillised using ethand as a solvent to obtain the product.

V. Attention or thinking questions

i. Attention

Try to write the reaction mechanism of this reaction.

ii. Thinking questions

(i) What does it mean that the color of potassium permanganate in this reaction fades?

(ii) Why should potassium permanganate in this reaction be added in batches?

实验二十七　硝基苯的合成

一、实验目的

1. 掌握硝基苯的亲电取代反应的机理。
2. 熟悉硝基苯的实验操作流程。
3. 了解硝基苯的理化性质及用途。

二、实验原理

（一）化合物简介

硝基苯是一种重要的有机中间体，化学式为 $C_6H_5NO_2$，呈无色或微黄色具苦杏仁味的油状液体。难溶于水，易溶于乙醇、乙醚、苯和油。密度比水大。遇明火、高热会燃烧、爆炸。

2017 年 10 月 27 日，世界卫生组织国际癌症研究机构公布的致癌物清单，经初步整理，硝基苯在 2B 类致癌物清单中。

硝基苯用三氧化硫磺化，得 3-硝基苯磺酸。硝基苯用氯磺酸磺化，得间硝基苯磺酰氯，用作染料、医药等中间体。硝基苯经氯化，得 3-硝基氯苯，广泛用于染料、农药的生产，经还原后，可得到间氯苯胺，是医药、农药、荧光增白剂、有机颜料等的中间体。硝基苯再硝化可得间二硝基苯，经还原可得间苯二胺，用作染料中间体、环氧树脂固化剂、石油添加剂、水泥促凝剂。间二硝基苯用硫化钠进行部分还原，则得间硝基苯胺，是偶氮染料和有机颜料等的中间体。

（二）合成工艺路线

硝基苯的合成，主要通过苯在混酸作用下制备，其合成工艺路线如下：

$$\underset{\text{浓}HNO_3}{\overset{\text{浓}H_2SO_4}{\longrightarrow}}$$ 苯 → 苯-NO_2 + H_2O

三、主要仪器和试剂

1. 主要仪器：水浴锅、三颈烧瓶、分析天平、烧杯、冷凝管、干燥管、分液漏斗、量筒、温度计、锥形瓶、铁架台、真空泵、沸石、pH 试纸等。
2. 主要试剂：浓硝酸、浓硫酸、苯、氯化钠、无水硫酸镁等。

四、实验步骤

在配有磁力搅拌器、温度计、冷凝管、干燥管的 100 mL 三颈烧瓶中，加入 3.6 mL 浓硝酸、5 mL 浓硫酸，快速搅拌，随后加入 4.5 mL 苯，将烧瓶升温至 60℃，保持回流反应 30 min。反应结束后，将反应液倒入分液漏斗中分液，然后放入烧杯中，用等体积水充分洗涤，直至不显酸性为止,最后用水洗至中性,将有机层放入干燥烧杯中,用无水硫酸镁充分干燥,过夜,过滤,得产物。

五、注意事项与思考题

(一)注意事项

试写出本反应的反应机理。

(二)思考题

1. 浓硫酸在反应中的作用是什么?
2. 反应过程中,如温度过高对反应有何影响?

Experiment 27 Synthesis of nitrobenzene

Ⅰ. Purpose of the experiment

i. To master the mechanism of nitrobenzene electrophilic substitution reaction.

ii. To familiar with the experimental operation process of nitrobenzene.

iii. To understand the physicochemical properties and uses of nitrobenzene.

Ⅱ. Experimental principle

i. Compound introduction

Nitrobenzene is an important organic intermediate with the chemical formula $C_6H_5NO_2$. It is a colorless or yellowish oily liquid with a bitter almond flavor. It is insoluble in water, soluble in ethanol, ether benzene and oil. Its density is greater than that of water. It can burn and explode when exposed to open flames or high heat.

On October 27, 2017, the World Health Organization's International Agency for Research on Cancer published a list of carcinogens, nitrobenzene was preliminarily classified as a Group 2B carcinogen.

Nitrobenzene is sulfonated with sulfur trioxide to obtain 3-nitrobenzene sulfonic acid. Nitrobenzene is sulfonated by chlorosulfonic acid to obtain nitrobenzene sulfonyl chloride, which can be used as intermediates in dyes and medicine. Nitrobenzene is chlorinated to obtain 3-nitrochlorobenzene, which is widely used in the production of dyes and pesticides. After reduction, *m*-chloroaniline can be obtained, which is an intermediate for medicine, pesticides, fluorescent brightening agents, organic pigments and so on. Nitrobenzene renitrification can obtain m-dinitrobenzene, reduction can get *m*-phenylenediamine, used as dye intermediates, epoxy resin curing agent, petroleum additives, cement coagulant, *m*-dinitrobenzene, sodium sulfide for partial reduction. *m*-nitroaniline is the intermediate of azo dyes and organic pigments.

ii. Synthetic process route

Nitrobenzene is synthesized mainly through the preparation of benzene under the action of mixed acid. The synthesis process is as follows:

$$\text{C}_6\text{H}_6 \xrightarrow[\text{concentrated } HNO_3]{\text{concentrated } H_2SO_4} \text{C}_6\text{H}_5-NO_2 + H_2O$$

Ⅲ. Main instruments and reagents

i. Main instruments: water bath, round bottoming flask, analytical balance, beaker, drying tube, condensate tube, liquid separation funnel, measuring cylinder, thermometer, conical flask, iron frame, vacuum pump, zeolite, pH test paper,*etc*.

ii. Main reagents: concentrated nitric acid, concentrated sulfuric acid, benzene, sodium chloride, s anhydrous magnesium sulfate, *etc*.

Ⅳ. Experimental steps

Add 3.6 mL of concentrated nitric acid and 5 mL of concentrated sulfuric acid into a 100 mL three-necked round-bottom flask equipped with a magnetic agitator, a thermometer, a condensing tube and a drying tube. Stir quickly. Then add 4.5 mL of benzene, and heat the flask to 60℃. Maintain the reflux reaction for 30 min. After the reaction, pour the reaction liquid into the separating funnel. Put it into the beaker. Wash it fully with equal volume of water until it is not acidi. Finally wash it with water until it is neutral. Put the organic layer into the drying beaker. Dry it fully with anhydrous magnesium sulfate, and filter it overnight to obtain the product.

Ⅴ. Attention or thinking questions

i. Attention

Try to write the reaction mechanism of this reaction.

ii. Thinking questions

(i) What is the function of concentrated sulfuric acid in the experiment?

(ii) In the process of reaction, what is the effect of excessive temperature on the reaction?

实验二十八　对硝基肉桂酸的合成

一、实验目的

1. 掌握对硝基肉桂酸缩合的反应机理。
2. 熟悉对硝基肉桂酸的合成操作过程。
3. 熟悉重结晶的操作过程。
4. 了解对硝基肉桂酸的理化性质及用途。

二、实验原理

（一）化合物简介

对硝基肉桂酸是一种重要的医药、化工中间体，分子式为 $C_9H_7NO_4$，又名 3-（4-硝基苯基）-2-丙烯酸、4-硝基肉桂酸、对硝基丙烯酸、4-硝基桂皮酸。

（二）合成工艺路线

对硝基肉桂酸的合成，以丙二酸与对硝基苯甲醛为原料制得，其合成工艺路线如下：

$$\underset{NO_2}{C_6H_4}-CHO + \underset{COOH}{\overset{COOH}{CH_2}} \xrightarrow{Py} \underset{NO_2}{C_6H_4}-CH=CHCOOH + CO_2 + H_2O$$

三、主要仪器和试剂

1. 主要仪器：三颈烧瓶、回流冷凝器、干燥管、分析天平、温度计、水浴锅、抽滤瓶、布氏漏斗、烧杯、玻璃棒、真空泵等。
2. 主要试剂：丙二酸、无水吡啶、哌啶、对硝基苯甲醛、浓盐酸等。

四、实验步骤

在配有磁力搅拌器、温度计、回流冷凝管、干燥管以及恒压滴液漏斗的 100 mL 三颈烧瓶中，加入无水吡啶 18 mL、丙二酸 8 g，快速搅拌，再加入哌啶 0.7 mL 以及对硝基苯甲醛 9 g，升温至 80℃后，保温反应 3 h。反应结束后，将反应体系冷却至室温，将反应液倒入约 100 mL 的冰水中，用浓盐酸调节 pH 至 2，抽滤，用水洗充分滤饼至中性，得到粗产物，可以用乙醇重结晶，得到纯品。

五、注意事项与思考题

（一）注意事项
试写出本反应的反应机理。

（二）思考题
1. 在重结晶的过程中，要注意什么？
2. 本反应中加入吡啶和哌啶的目的是什么？

Experiment 28　Synthesis of *p*-nitro cinnamic acid

Ⅰ. Purpose of the experiment

  i. To master the condensation reaction mechanism of *p*-nitro cinnamic acid.
  ii. To familiar with the synthetic process of *p*-nitro cinnamic acid.
  iii. To familiar with the process of recrystallization.
  iv. To understand the physicochemical properties and applications of *p*-nitro cinnamic acid.

Ⅱ. Experimental principle

i. Compound introduction

  p-nitro cinnamic acid is an important intermediate in medicine and chemical industry with the molecular formula $C_9H_7NO_4$, also known as 3-(4-nitrophenyl)-2-acrylic acid, 4-nitrocinnamic acid, *p*-nitro cinnamic acid, 4-nitrocinnamic acid.

ii. Synthetic process route

  The synthesis of *p*-nitro cinnamic acid is mainly made from malonic acid and *p*-nitro benzaldehyde. The synthesis process is as follows:

Ⅲ. Main instruments and reagents

  i. Main instruments: round-bottoming flask, reflux condenser, analytical balance, drying tube, thermometer, water bath, filter flask, Brinell funnel, beaker, glass rod, vacuum pump, *etc*.
  ii. Main reagents: malonic acid, pyridine, piperidine, *p*-nitro benzaldehyde, concentrated hydrochloric acid, *etc*.

Ⅳ. Experimental steps

  Add 18 mL of anhydrous pyridine, 8 g of malonic acid into a 100 mL three-necked round-bottom flask equipped with magnetic agitators, a thermometer, condensing tubes, drying tubes and constant pressure drip funds. Stir quickly. Then add 0.7 mL of piperidine and 9 g of *p*-nitro benzaldehyde. Heat to 80 ℃ and maintain the reaction for 3 h. After the reaction, cool the reaction system to room temperature. Pour the reaction solution into about 100 mL of ice water. Adjust the pH to 2 with concentrated hydrochloric acid. Filter by suctim. Wash the filter cake thoroughly with water until neutral to obtain the crude product, which could be recrystallized with ethanol to obtain the pure product.

V. Attention or thinking questions

i. Attention

Try to write the reaction mechanism of this reaction.

ii. Thinking questions

(i) What should we pay attention to in the process of recrystallization?

(ii) What is the purpose of adding pyridine and piperidine in this reaction?

实验二十九 环己酮的合成

一、实验目的

1. 掌握环己酮的合成反应机理。
2. 熟悉环己酮的合成实验操作流程。
3. 了解环己酮的理化性质及用途。

二、实验原理

（一）化合物简介

环己酮是一种重要有机化合物，化学式是 $C_6H_{10}O$，为羰基碳原子包括在六元环内的饱和环酮。无色透明液体，带有泥土气息，含有痕迹量的酚时，则带有薄荷味。不纯物为浅黄色，随着存放时间过长，生成杂质而显色，呈水白色到灰黄色，具有强烈的刺鼻臭味。

2017 年 10 月 27 日，世界卫生组织国际癌症研究机构公布的致癌物清单，经初步整理，环己酮在 3 类致癌物清单中。

环己酮是重要化工原料，是制造尼龙、己内酰胺和己二酸的主要中间体。也是重要的工业溶剂，如用于油漆，特别是用于那些含有硝化纤维、氯乙烯聚合物及其共聚物或甲基丙烯酸酯聚合物油漆等。用于有机磷杀虫剂及许多类似物等农药的优良溶剂。用作染料的溶剂。作为活塞型航空润滑油的黏滞溶剂。脂、蜡及橡胶的溶剂。也用作擦亮金属的脱脂剂，木材着色涂漆。用作指甲油等化妆品的高沸点溶剂。通常与低沸点溶剂和中沸点溶剂配制成混合溶剂，以获得适宜的挥发速度和黏度。

（二）合成工艺路线

环己酮的合成，以环己醇为原料，通过氧化反应制得。其合成工艺路线如下：

$$\text{环己醇} \xrightarrow{[O]} \text{环己酮}$$

三、主要仪器和试剂

1. 主要仪器：恒压滴液漏斗、200℃温度计、三颈烧瓶、分析天平、三口连接管、水浴锅、量筒、加热套、石棉网、冷凝管、尾接管、锥形瓶、分液漏斗、磁力加热搅拌器、真空泵等。

2. 主要试剂：次氯酸钠、碘化钾、碘化钾淀粉试纸、饱和亚硫酸氢钠溶液、氯化铝、沸石、无水碳酸钠、无水硫酸镁、氯化钠等。

四、实验步骤

在配有磁力搅拌器、温度计、冷凝管以及恒压滴液漏斗的 250 mL 三颈烧瓶中，分别加入环己醇 5 g、乙酸 25 mL，快速搅拌，在冰水浴冷却下，缓慢滴加次氯酸钠水溶液（约 1.8 mol/L）38 mL，升温至 30~35 ℃，滴加结束后，保温反应 5 min，用碘化钾淀粉试纸检验，有蓝色出现，若无，应再补加次氯酸钠溶液 5 mL，保证有过量次氯酸钠存在，使氧化反应完全。在室温下继续搅拌 30 min，加入饱和亚硫酸氢钠溶液，直至反应液对碘化钾淀粉试纸不显蓝色。

向反应混合物中加入水 30 mL、氯化铝 3 g 和几粒沸石，在石棉网上加热蒸馏，直至馏出液无油珠滴出为止。

在搅拌下，将馏出液分批加入无水碳酸钠，至反应液呈中性为止，然后加入精制氯化钠，使之变成饱和溶液，将混合液倒入分液漏斗中，分出上层有机层，用无水硫酸镁充分干燥，过夜，过滤后得到产物。

五、注意事项与思考题

（一）注意事项

试写出本反应的反应机理。

（二）思考题

1. 实验中使用精制氯化钠有何作用？
2. 第一次蒸馏所得到馏分的成分是什么？

Experiment 29 Synthesis of cyclohexanone

Ⅰ. Purpose of the experiment

i. To master the synthesis mechanism of cyclohexanone.

ii. To familiar with the experimental operation process of synthesis of cyclohexanone.

iii. To understand the physical and chemical properties and uses of cyclohexanone.

Ⅱ. Experimental principle

i. Compound introduction

Cyclohexanone is an important organic compound with the chemical formula is $C_6H_{10}O$. It is carbonyl carbon atoms included in the six membered ring saturated cycloketone. It is a colorless transparent liquid, with an earthy smell. When containing traces of phenol, it has a mint smell. Impurity is light yellow. While prolonged storage, impurities develop, resulting in a colour ranging from water white to gray yellow, with a strong pungent smell.

On October 27, 2017, the World Health Organization's International Agency for Research on Cancer published a list of carcinogens, cyclohexanone was preliminarily classified in Group 3.

Cyclohexanone is an important chemical raw material, which is the main intermediate in the manufacture of nylon, caprolactam and adipic acid. It is also an important industrial solvent, such as for paints, especially for those containing nitrocellulose, vinyl chloride polymer and its copolymer or methacrylate polymer paints. It serves as an excellent solvent for organophosphate pesticides and many similar agricultural chemicals. It is used as a solvent for dyes and as a viscous solvent in piston-type aviation lubricants. It acts as a solvent for fats, waxes, and rubber. Additionally, it is used as a degreaser for polishing metals and in wood staining and finishing. It serves as a high-boiling solvent in cosmetics such as nail polish. Typically, it is formulated with low-boiling and medium-boiling solvents to achieve suitable evaporation rates and viscosity.

ii. Synthetic process route

The synthesis of cyclohexanone is mainly produced by the oxidation reaction of cyclohexanol. The synthesis process is as follows:

$$\underset{}{\text{cyclohexanol}} \xrightarrow{[O]} \underset{}{\text{cyclohexanone}}$$

Ⅲ. Main instruments and reagents

i. Main instruments: constant pressure dripping funnel, 200 ℃ thermometer, flask, analysis balance, three-port connecting pipe, water bath, measuring cylinder, heating sleeve, asbestos net, condensation tube, tail nozzle, conical bottle, liquid separation funnel, magnetic heating agitator, vacuum pump, *etc.*

ii. Main reagents: sodium hypochlorite, potassium iodide, potassium iodide starch test paper, sodium bisulfite, aluminum chloride, zeolite, anhydrous sodium carbonate, anhydrous magnesium sulfate, sodium chloride, *etc.*

IV. Experimental steps

Add 5 g of cyclohexanol and 25 mL of acetic acid respectively into a 250 mL three-neck round-bottom flask equipped with a magnetic stirrer, a thermometer, a condensing tube, a drying tube and a constant pressure drip funnel. Stir quickly. Then slowly add 38 mL of sodium hypochlorite aqueous solution (about 1.8 mol/L) under the cooling ice water bath, and heat up to 30-35 ℃. After the addition, maintain the temperature for 5 min and test with potassium iodide starch paper. If no blue color appears, add 5 mL of sodium hypochlorite solution to ensure the existence of excessive sodium hypochlorite and complete oxidation reaction. Continue stirring at room temperature for 30 min and add saturated sodium bisulfite solution until the reaction solution no longer shows blue color to the potassium iodide starch test paper.

Add 30 mL of water, 3 g of aluminum chloride and a few grains of zeolite to the reaction mixture. Heat the asbestos mesh and distillation until the distillate drops out of the oil-free beads.

Under stirring, add anhydrous sodium carbonate in batches to the distillate, until the reaction liquid is neutral, and then add refined sodium chloride, so that it becomes a saturated solution. Pour the mixture into the separator funnel, separate out the upper organic layer, dry thoroughly with anhydrous magnesium sulfate overnight, and filter to obtain the product.

V. Attention or thinking questions

i. Attention

Try to write the reaction mechanism of this reaction.

ii. Thinking questions

(i) What is the effect of using refined sodium chloride in the experiment?

(ii) What is the composition of the fraction obtained by the first distillation?

实验三十 扁桃酸乙酯的合成

一、实验目的

1. 掌握扁桃酸乙酯合成酯化反应机理。
2. 熟悉扁桃酸乙酯的合成操作过程。
3. 了解扁桃酸乙酯的理化性质及用途。
4. 了解甲苯共沸带水的基本原理。

二、实验原理

（一）化合物简介

扁桃酸乙酯是一种重要的有机化合物，分子式为 $C_{10}H_{12}O_3$，又名苯乙醇酸乙酯、苦杏仁酸乙酯、羟基苯基乙酸乙酯、α-羟基苯乙酸乙酯，无色透明液体，其作为制备匹莫林的中间体。

（二）合成工艺路线

扁桃酸乙酯的合成，以扁桃酸与乙醇为原料，进行酯化反应制得，其合成工艺路线如下：

$$\text{PhCH(OH)COOH} \xrightarrow[\text{H}_2\text{SO}_4]{\text{C}_2\text{H}_5\text{OH}} \text{PhCH(OH)COOC}_2\text{H}_5$$

三、主要仪器和试剂

1. 主要仪器：三颈烧瓶、回流冷凝器、分析天平、滴液漏斗、温度计、水浴锅、烧杯、玻璃棒、真空泵等。
2. 主要试剂：扁桃酸、无水乙醇、浓硫酸、饱和碳酸钠溶液、饱和氯化钠溶液、无水硫酸镁、甲苯等。

四、实验步骤

在配有磁力搅拌器、温度计、冷凝管、干燥管以及恒压滴液漏斗的 100 mL 三颈烧瓶中，加入扁桃酸 10 g 以及无水乙醇 40 mL，快速搅拌，缓慢滴加浓硫酸 2 mL，升温至回流，保温反应 2 h。减压浓缩，冷却至室温，将浓缩物倒入 50 mL 冰水中，随后，用饱和碳酸钠溶液调节 pH 至 8，接着用 100 mL 二氯甲烷萃取 3 次，用饱和氯化钠溶液洗涤有机层，然后用无水硫酸镁充分干燥，过夜，抽滤，浓缩，向浓缩物中加入甲苯 50 mL，共沸带水，得到浅黄色油状物。

五、注意事项与思考题

(一)注意事项

试写出本反应的反应机理。

(二)思考题

1. 本反应中加入少量浓硫酸的作用是什么?
2. 简述甲苯共沸带水的机理是什么?

Experiment 30 Synthesis of ethyl mandelate

I. Purpose of the experiment

i. To master the esterification reaction mechanism of ethyl mandelate synthesis.
ii. To familiar with the synthetic operation process of ethyl mandelate.
iii. To understand the physicochemical properties and applications of ethyl mandelate.
iv. To understand the basic principle of toluene azeotrope water.

II. Experimental principle

i. Compound introduction

Ethyl mandelate is an important organic compound with the molecular formula $C_{10}H_{12}O_3$, also known as ethyl phenyl glycolate, ethyl amygdalic acid, ethyl hydroxy phenylacetic acid, ethyl α-hydroxy ethyl phenylacetate acid. It is a colorless and transparent liquid, which is used as an intermediate in the preparation of pemoline.

ii. Synthetic process route

Ethyl mandelate is synthesized mainly through esterification of mandelic acid and ethanol. The synthesis process is as follows:

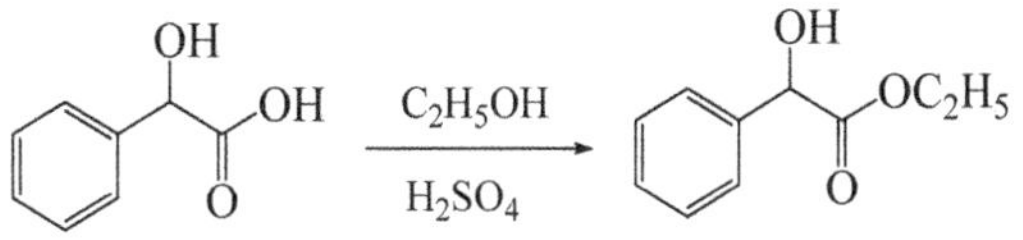

III. Main instruments and reagents

i. Main instruments: round-bottoming flask, reflux condenser, analytical balance, drip funnel, thermometer, water bath, beaker, glass rod, vacuum pump, *etc.*
ii. Main reagents: mandelic acid, anhydrous ethanol, concentrated sulfuric acid, sodium carbonate, anhydrous magnesium sulfate, toluene, *etc.*

IV. Experimental steps

Add 10 g of mandelic acid and 40 mL of anhydrous ethanol into a 100 mL three-neck round-bottoming flask equipped with a magnetic stirrer, a thermometer, a condensing tube, a drying tube and a constant pressure drip funnel. Stir quickly. Slowly add 2 mL of concentrated sulfuric acid. Heat up to reflux, and maintain the reaction for 2 h. Concentrate under reduced pressure, and cool to room temperature. Pour the concentrate into 50 mL of ice water, then adjust the pH to 8 using saturated sodium carbonate solution. Next, extract three times with 100 mL of dichloromethane.

Wash the organic layer with saturated sodium chloride solution, then dry thoroughly with anhydrous magnesium sulfate overnight. Filter under vacuum and concentrate. Add 50 mL of toluene to the concentrate, performing azeotropic distillation with water to obtain a light yellow oily substance.

V. Attention or thinking questions

i. Attention

Try to write the reaction mechanism of this reaction.

ii. Thinking questions

(i) What is the effect of adding a small amount of concentrated sulfuric acid in this reaction?

(ii) What is the mechanism of toluene azeotrope water?

实验三十一　苯甲酸的合成

一、实验目的

1. 掌握苯甲酸合成反应机理。
2. 熟悉苯甲酸合成的实验操作过程。
3. 熟悉苯环侧链上的氧化反应。
4. 了解苯甲酸的理化性质及用途。

二、实验原理

（一）化合物简介

苯甲酸是一种重要的医药、化工中间体，分子式为 $C_7H_6O_2$，作为一种芳香酸类有机化合物，也是最简单的芳香酸，其最初由安息香胶制得，故称安息香酸。外观为白色针状或鳞片状结晶。100℃以上时会升华。微溶于冷水、己烷，溶于热水、乙醇、乙醚、氯仿、苯、二硫化碳和松节油等。

苯甲酸以游离酸、酯或其衍生物的形式广泛存在于自然界中。主要用于制备苯甲酸钠防腐剂，并用于合成药物、染料。还用于制作增塑剂、媒染剂、杀菌剂和香料等。可由甲苯在二氧化锰存在时，直接氧化制得，或用水蒸气将苯二甲酸酐脱羧制备。

（二）合成工艺路线

苯甲酸的合成，主要是通过甲苯在强氧化剂高锰酸钾的作用下制得，其合成工艺路线具体如下：

$$\text{C}_6\text{H}_5\text{CH}_3 \xrightarrow{\text{KMnO}_4} \text{C}_6\text{H}_5\text{COOH}$$

三、主要仪器和试剂

1. 主要仪器：分析天平、三颈烧瓶、锥形瓶、加热套、磁力搅拌器、分水器、蒸馏头、回流冷凝器、分液漏斗、温度计、水浴锅、烧杯、玻璃棒、沸石、真空泵等。
2. 主要试剂：甲苯、高锰酸钾、浓盐酸等。

四、实验步骤

在配有磁力搅拌器、温度计、回流冷凝管的 250 mL 三颈烧瓶中，加水 100 mL、甲苯 2.7 mL，升温至沸腾，随后，从冷凝管上端管口，分批 3 批加入 8.5 g 高锰酸钾，然后用约 25 mL 水，冲洗在冷凝管上附着的高锰酸钾，继续回流反应，直至甲苯层消失，在回流液中无油珠时停止加热，反应混合物趁热抽滤，用少许热水充分洗涤滤饼，合并滤液及洗涤液，冷却至室温，再用冰水冷却，随后用浓盐酸调节酸碱度，直至固体完全析出，抽滤，得苯甲酸粗品。

五、注意事项与思考题

（一）注意事项

试写出本反应的反应机理。

（二）思考题

1. 反应完毕后，如果滤液呈紫色，加入亚硫酸钠有何作用？
2. 加入高锰酸钾时，一定要分批次缓慢加入，否则会从冷凝管中，将反应液喷出。
3. 注意在最后调节酸碱度时，注意调节速度，保证晶体完全析出。

Experiment 31   Synthesis of benzoic acid

I. Purpose of the experiment

i. To master the synthesis mechanism of benzoic acid.

ii. To familiar with the experimental operation process of benzoic acid synthesis.

iii. To familiar with the oxidation reaction on the side chain of the benzene ring.

iv. To understand the physicochemical properties and uses of benzoic acid.

II. Experimental principle

i. Compound introduction

Benzoic acid is an important intermediate in the pharmaceutical and chemical industry with the molecular formula $C_7H_6O_2$. As an aromatic acid organic compound, it is originally prepared from benzoin gum, so it is called benzoin acid. It appears as white acicular or scaly crystal and subimes above 100℃. It is slightly soluble in cold water, hexane, and soluble in hot water, ethanol, ether, chloroform, benzene, carbon disulfide and turpentine, *etc*.

Benzoic acid exists widely in nature in the form of free acid, ester or its derivatives. It is mainly used for the preparation of sodium benzoate preservatives and for the synthesis of drugs and dyes. It is also used to make plasticizer, mordant, fungicide and spices. It can be prepared by direct oxidation of toluene in the presence of manganese dioxide or by decarboxylation of phthalic anhydride with steam.

ii. Synthetic process route

The synthesis of benzoic acid is mainly prepared by toluene under the action of the strong oxidant potassium permanganate. The synthesis process is as follows:

$$\text{C}_6\text{H}_5\text{CH}_3 \xrightarrow{\text{KMnO}_4} \text{C}_6\text{H}_5\text{COOH}$$

III. Main instruments and reagents

i. Main instruments: analytical balance, conical flask, heating set, magnetic stirrer, water separator, three-necked flask, distillation head, reflux condenser, liquid separation funnel, thermometer, water bath, beaker, vacuum pump, glass rod, zeolite, *etc*.

ii. Main reagents: toluene, potassium permanganate, concentrated hydrochloric acid, *etc*.

IV. Experimental steps

Add 100 mL of water and 2.7 mL of toluene into a 250 mL three-neck round-bottom flask equipped with magnetic agitators, thermometers and condensing tubes, and heat up to boiling. Then, add 8.5 g of potassium permanganate in three batches from the upper end of the condensing tube. And rinse the potassium permanganate attached to the condensing tube with about 25 mL of water to continue the reflux reaction. Until the toluene layer disappears, stop heating when there are no oil beads in the reflux liquid. Filter the reaction mixture while hot, wash the filter cake thoroughly with a little hot water, and combine the filtrate and wash liquid. Cool to room temperature. Then cool with ice water. A just the pH with concentrated hydrochloric acid until the solid completely precipitates. And filter to obtain the crude benzoic acid.

V. Attention or thinking questions

i. Attention

Try to write the reaction mechanism of this reaction.

ii. Thinking questions

(i) After the reaction is completed, if the filtrate is purple, what is the effect of adding sodium sulfite?

(ii) When adding potassium permanganate, be sure to add it slowly in batches, otherwise the reaction liquid will be ejected from the condensing tube.

(iii) Pay attention to the final adjustment of pH, pay attention to the adjustment of speed, to ensure that the crystal is completely precipitation.

实验三十二　4-溴代丁酸乙酯的合成

一、实验目的

1. 掌握 4-溴丁酸乙酯内酯环开环卤代的反应机理。
2. 熟悉 4-溴丁酸乙酯的实验操作过程。
3. 了解 4-溴丁酸乙酯的理化性质及用途。

二、实验原理

（一）化合物简介

4-溴丁酸乙酯为羧酸酯类衍生物，是一种非常重要的有机化合物，分子式为 $C_6H_{11}BrO_2$，呈无色透明至黄色液体状，其可用作农药、医药中间体。

（二）合成工艺路线

4-溴丁酸乙酯的合成，以 γ-丁内酯与无水乙醇作为原料制得，其合成工艺路线具体如下：

$$\text{γ-丁内酯} + C_2H_5OH \xrightarrow{HBr} Br(CH_2)_3COOC_2H_5$$

三、主要仪器和试剂

1. 主要仪器：三颈烧瓶、回流冷凝器、干燥管、分析天平、滴液漏斗、温度计、水浴锅、烧杯、玻璃棒、布氏漏斗、抽滤瓶、分液漏斗、真空泵等。
2. 主要试剂：γ-丁内酯、无水乙醇、乙酸乙酯、无水硫酸镁等。

四、实验步骤

在配有磁力搅拌器、温度计、冷凝管、干燥管、HBr 导气管及气体吸收装置的 250 mL 三颈烧瓶中，加入 50 mL 无水乙醇和 26 g γ-丁内酯，冰水浴冷却 0℃，快速搅拌，持续通入 HBr 气体 1 h。反应结束后，将反应液缓慢倒入冰水中，用 75 mL 乙酸乙酯萃取 3 次，合并有机相，随后，用无水硫酸镁进行充分干燥。抽滤，浓缩滤液，蒸馏，收集 97~99℃的馏分，得油状产物。

五、注意事项与思考题

（一）注意事项

试写出本反应的反应机理。

（二）思考题

本反应中使用溴化氢作为溴化剂，还有哪些溴化剂可以使用？

Experiment 32 Synthesis of 4-bromo-ethyl butyrate

Ⅰ. Purpose of the experiment

i. To master the reaction mechanism of ring opening halogenation of 4-bromobutyrate ethyl lactone ring.

ii. To familiar with the experimental operation process of ethyl 4-bromobutyrate.

iii. To understand the physicochemical properties and applications of ethyl 4-bromobutyrate.

Ⅱ. Experimental principle

i. Compound introduction

Ethyl 4-bromobutyrate is a carboxylate derivative and is a very important organic compound. With the molecular formula $C_6H_{11}BrO_2$. It appears as a colorless to yellow transparent liquid. It can be used as a pesticide, medicine intermediate.

ii. Synthetic process route

The synthesis of 4-bromobutyrate ethyl ester is mainly prepared by γ-butyrolactone and anhydrous ethanol as raw materials. The synthesis process is as follows:

$$\text{γ-butyrolactone} + C_2H_5OH \xrightarrow{HBr} Br(CH_2)_3COOC_2H_5$$

Ⅲ. Main instruments and reagents

i. Main instruments: three-neck flask, reflux condenser, drying tube, analytical balance, drip funnel, thermometer, water bath, beaker, vacuum pump, glass rod, Brinell funnel, filter bottle, separator funnel, *etc*.

ii. Main reagents: γ-butyrolactone, anhydrous ethanol, ethyl acetate, anhydrous magnesium sulfate, *etc*.

Ⅳ. Experimental steps

Add 50 mL of anhydrous ethanol and 26 g of γ-butyrolactone into a 250 mL three-neck bottling flask equipped with magnetic agitators, thermometers, condensing tubes, drying tubes, HBr guide tubes and absorbers. Cool the mixture at 0℃ in an ice water bath. Stir quickly, and continuously introduce HBr gas for 1h. At the end of the reaction, slowly pour the reaction solution into ice water, and extract three times with 75 mL ethyl acetate. Combin the organic phase. Then thoroughly dry with anhydrous magnesium sulfate. At last extract filtration, concentrate the filtrate, distillate, and collect the fraction at 97-99℃ to obtain an oily products.

Ⅴ. Attention or thinking questions

i. Attention

Try to write the reaction mechanism of this reaction.

ii. Thinking questions

Hydrogen bromide is used as the brominating agent in this reaction. What other brominating agents can be used?

实验三十三　香豆素-3-羧酸的合成

一、实验目的

1. 掌握香豆素-3-羧酸的合成反应机理。
2. 熟悉香豆素-3-羧酸合成操作过程。
3. 熟悉本反应的中重结晶方法。
4. 了解香豆素-3-羧酸的理化性质及用途。

二、实验原理

（一）化合物简介

香豆素-3-羧酸是一种重要的医药、化工中间体，分子式为 $C_{10}H_6O_4$。又名为 2-氧代-2H-苯并吡喃-3-羧酸、香豆素-3-羧酸、3-羧酸香豆素、2-氧代-2H-1-苯并吡喃-3-羧酸。

（二）合成工艺路线

香豆素-3-羧酸的合成，主要是通过水杨醛和丙二酸酯在有机碱的催化下，低温合成香豆素的衍生物，这种合成方法称为 Knovenagel 反应。其水杨醛与丙二酸酯，在六氢吡啶催化下，缩合生成中间体香豆素-3-甲酸乙酯，随后加碱水解，酯基和内酯均被水解，然后酸化，再次闭环内酯化，即生成香豆素-3-羧酸。其合成工艺路线具体如下：

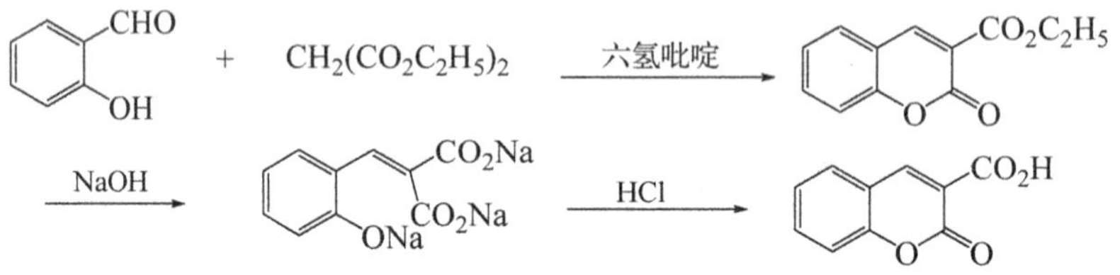

三、主要仪器和试剂

1. 主要仪器：锥形瓶、分析天平、加热套、磁力搅拌器、分水器、三颈烧瓶、蒸馏头、回流冷凝管、干燥管、分液漏斗、温度计、恒压滴液漏斗、水浴锅、烧杯、玻璃棒、真空泵等。
2. 主要试剂：水杨醛、丙二酸二乙酯、无水乙醇、六氢吡啶、乙酸、氢氧化钠、浓盐酸、无水氯化钙等。

四、实验步骤

（一）香豆素-3-甲酸乙酯的合成

在配有磁力搅拌器、温度计、冷凝管、干燥管以及恒压滴液漏斗的 100 mL 三颈烧瓶中，加入水杨醛 4.9 g、丙二酸二乙酯 7.2 g、无水乙醇 25 mL、六氢吡啶以及 1 滴乙酸，升温至回流,保温反应 2 h,反应结束后,冷却至室温,向反应体系中加水 35 mL,放置冰箱冷却,有固体析出,抽滤,滤饼用少量冷的乙醇洗涤 3 次,干燥,得香豆素-3-甲酸乙酯粗产物,待用。亦可用 25%乙醇进行重结晶。

（二）香豆素-3-羧酸的合成

在配有磁力搅拌器、温度计、冷凝管以及恒压滴液漏斗的 100 mL 三颈烧瓶中,加入上步待用香豆素-3-甲酸乙酯 4 g、氢氧化钠 3 g、乙醇 20 mL 以及水 10 mL,快速搅拌,升温至回流,保温反应 30 min,反应结束后,趁热将反应体系中的反应液,倒至装入盐酸溶液中(浓盐酸 10 mL 与水 50 mL),注意在倒入过程中,一边倒入,一边快速搅拌,有白色固体析出,用冰水浴冷却,使得晶体析出完全,减压抽滤,用少许冷水充分洗涤,抽干,干燥,得到香豆素-3-羧酸产物。

五、注意事项与思考题

（一）注意事项

试写出本反应的反应机理。

（二）思考题

本反应中用六氢吡啶,为有机碱催化剂,还可以用哪些催化剂?

Experiment 33   Synthesis of coumarin-3-carboxylic acid

I. Purpose of the experiment

i. To master the synthesis mechanism of coumarin-3-carboxylic acid.

ii. To familiar with the synthesis process of coumarin-3-carboxylic acid.

iii. To familiar with the medium recrystallization method of this reaction.

iv. To understand the physicochemical properties and applications of coumarin-3-carboxylic acid.

II. Experimental principle

i. Compound introduction

Coumarin-3-carboxylic acid is an important intermediate in medicine and chemical industry with the molecular formula $C_{10}H_6O_4$. It is also known as 2-oxy-2h-benzopyrane-3-carboxylic acid, coumarin-3-carboxylic acid, 3-carboxylic coumarin, 2-oxy-2h-1-benzopyrane-3-carboxylic acid.

ii. Synthetic process route

The synthesis of coumarin-3-carboxylic acid is mainly through the low temperature synthesis of coumarin derivatives with salicylaldehyde and malonate under the catalysis of organic base. This synthesis method is called Knovenagel reaction. Under the catalysis of hexahydropyridine, the salicylaldehyde and malonate are condensation to form the intermediate coumarin-3-ethyl-formate, which is then hydrolyzed by alkali. The ester group and the lactone are hydrolyzed. Then acidize, and again esterifie in the closed loop. Coumarin-3-carboxylic acid is produced. The synthesis process is as follows:

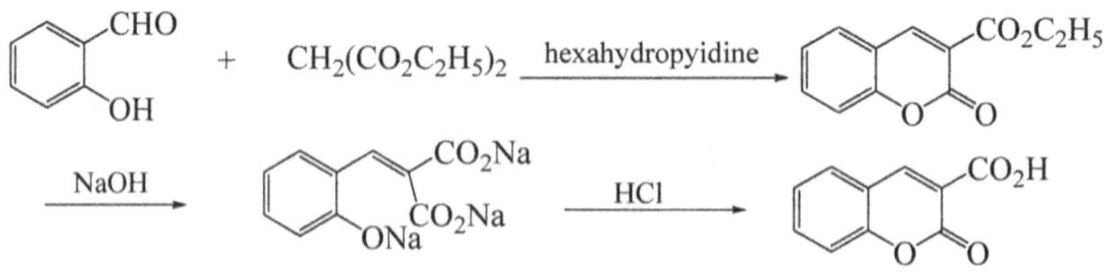

III. Main instruments and reagents

i. Main instruments: conical flask, analysis balance, heating set, magnetic stirrer, water separator, three-neck flask, distillation head, reflux condenser, drying tube, liquid separation funnel, thermometer, water bath, beaker, glass rod, vacuum pump, *etc*.

ii. Main reagents: salicylaldehyde, diethyl malonate, anhydrous ethanol, hexa-pyridine, acetic acid, sodium hydroxide, concentrated hydrochloric acid, anhydrous calcium chloride, *etc*.

Ⅳ. Experimental steps

i. Synthesis of coumarin-3-ethyl-formate

Add 4.9 g of salicylaldehyde, 7.2 g of diethyl malonate, 25 mL of anhydrous ethanol, hexahydropyridine and 1 drop of acetic acid into a 100 mL three-necked round-bottom flask equipped with a magnetic agitator, a thermometer, a condensing tube, a drying tube and a constant pressure drip fund. Heat up to reflux and maintain for 2 h. After the reaction, cool to room temperature. Add 35 mL of water into the reaction system. Place in the refrigerator for cooling, resulting in solid precipitation. Filter the solid and wash the filter cake 3 times with a small amount of cold ethanol. Dry to yield to crude product of Coumarin-3-ethyl formate for further use. Recrystallization can also be performed using 25% ethanol.

ii. Synthesis of coumarin-3-carboxylic acid

Add 4 g of coumarin-3-ethyl-formate, 3 g of sodium hydroxide, 20 mL of ethanol and 10 mL of water into a 100 mL three-neck round-bottoming flask equipped with a magnetic stirrer, a thermometer, a condensing tube and a constant pressure drip funnel. Stir quickly. Heat up to reflux, and maintain for 30 min. After the reaction, while hot, pour the reaction liquid in the reaction system into a hydrochloric acid solution (10 mL of concentrated hydrochloric acid and 50 mL of water), ensuring to stir quickly while pouring. A white solid precipitates, and cool with an ice water bath to ensure complete crystallisation. Then filter under reduced pressure, thoroughly wash with a small amount of cold water. Dry to yield the product of coumarin-3-carboxylic acid.

Ⅴ. Attention or thinking questions

i. Attention

Try to write the reaction mechanism of this reaction.

ii. Thinking questions

In this reaction, hexahydro pyridine is an organic base catalyst. What other catalysts can be used?

实验三十四　对甲基苯乙酮的合成

一、实验目的

1. 掌握对甲基苯乙酮合成中 Fridel-Crafts 酰基化反应的机理。
2. 熟悉对甲基苯乙酮的合成实验操作流程。
3. 熟悉实验中涉及的无水操作方法。
4. 了解对甲基苯乙酮的理化性质及用途。

二、实验原理

（一）化合物简介

对甲基苯乙酮是一种重要的有机化合物，分子式为 $C_9H_{10}O$。为无色针状结晶体或近似无色液体，在稍低温度下，便会凝固，易溶于乙醚、苯、氯仿与丙二醇，几乎不溶于水。溶于乙醇和大多数非挥发性油，微溶于丙二醇与矿物油，不溶于甘油和水。天然对甲基苯乙酮存在于玫瑰木精油、含羞草花油等中。

对甲基苯乙酮具有类似山楂花的芳香，呈强烈的山楂似香气、水果和花香。可用于配制金合欢型、皂用紫丁香型香精等，亦可作果实食品香精。可用于含羞花、山楂花、金合欢、刺槐、报春花、葵花、风信子、紫丁香、百合、木香、苔香等类型中。可与香豆素、大茴香醛、洋茉莉醛共用于皂用薰衣草、香薇、素心兰、新刈草型中。可微量用于杏仁、香荚兰豆香型的食用香精中，还可少量用于烟草香精中。

（二）合成工艺路线

对甲基苯乙酮的合成，以甲苯与乙酸酐为原料，进行傅克酰基化反应制得，其合成工艺路线具体如下：

主反应：

$$H_3C-C_6H_4-H + (CH_3CO)_2O \xrightarrow{AlCl_3} H_3C-C_6H_4-CO-CH_3 + CH_3CO_2H$$

副反应：

$$H_3C-C_6H_4-H + (CH_3CO)_2O \xrightarrow{AlCl_3} H_3C-C_6H_3(CH_3)-CO-CH_3 + CH_3CO_2H$$

三、主要仪器和试剂

1. 主要仪器：加热套、分析天平、三颈烧瓶、蒸馏头、回流冷凝管、干燥管、恒压滴液漏斗、分液漏斗、温度计、油浴锅、烧杯、玻璃棒、布氏漏斗、抽滤瓶、真空泵等。

2. 主要试剂：无水甲苯、乙酸酐、三氧化铝、浓盐酸、10%氢氧化钠、无水硫酸镁等。

四、实验步骤

在配有磁力搅拌器、温度计、回流冷凝管、干燥管以及恒压滴液漏斗的 250 mL 三颈烧瓶中，加入三氯化铝 11 g、无水甲苯 15 mL，快速搅拌，缓慢滴加乙酸酐 4.5 mL 与 3 mL 混合溶剂，在 15 min 内滴加结束，升温至 100℃，保温反应 30 min，反应结束后，用冰水浴冷却反应体系，快速搅拌下，缓慢滴加稀盐酸 50 mL（浓盐酸 25 mL+水 25 mL），待铝盐完全溶解后，倒入分液漏斗，分出甲苯有机层，分别用水、10%氢氧化钠溶液进行充分洗涤，再次分出甲苯层，用无水硫酸镁，进行充分干燥过夜。

对无水硫酸镁干燥后的甲苯有机层，进行抽滤，得干燥甲苯层，随后，进行蒸馏处理，收集 224～226℃的馏分，得产物。

五、注意事项与思考题

（一）注意事项

1. 试写出本反应的反应机理。
2. 本反应对无水要求非常严格，务必将反应中所用的试剂、仪器进行干燥处理。
3. 本反应中，所用的催化剂三氯化铝，容易吸水潮解，避免过长时间暴露在空气中。

（二）思考题

1. 本反应涉及的试剂、仪器的无水要求非常严格，在操作过程中有哪些注意事项？
2. 本反应后为什么要加入稀盐酸？
3. 本反应中用乙酸酐作为酰化剂，还有哪些酰化剂可以用在本反应中？

Experiment 34 Synthesis of *p*-methyl acetophenone

I. Purpose of the experiment

i. To mastered the acylation mechanism of Frdel-Crafts in *p*-methyl acetophenone synthesis.
ii. To familiar with the experimental operation process of synthesis of *p*-methyl acetophenone.
iii. To familiar with the water-free operation methods involved in the experiment.
iv. To understand the physicochemical properties and uses of *p*-methyl acetophenone.

II. Experimental principle

i. Compound introduction

p-methyl acetophenone is an important organic compound with the formula $C_9H_{10}O$. It is a colorless acicular crystal or nearly colorless liquid that solidifies at slightly lower temperatures. It is easily soluble in ether, benzene, chloroform and propylene glycol, and almost insoluble in water. It is soluble in ethanol and most non-volatile oils, slightly soluble in propylene glycol and mineral oils, insoluble in glycerol and water. Natural *p*-methyl acetophenone is found in rosewood essential oil, mimosa flower oil, *etc*.

p-methyl acetophenone has an aroma similar to Shanzha flower, with a strong hawthorn like aroma along with fruity and floral notes. It can be used to prepare acacia flavor, soap lilac flavor and fruit food flavor. It can be used in shy flower, hawthorn flower, acacia, robinia, primrose, sunflower, hyacinth, lilac, wood, moss and other types. It can be used together with coumarin, anisaldehyde for soap lavender, sweet myrtle, plain heart orchid and new grass mower. It can be used in a trace amount of almond, vanilla flavor food, and also can be used in a small amount of tobacco flavor.

ii. Synthetic process route

The synthesis of *p*-methyl acetophenone is mainly prepared by the Frdel-Crafts acylation reaction of toluene and acetic anhydride as raw materials. The synthesis process is as follows:

Main reaction:

$$H_3C-C_6H_4-H + (CH_3CO)_2O \xrightarrow{AlCl_3} H_3C-C_6H_4-\underset{\underset{O}{\|}}{C}-CH_3 + CH_3CO_2H$$

Side effects:

$$H_3C-C_6H_4-H + (CH_3CO)_2O \xrightarrow{AlCl_3} H_3C-C_6H_3(CH_3)-\underset{\underset{O}{\|}}{C}-CH_3 + CH_3CO_2H$$

III. Main instruments and reagents

i. Main instruments: heating set, analysis balance, three-necked flask, distillation head, drying tube, reflux condenser, liquid separation funnel, thermometer, oil bath, beaker, glass rod, Brinell funnel, suction bottle, vacuum pump, *etc.*

ii. Main reagents: toluene, acetic anhydride, aluminium trioxide, concentrated hydrochloric acid, sodium hydroxide, anhydrous magnesium sulfate, *etc.*

IV. Experimental steps

Add 11 g of aluminum trichloride and 15 mL of anhydrous toluene into a 250 mL three-necked round-bottom flask equipped with a magnetic agitator, a thermometer, a condensing tube, a drying tube and a constant pressure drip funnel. Stir quickly. Slowly add 4.5 mL of acetic anhydride and 3 mL of mixed solvent, completing the addition within 15 min. Heat up to 100 ℃ and maintain for 30 min. After the reaction, cool the reaction system with an ice water bath. Under rapidly stirring, slowly add 50 mL diluted hydrochloric acid (25 mL of concentrated hydrochloric acid+25 mL of water). After the aluminum salt is completely dissolved, pour the liquid separator into the hopper to separate out the toluene organic layer, which is thoroughly washed with water and 10% sodium hydroxide solution respectively. Dry thoroughly with anhydrous magnesium sulfate overnight.

After drying the toluene organic layer of anhydrous magnesium sulfate, perform suction filtration to obtain the dry toluene layer. Then, carry out the distillation process, collecting the fraction at 224-226 ℃ to obtain the product.

V. Attention or thinking questions

i. Attention

(i) Try to write the reaction mechanism of this reaction.

(ii) This reaction is very strict on anhydrous requirements, it must be used in the reaction reagents, instruments for drying treatment.

(iii) In this reaction, the catalyst aluminum trichloride is easy to absorb water and dehydrate, so as to avoid prolonged exposure to the air.

ii. Thinking questions

(i) The reagent and instrument involved in this reaction have very strict anhydrous requirements. What should be noted in the operation process?

(ii) Why should diluted hydrochloric acid be added after the reaction?

(iii) Acetic anhydride is used as an acylation agent in this reaction. What other acylation agents can be used in this reaction?

实验三十五　Cannizzaro 反应合成苯甲酸和苯甲醇

一、实验目的

1. 掌握 Cannizzaro 反应合成苯甲酸及苯甲醇的反应机理。
2. 熟悉实验过程中的各种常见操作。
3. 了解苯甲酸、苯甲醇的理化性质及用途。

二、实验原理

（一）化合物简介

苯甲酸是一种芳香酸类有机化合物，分子式为 $C_7H_6O_2$，也是最简单的芳香酸。最初由安息香胶制得，故称安息香酸。熔点 122.13℃，沸点 249.2℃，相对密度 1.2659。外观为白色针状或鳞片状结晶。100℃以上时会升华。微溶于冷水、己烷，溶于热水、乙醇、乙醚、氯仿、苯、二硫化碳和松节油等。

苯甲酸以游离酸、酯或其衍生物的形式广泛存在于自然界中。主要用于制备苯甲酸钠防腐剂，并用于合成药物、染料。还用于制作增塑剂、媒染剂、杀菌剂和香料等。最初苯甲酸是由安息香胶干馏或碱水水解制得，也可由马尿酸水解制得。工业上常用甲苯、邻二甲苯或萘为原料，制备苯甲酸，上述原料可从煤焦油或石油中获得。此外，由甲苯生产苯甲醛时，可副产苯甲酸。苯甲酸的工业生产方法，主要有甲苯液相空气氧化法、三氯甲苯水解法、邻苯二甲酸酐脱羧法，此外还有苄卤氧化法。

苯甲醇是一种有机化合物，分子式是 C_7H_8O，其是最简单的芳香醇之一，可看作是苯基取代的甲醇。在自然界中，多数以酯的形式，存在于香精油中，例如茉莉花油、风信子油和秘鲁香脂中都含有此成分。

苯甲醇是极有用的定香剂，是茉莉、月下香、伊兰等香精调配时不可缺少的香料。用于配制香皂、日用化妆香精。苯甲醇能缓慢的自然氧化，一部分生成苯甲醛和苄醚，使市售产品常带有杏仁香味，故不宜久贮。苯甲醇在工业化学品生产中用途广泛，用于涂料溶剂、照相显影剂、聚氯乙烯稳定剂、合成树脂溶剂、维生素 B 注射液的溶剂、药膏或药液的防腐剂等。可用作尼龙丝、纤维及塑料薄膜的干燥剂、染料、纤维素酯、酪蛋白的溶剂、制取苄基酯或醚的中间体。同时，广泛用于制笔（圆珠笔油）、油漆溶剂等方面。储存于阴凉、通风的库房，远离火种、热源，应与氧化剂、食用化学品分开存放，切忌混储，配备相应品种和数量的消防器材。

（二）合成工艺路线

Cannizzaro 反应（坎尼扎罗反应），是无 α-活泼氢的醛，在强碱作用下，发生分子间氧化还原反应，生成一分子羧酸和一分子醇的有机歧化反应。意大利化学家斯坦尼斯劳·坎尼扎罗在 1895 年，通过用草木灰处理苯甲醛，得到了苯甲酸和苯甲醇，首先发现了这个反应，因此，称坎尼扎罗反应。不含 α-氢原子的脂肪醛、芳醛或杂环醛类，在浓碱作用下，醛分子自身，同时发生氧化与还原反应，生成相应的羧酸(在碱溶液中生成羧酸盐)和醇的有机歧化反应。

本反应通过苯甲醛，在浓碱的作用下，发生 Cannizzaro 反应合成苯甲酸与苯甲醇，其合成工艺路线具体如下：

$$PhCHO \xrightarrow{KOH} PhCH_2OH + PhCO_2K$$

$$PhCO_2K \xrightarrow{H^+} PhCOOH$$

三、主要仪器和试剂

1. 主要仪器：三颈烧瓶、锥形瓶、分析天平、加热套、恒压滴液漏斗、磁力搅拌器、干燥管、蒸馏头、回流冷凝器、分液漏斗、温度计、油浴锅、烧杯、玻璃棒、真空泵等。

2. 主要试剂：苯甲醛、氢氧化钾、乙醚、10%碳酸钠、浓盐酸、无水硫酸镁、无水碳酸钾、饱和亚硫酸氢钠、甲醇等。

四、实验步骤

在配有磁力搅拌器、温度计、冷凝管的 100 mL 三颈烧瓶中，加入氢氧化钾溶液（氢氧化钾 9 g 与水 9 mL），快速搅拌，冰水冷却至室温，缓慢滴加新蒸苯甲醛 10 mL，反应体系变成白色糊状物，放置过夜，向反应体系中加水 30 mL，剧烈搅拌，使得固体全部溶解。用乙醚进行萃取 3 次，合并有机相，分别用饱和亚硫酸氢钠 2.5 mL、10%碳酸钠溶液 5 mL 以及水 5 mL 洗涤，然后用无水硫酸镁进行充分干燥，过滤，将滤液中乙醚蒸去，再进行蒸馏苯甲醇，收集 204~206℃的馏分，即为苯甲醇。

在上述乙醚萃取后的水溶液中，用浓盐酸调节酸碱度，使刚果红试纸变蓝。放置冰箱，使之充分冷却，有固体析出，即为苯甲酸粗产物。粗产物可以用水进行重结晶得苯甲酸纯品。

五、注意事项与思考题

（一）注意事项

试写出本反应的反应机理。

（二）思考题

本反应中的两种产物，是根据什么原理分离提纯的？

Experiment 35 Synthesis of benzoic acid and benzyl alcohol by Cannizzaro reaction

I. Purpose of the experiment

i. To study the reaction mechanism of synthesis of benzoic acid and benzyl alcohol by Cannizzaro reaction.

ii. To familiar with various common operations during the experiment.

iii. To understand the physicochemical properties and applications of benzoic acid and benzyl alcohol.

II. Experimental principle

i. Compound introduction

Benzoic acid is an organic compound of aromatic acids with the molecular formula $C_7H_6O_2$, which is also the simplest aromatic acid. It was originally made from benzoin gum, hence the name benzoin acid. Its melting point is 122.13℃ and boiling point is 249.2℃. Its relative density is 1.2659. It appears as white acicular or scaly crystal. It sublimates above 100℃. It is slightly soluble in cold water, hexane, and soluble hot water, ethanol, ether, chloroform, benzene, carbon disulfide and turpentine, *etc*.

Benzoic acid exists widely in nature in the form of free acid, ester or its derivatives. It is mainly used for the preparation of sodium benzoate preservatives and for the synthesis of drugs and dyes. It is also used to make plasticizer, mordant, fungicide and spices. Benzoic acid is initially prepared by benzoin distillation or alkaline hydrolysis, but also by horse uric acid hydrolysis. Benzene, o-xylene or naphthalene are commonly used in industry as raw materials to prepare benzoic acid, which can be obtained from coal tar or petroleum. In addition, benzoic acid can be produced byproduct when benzaldehyde is produced from toluene. The industrial production methods of benzoic acid mainly include the toluene liquid phase air oxidation method, trichloro toluene hydrolysis method, phthalic anhydride decarboxylation method, and the benzyl halide oxidation method.

Benzyl alcohol is an organic compound with the molecular formula C_7H_8O. It is one of the simplest aromatic alcohols, which can be regarded as phenyl-substituted methanol. In nature, it mostly exists in-the form of esters and is found in essential oils, such as jasmine oil, hyacinth oil, and Peruvian balm.

Benzyl alcohol is a very useful fixed fragrance agent, as jasmine, incense under the moon, Ilan and other fragrances in the deployment of indispensable spices. It is used for preparing soap, daily cosmetic flavor. Benzyl alcohol can be slowly natural oxidation, part of the formation of benzaldehyde and benzyl ether, so that commercially available products often with almond flavor, so it is not suitable for long storage. Benzyl alcohol is widely used in the production of industrial chemicals. It is used in coating solvent, photographic developer, polyvinyl chloride stabilizer, medicine, synthetic resin solvent, vitamin B injection solvent, ointment or liquid preservative. It can be used as a desiccant for nylon silk, fiber and plastic film, solvent of cellulose ester and casein, intermediate of benzyl ester or ether. At the same time, it is widely used in pens (ballpoint pen oil), paint solvent and other aspects.

It should be stored in a cool, ventilated warehouse, away from fire and heat sources, and should be separated from oxidizer and edible chemicals, avoiding mixed storage, equipped with the appropriate variety and quantity of fire equipment.

ii. Synthetic process route

Cannizzaro reaction is an aldehyde without α-hydrogen. Under the action of a strong base, an intermolecular redox reaction occurs, resulting in the organic disproportionation of one molecule of carboxylic acid and one molecule of alcohol. The Italian chemist Stanislaw Cannizzaro dicovered this reaction in 1895 by treating benzaldehyde with plant ash, obtained benzoic acid and benzyl alcohol, first discovered this reaction, therefore, hence the name Cannizzaro reaction. For aliphatic aldehydes, aromatic aldehydes or heterocyclic aldehydes that do not contain α-hydrogen atoms, under the action of concentrated alkali, the aldehyde molecules themselves undergo oxidation and reduction reactions at the same time to generate the corresponding carboxylic acid (carboxylate is formed in alkali solution) and the organic disproportionation reaction of alcohol.

In this reaction, benzoic acid and benzyl alcohol can be synthesized by the Cannizzaro reaction through benzaldehyde under the action of concentrated alkali. The synthesis process is as follows:

$$\text{PhCHO} \xrightarrow{\text{KOH}} \text{PhCH}_2\text{OH} + \text{PhCO}_2\text{K}$$

$$\text{PhCO}_2\text{K} \xrightarrow{\text{H}^+} \text{PhCOOH}$$

III. Main instruments and reagents

i. Main instruments: conical flask, analysis balance, heating set, constant pressure dripping funnel, magnetic stirrer, drying tube, three-necked flask, distillation head, reflux condenser, liquid separation funnel, thermometer, oil bath, beaker, glass rod, vacuum pump, *etc.*

ii. Main reagents: benzaldehyde, potassium hydroxide, ether, 10% sodium carbonate, concentrated hydrochloric acid, anhydrous magnesium sulfate, anhydrous potassium carbonate, sodium bisulfite, methanol, *etc.*

IV. Experimental steps

Add potassium hydroxide solution (9 g of potassium hydroxide and 9 mL of water) into a 100 mL three-necked round-bottomed flask equipped with a magnetic agitator, a thermometer and a condensing tube, and stir quickly. Cool to room temperature with the ice water. Slowly add 10 mL of freshly steamed benzaldehyde. The reaction system becomes a white paste, which is left overnight. Add 30 mL of water into the reaction system and stir vigorously until all solids dissolve. Carry out the extraction with ethyl ether 3 times, combin the organic phase, and wash with 2.5 mL of saturated sodium bisulfite, 5 mL of 10% sodium carbonate solution and 5 mL of water. Then, dry thoroughly anhydrous magnesium sulfate and filter. Evaporate the ether from the filtrate. Followed by distillation of benzyl alcohol, collect the fraction at 204-206℃, which is benzyl alcohol.

In the aqueous solution after ether extraction, adjust the pH with concentrated hydrochloric acid until Congo red test paper turns blue. Place it in the refrigerator to cool, and solid will precipitate which is the crude product of benzoic acid. The crude product can be recrystallized from water to obtain pure benzoic acid.

V. Attention or thinking questions

i. Attention

Try to write the reaction mechanism of this reaction.

ii. Thinking questions

According to what principle are the two products in this reaction separated and purified?

实验三十六　二苯甲醇的合成（2）

一、实验目的

1. 掌握二苯甲醇合成反应机理。
2. 熟悉二苯甲醇的合成实验操作过程。
3. 了解二苯甲醇的理化性质及用途。

二、实验原理

（一）化合物简介

二苯甲醇又名二苯基甲醇、双苯甲醇、α-苯基苯甲醇、羟基-二苯基甲烷，分子式为 $C_{13}H_{12}O$。常温下白色至浅米色结晶固体，易溶于乙醇、醚、氯仿和二硫化碳，20℃水中溶解度仅为 0.5 g/L。低毒，避免与皮肤和眼睛接触，缺乏有关毒性数据，毒性可参照甲醇。遇明火、高温、遇强氧化剂可燃烧，释放出有毒气体。

二苯甲醇主要用于有机合成、医药工业中间体。作为苯甲托品、苯海拉明及乙酰唑胺的中间体。

（二）合成工艺路线

二苯甲醇的合成，以二苯甲酮为原料，在锌粉的条件下还原制得，其合成工艺路线具体如下：

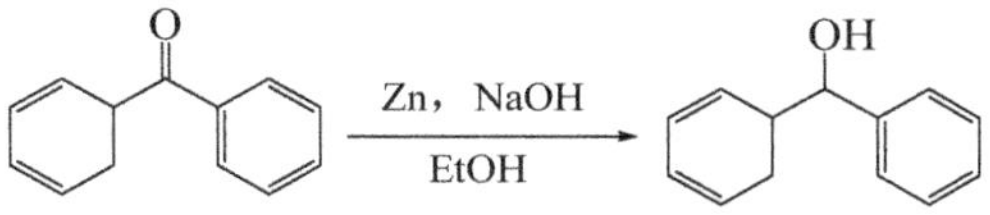

三、主要仪器和试剂

1. 主要仪器：加热套、磁力搅拌器、分析天平、干燥管、三颈烧瓶、蒸馏头、回流冷凝器、滴液漏斗、温度计、油浴锅、烧杯、玻璃棒、布氏漏斗、抽滤瓶、真空泵等。
2. 主要试剂：二苯甲酮、锌粉、95%乙醇、石油醚、浓盐酸、氢氧化钠等。

四、实验步骤

在配有磁力搅拌器、温度计、冷凝管、干燥管的 100 mL 三颈烧瓶中，加入氢氧化钠 1 g、二苯甲酮 1 g、锌粉 1 g 以及 95%乙醇 10 mL，在室温下搅拌 30 min，升温至 80℃，保温反应 20 min。反应结束后，抽滤，滤饼用乙醇充分洗涤，合并滤液，倒至盛有 60 mL 冰水的烧杯中，快速搅拌，用浓盐酸调 pH 至 5~6，抽滤，干燥，得粗产物。粗产物可以用石油醚进行重结晶，得到二苯甲醇纯品。

五、注意事项与思考题

（一）注意事项

试写出本反应的反应机理。

（二）思考题

1. 设计一条新的二苯甲醇的合成工艺路线？
2. 本反应中用石油醚作为重结晶溶剂，还有哪些溶剂可以用作此产物的重结晶？
3. 本反应中用金属锌作为催化剂，还有哪些催化剂可以用在本反应中？

Experiment 36 Synthesis of diphenyl carbinol (2)

I. Purpose of the experiment

i. To master the synthesis mechanism of diphenyl carbinol.
ii. To familiar with the experimental operation process of diphenyl carbinol synthesis.
iii. To understand the physicochemical properties and uses of diphenyl carbinol.

II. Experimental principle

i. Compound introduction

Diphenyl carbinol, also known as diphenyl carbinol, α-phenylcarbinol, hydroxy-diphenylmethane, has the molecular formula $C_{13}H_{12}O$. It is a white to light beige crystalline solid at room temperature. It is easily solubled in ethanol, ether, chloroform and carbon disulfide, with a solubility of only 0.5 g/L at 20 ℃. It is low toxicity. Pay attention to avoid contact with skin and eyes. With lacking of toxicity data, it can refer to methanol toxicity. It is flammable when exposed to open flame, high temperature, strong oxidant, and release toxic gas.

Diphenyl carbinol is mainly used as an intermediate in organic synthesis and in the pharmaceutical industry. It serves as an intermediate of benzatropine, Diphenhydramine and acetazolamide.

ii. Synthetic process route

The synthesis of diphenyl carbinol is mainly prepared by the reduction of benzophenone under the condition of zinc powder. The synthesis process is as follows:

$$\text{Ph-CO-Ph} \xrightarrow[\text{EtOH}]{\text{Zn, NaOH}} \text{Ph-CH(OH)-Ph}$$

III. Main instruments and reagents

i. Main instruments: heating sleeve, magnetic stirrer, analysis balance, drying tube, three-necked flask, distillation head, reflux condenser, drip funnel, thermometer, oil bath, beaker, glass rod, Brinell funnel, suction bottle, vacuum pump, *etc*.

ii. Main reagents: benzophenone, zinc powder, 95% ethanol, petroleum ether, concentrated hydrochloric acid, sodium hydroxide, *etc*.

IV. Experimental steps

Add 1 g of sodium hydroxide, 1 g of benzophone, 1 g of zinc powder and 10mL of 95% ethanol into a 100 mL three-necked round-bottom flask equipped with a magnetic agitator, a thermometer, a condensing tube and a drying tube. Stir at room temperature for 30 min, then heat to 80℃, and maintain the reaction for 20 min. After the reaction, filter under vacuum, wash the filter cake with ethanol, combin filtrate, and pour it into a beaker containing 60 mL of ice water. Stir quickly. Adjust pH to 5-6 with concentrated hydrochloric acid. Filter and dry. Obtain the crude product. The crude product can be recrystallized with petroleum ether to obtain pure diphenyl carbinol.

V. Attention or thinking questions

i. Attention

Try to write the reaction mechanism of this reaction.

ii. Thinking questions

(i) Design a new synthesis process of diphenyl carbinol?

(ii) Petroleum ether is used as the recrystallization solvent in this reaction. What other solvents can be used as the recrystallization of this product?

(iii) Zinc metal is used as the catalyst in this reaction. What other catalysts can be used in this reaction?

第三部分 药物合成实验

Part Ⅲ Drug synthesis experiment

实验一　青霉素 G 钾盐的氧化

一、实验目的

1. 掌握青霉素的合成反应机理。
2. 熟悉青霉素的合成实验操作过程。
3. 了解青霉素的理化性质以及临床用途。

二、实验原理

（一）药物简介

青霉素钾又名苄青霉素、苄青霉素 G 钾盐，白色晶体性粉末，无臭或微有特异性臭，有吸湿性。易溶于水、生理盐水、葡萄糖溶液。水溶液在室温放置易失效，遇酸、碱、氧化剂等迅速失效。

青霉素对溶血性链球菌等链球菌属、肺炎链球菌和不产青霉素酶的葡萄球菌具有良好抗菌作用，对肠球菌有中等度抗菌作用；对淋病奈瑟菌、脑膜炎奈瑟菌、白喉棒状杆菌、炭疽芽孢杆菌、牛型放线菌、念珠状链杆菌、李斯特菌、钩端螺旋体和梅毒螺旋体敏感；对流感嗜血杆菌和百日咳鲍特氏菌亦具有一定抗菌活性，其他革兰氏阴性需氧或兼性厌氧菌对本药物敏感性差。

青霉素药物对梭状芽孢杆菌属、消化链球菌厌氧菌以及产黑色素拟杆菌等具良好抗菌作用，对脆弱拟杆菌的抗菌作用差。青霉素通过抑制细菌细胞壁合成，发挥杀菌作用。

化学结构式为：

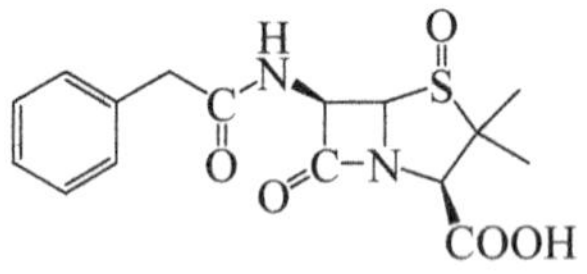

($2S$，$5R$，$6R$)-3,3-二甲基-6-(2-苯乙酰氨基)-7-氧代-4-硫杂-1-氮杂双环[3.2.0]庚烷-2-甲酸

（二）合成工艺路线

青霉素的合成，以过氧化氢氧化青霉素钾为原料制得，其合成工艺路线如下：

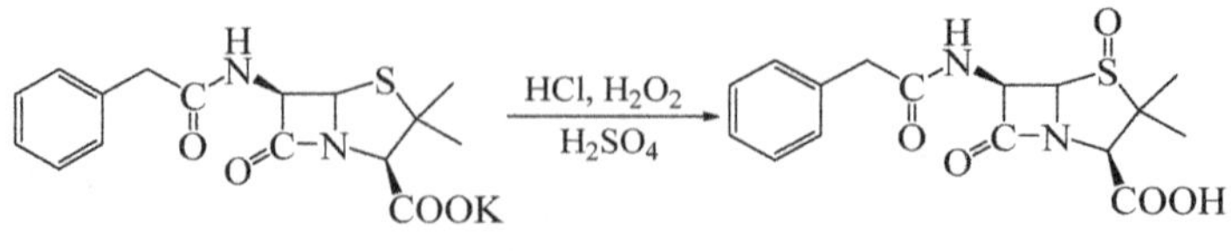

三、主要仪器和试剂

1. 主要仪器：三颈烧瓶、磁力搅拌器、分析天平、温度计、恒压滴液漏斗、冷凝管、分液漏斗、烧杯、橡胶管、锥形瓶、抽滤瓶、布氏漏斗、滤纸、真空泵、淀粉-KI 试纸等。

2. 主要试剂：50%过氧化氢、青霉素 G 钾盐、浓硫酸、稀盐酸等。

四、实验步骤

在配有磁力搅拌器、温度计、冷凝管以及恒压滴液漏斗的 100 mL 三颈烧瓶中，加入青霉素 G 钾盐 7.5 g、水 7 mL，随后用稀盐酸（0.5 mol/L）调节 pH 至 5，用冰水浴冷却至 0～5℃，随后，缓慢滴加 2 g H_2O_2（50%），继续搅拌反应，通过点薄层板，跟踪反应情况，当反应结束后，缓慢滴加硫酸（1 mol/L），当反应体系出现白色混浊后，改用稀硫酸（0.1 mol/L）调节 pH 至 2，冰水浴冷却，静置 2 h，抽滤，将冰水充分洗涤滤饼，至淀粉-KI 试纸不显色为止，干燥，得产物。

五、注意事项与思考题

（一）注意事项

试写出本反应的反应机理。

（二）思考题

1. 简述青霉素 G 的临床用途？
2. 解释青霉素 G 化学性质不稳定的原因？

Experiment 1 Oxidation of penicillin G potassium salts

I. Purpose of the experiment

i. To master the synthesis mechanism of penicillin.

ii. To mamiliar with the experimental operation process of penicillin synthesis.

iii. To understand the physicochemical properties and clinical uses of penicillin.

II. Experimental principle

i. Drug introduction

Penicillin potassium, also known as benzyl penicillin, benzylpenicillin G potassium salt, is a white crystalline powder. It is odorless or slightly specific odor, hygroscopicity. It is soluble in water, saline, and glucose solution. The aqueous solution is prone to inactivation when placed at room temperature, and rapidly loses efficacy upon contact with acid, base, oxidant, *etc*.

Penicillin has good antibacterial action against Streptococcus hemolyticus, Streptococcus pneumoniae and staphylococcus, which does not produce penicillase, and has moderate antibacterial action against enterococcus. It has antibacteriol action against Neisseria gonorrhoeae, Neisseria meningitidis, Corynebacterium diphtheria, Bacillus anthracis, actinomyces bovis, Streptobacillus candida, listeria, leptospirosis and Treponema pallidum. It also has certain antibacterial activity against Haemophilus influenzae and Bordetella pertussis. Other Gram-negative aerobic or facultative anaerobes have poor sensitivity to this drug.

Penicillin has good antibacterial action against Clostridium, anaerobic bacteria of digestive streptococcus and Bacteroides melanin, but poor antibacterial action against bacteroides fragile. Penicillin acts as a bactericidal agent by inhibiting cell wall synthesis.

The chemical structure formula is:

(2S, 5R, 6R) -3, 3-dimethyl-6-(2-acetyl amino benzene)-7-4-Sulfur impurity oxygen generation-1-nitrogen mixed double loop [3.2.0] heptane-2-formic acid

ii. Synthetic process route

The synthesis of penicillin is mainly made by oxidizing penicillin potassium with hydrogen peroxide as raw material. The synthesis process is as follows:

III. Main instruments and reagents

i. Main instruments: three-necked flask, magnetic stirrer, analytical balance, thermometer, constant pressure drip funnel, condensing tube, separation funnel, beaker, rubber tube, conical bottle, suction bottle, Brinell funnel, filter paper, vacuum pump, starch-KI test paper, *etc*.

ii. Main reagents: 50% hydrogen peroxide, penicillin G potassium salt, concentrated sulfuric acid, hydrochloric acid, *etc*.

IV. Experimental steps

Add 7.5 g of Penicillin G potassium salt and 7 mL of water into a 100 mL three necked bottom flask equipped with a magnetic agitator, a thermometer, a condensation tube and a constant-pressure drip funnel. Adjust pH to 5 with dilute hydrochloric acid (0.5 mol/L) and cool to 0-5 ℃ with an ice water bath. Slowly add 2 g of H_2O_2 (50%), and continue to stir the reaction. Monitor the reaction using thin-layer chromatography. When the reaction is over, slowly add sulfuric acid (1 mol/L). When the reaction system appears white turbidity, use dilute sulfuric acid (0.1 mol/L) to adjust the pH to 2. Cool in an ice water bath, and stand for 2 h. Filter, and wash the filter cake with ice water fully until the starch -KI test paper shows no color. Dry to obtain the product.

V. Attention or thinking questions

i. Attention

Try to write the reaction mechanism of this reaction.

ii. Thinking questions

(i) Describe the clinical use of penicillin G?

(ii) Explain the chemical instability of penicillin G?

实验二 盐酸胍法辛的合成

一、实验目的

1. 掌握盐酸胍法辛药物的合成反应机理。
2. 熟悉盐酸胍法辛药物的合成实验操作过程。
3. 了解盐酸胍法辛药物的理化性质及临床用途。

二、实验原理

（一）药物简介

盐酸胍法辛，化学名称为 N-脒基-2-（2,6-二氯苯基）乙酰胺单盐酸盐，其分子式为 $C_9H_9Cl_2N_3O$，分子量为 282.55，其为白色至类白色结晶粉末，难溶于水和乙醇，在甲醇中有相对较高的溶解度（>30 mg/mg）。

盐酸胍法辛是在全球范围内获准的第一个选择性 α-2A 肾上腺能受体激动剂，其对 α-2A 受体亚型的亲和力与 α-2B 或 α-2C 亚型相比高 15~20 倍。在治疗注意缺陷与多动障碍过程中，盐酸胍法辛的作用机制尚未得到充分确定。

临床前研究表明，盐酸胍法辛在前额皮质及基底神经节中，通过对 α-2 肾上腺素能受体中突触去甲肾上腺素传递的直接改性而调节信号。盐酸胍法辛属于抗高血压药物，用于治疗中度至重度高血压。

化学结构式为：

N-脒基-2-(2,6-二氯苯基）乙酰胺单盐酸盐

（二）合成工艺路线

盐酸胍法辛的合成，以 2,6-二氯氯苄为原料，经过亲核取代反应生成 2,6-二氯苯乙腈，在酸性条件下醇解成酯后，与胍反应成胍法辛，最后与盐酸成盐制备，其合成工艺路线如下：

三、主要仪器和试剂

1. 主要仪器：三颈烧瓶、磁力搅拌器、分析天平、温度计、恒压滴液漏斗、冷凝管、干燥管、分液漏斗、烧杯、橡胶管、锥形瓶、抽滤瓶、布氏漏斗、滤纸、真空泵等。

2. 主要试剂：2,6-二氯氯苄、35%氰化钠、三乙胺、氯苯、无水甲醇、浓硫酸、盐酸胍、异丙醇、异丙醇钠、2,6-二氯苯乙腈、无水硫酸钠、乙醇、乙醚、10%碳酸钠等。

四、实验步骤

（一）2,6-二氯苯乙腈的合成

在配有磁力搅拌器、温度计、冷凝管以及恒压滴液漏斗的 100 mL 三颈烧瓶中，加入 2,6-二氯氯苄 11.8 g、35%氰化钠水溶液 9.2 g、三乙胺 0.1 g 以及氯苯 15 mL，快速搅拌，升温至 95℃，保温反应 3 h。反应结束后，倒入分液漏斗中，分出有机层，用水充分洗涤至 pH 7，然后，用无水硫酸钠充分干燥，过夜，过滤。滤液蒸去溶液，用冰水浴冷却，有固体析出，得粗产物。可以用乙醇进行重结晶，得到 2,6-二氯苯乙腈的纯品。

（二）2,6-二氯苯乙酸甲酯的合成

在配有磁力搅拌器、温度计、冷凝管、干燥管以及恒压滴液漏斗的 100 mL 三颈烧瓶中，加入 2,6-二氯苯乙腈 4.8 g、无水甲醇 50 mL 以及浓硫酸 2 mL，快速搅拌，升温至回流，保温反应 16 h。反应结束后，通过减压蒸去甲醇，将反应瓶的残余物，倒入冰水 30 mL 中，用乙醚充分萃取 3 次，合并有机层，分别用水、10%碳酸钠溶液、水洗涤，然后用无水硫酸钠，充分干燥，过夜，抽滤，滤液回收溶剂后，减压蒸馏，收集馏分，得 2,6-二氯苯乙酸甲酯无色液体。

（三）盐酸胍法辛的合成

在配有磁力搅拌器、温度计、冷凝管、干燥管以及恒压滴液漏斗的 100 mL 三颈烧瓶中，加入盐酸胍 1.1 g、异丙醇钠 0.9 g 以及异丙醇 20 mL，快速搅拌，室温反应 24 h，反应结束后，过滤，除去氯化钠，向滤液中加入 2,6-二氯苯乙酸甲酯 2.2 g、无水异丙醇 5 mL，室温搅拌 30 min 后，回收溶剂，冷却，有固体析出。向固体中加入适量的异丙醇，调成浆状后，用氯化氢的乙醇溶液，调节 pH 至 1～2，减压抽滤，除去不溶物，滤液浓缩后，加入适量乙醚，有白色晶体析出，过滤，得粗产物，可以用乙醇-乙醚进行重结晶，得到盐酸胍法辛纯品。

五、注意事项与思考题

1. 试写出本反应的反应机理。
2. 熟悉重结晶的操作过程。

Experiment 2 Synthesis of guanfacine hydrochloride

I. Purpose of the experiment

i. To master the synthetic reaction mechanism of guanfacine hydrochloride.

ii. To familiar with the experimental operation process of guanfacine hydrochloride synthesis.

iii. To understand the physicochemical properties and clinical applications of guanfacine hydrochloride.

II. Experimental principle

i. Drug introduction

Guanfacine hydrochloride, chemically known is *n*-amidine-2-(2,6-dichlorophenyl) acetamide monohydrochloride, has a molecular formula of $C_9H_9Cl_2N_3O$ and a molecular weight of 282.55. It appears as a white to off-white crystalline powder. It is poorly soluble in water and ethanol, and has a relatively high solubility in methanol (>30 mg/mg).

Guanfacine hydrochloride is the first selective α-2a adrenoceptor agonist licensed worldwide, with an affinity for the α-2a receptor subtype that is 15-20 times higher than that for the α-2b or α-2c subtypes. The mechanism of action of guanfacine hydrochloride in the treatment of attention deficit and hyperactivity disorder has not been fully established.

Preclinical studies have shown that guanfacine hydrochloride modulates signaling in the prefrontal cortex and basal ganglia by directly modifying synaptic norepinephrine transmission in α-2 adrenergic receptors. Guanfacine hydrochloride be longs to the class of antihypertensive drug and is used to treat moderate to severe hypertension.

The chemical structure formula is:

N-amidine-2 -(2, 6-dichlorophenyl) acetamide monohydrochloride

ii. Synthetic process route

The synthesis of guan facine hydrochloride begins with 2, 6-dichlorophenylbenzene as the starting material, which undergoes a nucleophilic substitution reaction, which can be alkylated into esters under acidic conditions, then react with guanidine to form guanidine, and finally form salt with hydrochloric acid. The synthesis process is as follows:

$$\underset{\text{Cl}}{\overset{\text{Cl}}{\bigcirc}}\text{CH}_2\text{Cl} \xrightarrow[\text{chlorobenzene}]{\text{NaCN,(C}_2\text{H}_5)_3\text{N}} \underset{\text{Cl}}{\overset{\text{Cl}}{\bigcirc}}\text{CH}_2\text{CN} \xrightarrow[\text{H}_2\text{SO}_4]{\text{CH}_3\text{OH}} \underset{\text{Cl}}{\overset{\text{Cl}}{\bigcirc}}\text{CH}_2\text{COOCH}_3$$

$$\xrightarrow[\text{(2) HCl, C}_2\text{H}_5\text{OH}]{\text{(1) H}_2\text{N-C(NH)-NH}_2 \cdot \text{HCl, (CH}_3)_2\text{CHOH}} \underset{\text{Cl}}{\overset{\text{Cl}}{\bigcirc}}\text{CH}_2-\overset{\text{O}}{\underset{}{\text{C}}}-\underset{\text{H}}{\text{N}}-\overset{\text{NH}}{\underset{}{\text{C}}}-\text{NH}_2 \cdot \text{HCl}$$

III. Main instruments and reagents

i. Main instruments: three-necked flask, magnetic stirrer, analytical balance, thermometer, drying tube, constant pressure drip funnel, condensing tube, separation funnel, beaker, rubber tube, conical bottle, suction bottle, Brinell funnel, filter paper, vacuum pump, *etc.*

ii. Main reagents: 2, 6-dichlorobenzene, 35% sodium cyanide, triethylamine, chlorobenzene, anhydrous methanol, concentrated sulfuric acid, guanidine hydrochloride, isopropanol, sodium isopropanol, 2, 6-dichlorophenonitrile, anhydrous sodium sulfate, ethanol, ethyl ether, 10% sodium carbonate, *etc.*

IV. Experimental steps

i. Synthesis of 2, 6-dichlorophenacetonitrile

Add 11.8 g of 2, 6-dichlorobenzene, 9.2 g of 35% sodium cyanide aqueous solution, 0.1 g of triethylamine and 15mL of chlorobenzene into a 100 mL three-necked flask equipped with a magnetic agitator, a thermometer, a condensate tube and a constant pressure drip hopper. Stir rapidly and heat to 95℃, maintaining the reaction for 3 h. After the reaction, pour into the separation funnel, separate the organic layer, wash with water to pH 7. Then fully dry with anhydrous sodium sulfate overnight, and filter. The filtrate is steamed to remove the solution, and upon cooling in an ice water bath, some solid precipitates, yielding the crude product. It can be recrystallized with ethanol to obtain 2,6-dichlorophenacetonitrile.

ii. Synthesis of 2, 6-dichlorophenylacetic acid methyl ester

Add 4.8 g of 2, 6-dichlorophenacetonitrile, 50 mL of anhydrous methanol and 2 mL of concentrated sulfuric acid into a 100 mL three-necked flask equipped with a magnetic agitator, a thermometer, a condensing tube, a drying tube and a constant pressure drip funnel. Stir quickly and heat to reflux, maintaining the reaction for 16 h. After the reaction, the methanol is steamed by decompression. Pour the residue of the reaction bottle into 30 mL of ice water. Fully extract with ethyl ether for 3 times. Combin with the organic layer. Wash with water, 10% sodium carbonate solution and water respectively. Dry over anhydrous sodium sulfate overnight and filter. Recover the solvent from the filtrate by reduced pressure distillation to yield colorless liquid methyl 2, 6-

dichlorophenylacetic.

iii. Synthesis of guanfacine hydrochloride

Add 1.1 g of guanidine hydrochloride, 0.9 g of sodium isopropyl alcohol and 20 mL of isopropyl alcohol into a 100 mL three-neck flask equipped with a magnetic agitator, a thermometer, a condensing tube, a drying tube and a constant pressure drip funnel. Stir quickly and respond at room temperature for 24 h. After the reaction, filter to remove sodium chloride. Add 2.2 g of 2,6-dichlorophenylacetic acid methyl ester and 5 mL of anhydrous isopropyl alcohol. After stirring at room temperature for 30 min, the solvent is recovered and upon cooling. Some solids are separated out. Add the appropriate amount of isopropyl alcohol to the solid. Adjust to the pH to 1-2 with hydrogen chloride ethanol solution. Filter the mixture under reduced pressure to remove insoluble matter. Concentrate the filtrate. Add the appropriate amount of ether resulting in the precipitation of white crystals, which are filtered to obtain the crude product. It can be recrystallized with ethanol-ethers to obtain guanfacine hydrochloride pure product.

V. Attention or thinking questions

i. Try to write the reaction mechanism of this reaction.

ii. Familiar with the operation process of recrystallization.

实验三 诺氟沙星的合成

一、实验目的

1. 掌握诺氟沙星药物合成过程中涉及的缩合等重要反应。
2. 熟悉诺氟沙星药物全合成工艺。
3. 了解诺氟沙星药物合成中的各步中间体产物处理的有效措施、方法。
4. 了解诺氟沙星药物合成过程中的常规基本操作。
5. 了解诺氟沙星药物的理化性质和临床用途。

二、实验原理

（一）药物简介

诺氟沙星为白色至淡黄色结晶性粉末，无臭，味微苦，具有吸湿性，在二甲基甲酰胺中略溶，在水或乙醇中极微溶解，在醋酸、盐酸或氢氧化钠溶液中易溶。熔点为218～224℃。

诺氟沙星为第三代喹诺酮类抗菌药，会阻碍消化道内致病细菌DNA旋转酶的作用，阻碍细菌DNA复制，对细菌有抑制作用，是治疗肠炎痢疾的常用药。但此药对未成年人骨骼形成有延缓作用，会影响发育，故禁止未成年人服用。

化学结构式为：

1-乙基-6-氟-1,4-二氢-4-氧-7-（1-哌嗪基）-3-喹啉羧酸

（二）合成工艺路线

以氟氯苯胺、乙氧基次甲基丙二酸二乙酯为原料，经过高温缩合反应、环合反应，得到6-氟-7-氯-1,4-二氢-4-氧喹啉-3-羧酸乙酯；接着用溴乙烷试剂进行乙基化反应，得到1-乙基-6-氟-7-氯-1,4-二氢-4-氧喹啉-3-羧酸乙酯，然后经过水解反应，再与乙酸酐以及硼酸反应生成硼螯合物，紧接着与哌嗪进行缩合，最后在碱性条件下，水解得到终产物诺氟沙星。

$$\xrightarrow[\text{(2) OH}^-]{\text{(1) H}^+}$$

[结构式：环丙沙星类似物 - 6-氟-7-哌嗪基-1-乙基-4-氧-1,4-二氢喹啉-3-羧酸]

三、主要仪器和试剂

1. 主要仪器：分析天平、烧杯、恒压滴液漏斗、量筒、加热套、玻璃棒、旋转蒸发仪、三颈烧瓶、回流冷凝管、抽滤瓶、真空泵、布氏漏斗、油泵等。

2. 主要试剂：乙氧基次甲基丙二酸二乙酯、氟氯苯胺、无水碳酸钾、溴乙烷、甲苯、丙酮、二甲亚砜、二甲基甲酰胺、硼酸、乙酸酐、氯化锌、乙酸、液体石蜡、哌嗪等。

四、实验步骤

（一）6-氟-7-氯-1,4-二氢-4-氟喹啉-3-羧酸乙酯的合成

在配有磁力搅拌器、回流冷凝管、温度计装置的三颈烧瓶中，依次加氟氯苯胺、乙氧基次甲基丙二酸二乙酯，在快速搅拌下，升温至120℃，保温反应2.5 h，反应结束后，将体系降温至室温，随后将反应的回流装置，转换成蒸馏装置，加入石蜡80 mL，加热到250℃，有乙醇生成，回收乙醇，30 min后，将反应体系冷却到60℃，进行抽滤，滤饼分别用甲苯、丙酮洗涤，直至滤饼颜色为灰白色，干燥处理。

注：

1. 本反应对无水的要求特别严格，反应中所涉及的仪器、试剂等均需做无水处理。

2. 本反应的环合温度应控制在250℃左右，为避免温度大于270℃，开始反应时，反应液相对黏稠，为防止反应体系受热不均匀，应适当加快搅拌速度。

3. 本环合反应为Gould-Jacobs反应（古尔德-雅各布斯反应），反应条件控制不当，会有反式产物的生成。

（二）1-乙基-6-氟-7-氯-1,4-二氢-4-氧喹啉-3-羧酸乙酯的合成

在配有磁力搅拌器、回流冷凝管、温度计以及恒压滴液漏斗的250 mL三颈烧瓶中，加入6-氟-7-氯-1,4-二氢-4-氟喹啉-3-羧酸乙酯、无水碳酸钾以及二甲基甲酰胺，剧烈搅拌下，升温至70～80℃时，滴加溴乙烷，在60 min内滴加完毕，然后继续升温至100～110℃，保温反应6～8 h，反应完成后，减压蒸出二甲基甲酰胺，自然降温至50℃左右，随后加入200 mL水，有固体析出，抽滤，滤饼用水洗多次，干燥，得到粗品，随后用乙醇进行重结晶。

注：

1. 反应对无水的要求特别严格，本反应中所用的试剂二甲基甲酰胺、无水碳酸钾要先经过无水干燥处理，即使较少的含水量亦对反应有较大的影响。

2. 反应试剂溴乙烷沸点的沸点相对较低，非常容易挥发，为了避免损失，小心加入，亦要注意反应体系的气密性。

3. 一般来说，滤饼洗涤要充分，注意碾碎结成块状的固体，要多次用水洗涤，除去多余的碳酸钾。

4. 终产物的具体重结晶步骤：取一定量的 1-乙基-6-氟-7-氯-1,4-二氢-4-氧喹啉-3-羧酸乙酯粗品，加入其 4~5 倍量的乙醇溶剂，回流状态下，加入适量活性炭，趁热抽滤，随后将滤液自然冷却至 10℃ 左右，有晶体析出，抽滤，将滤饼充分洗涤，干燥，得重结晶产物。

（三）硼螯合物的合成

在配有磁力搅拌器、冷凝管、温度计以及恒压滴液漏斗的三颈烧瓶中，加入氯化锌 1 g、硼酸 3.3 g 以及乙酸酐 17 g，剧烈搅拌，同时升温至 79℃，引发反应，随后将反应体系升温至 120℃，缓慢滴加乙酸酐，滴加完毕后，回流反应 1.5 h，冷却，加入 1-乙基-6-氟-7-氯-1,4-二氢-4-氧喹啉-3-羧酸乙酯，回流反应 3 h，自然冷却至室温，加适量水，抽滤，用少量冷的乙醇，进行洗涤，直至颜色变成灰白色，干燥，得产品。

注：

1. 反应中，硼酸与乙酸酐反应生成硼酸三乙酰酯产物。

2. 反应中，对产物用乙醇进行洗涤时，要保证产物的温度已经降下来，再用冷的乙醇进行洗涤，防止产品损失。

（四）诺氟沙星的合成

在装有磁力搅拌器、回流冷凝管以及温度计的三颈烧瓶中，加入上步反应的螯合物 10 g、无水哌嗪 8 g 以及溶剂二甲亚砜溶剂 30 g，升温至 110℃，保温反应 3 h，随后，将反应体系温度降至 90℃，加入 10% NaOH 溶液 20 mL，接着回流反应 2 h，结束后，将反应体系降温至室温，加水 50 mL，用乙酸调节 pH 至 7.1~7.2，抽滤，水洗，得到粗产物。在 250 mL 烧杯中，加入粗产物以及水 100 mL，用乙酸调节 pH 至 4~5，有产物析出，抽滤，充分水洗，干燥，得到产物诺氟沙星。

注：

1. 诺氟沙星能够溶于碱，判断反应是否完全时可以在反应液中再加入 NaOH 进行回流，观察溶液的颜色是否澄清透明，若有此现象发生，则表明完全反应。

2. 在过滤产物的粗品时，要将滤饼充分洗涤，除去其中的乙酸盐，避免带入至精制的步骤中。

五、注意事项与思考题

（一）注意事项

1. 试写出古尔德-雅各布斯反应的反应机理。
2. 合成中涉及高温反应，请写出高温反应操作过程中的安全注意事项。

（二）思考题

1. 本反应过程中涉及的副产物有哪些？请列举。
2. 如减压除去二甲基甲酰胺溶剂时，有什么注意事项？

Experiment 3　Synthesis of norfloxacin

Ⅰ. Purpose of the experiment

　　i. To master the condensation and other important reactions involved in the synthesis of norfloxacin.

　　ii. To familiar with the total synthesis process of norfloxacin.

　　iii. To understand the effective measures and methods for the treatment of intermediates in each step of norfloxacin drug synthesis.

　　iv. To understand the routine and basic operations in the process of norfloxacin synthesis.

　　v. To understand the physicochemical properties and clinical uses of norfloxacin.

Ⅱ. Experimental principle

i. Drug introduction

　　Norfloxacin is a white to pale yellow powder, odorless, with a slightly bitter taste, hygroscopic crystalline. It is slightly soluble in dimethylformamide. It is very slightly soluble in water or ethanol. It is easily soluble in acetic acid, hydrochloric acid or sodium hydroxide solutions. Its melting point is 218-224℃.

　　Norfloxacin is a third-generation quinolone antibacterial drug, which can hinder the action of DNA rotating enzyme of pathogenic bacteria in the digestive tract, hinder bacterial DNA replication, and has an inhibitory effect on bacteria. Norfloxacin is a common drug for the treatment of enteritis and dysentery. However, this drug has a delaying effect on the skeletal development of mintors, which can affect growth, hence it is prohibited for use by minters.

　　The chemical structure formula is:

1-ethyl-6-fuoro-1,4-dihydro-4-oxo-7-(1-piperazinyl)-3-quindinecarboxylic acid

ii. Synthetic process route

　　Using 4-chloroaniline and ethyl 2-ethoxy-2-methylpropanedioate as raw materials, a high-temperature condensation reaction and cyclisation reaction yield 6-fluoro-7-chloro-1,4-dihydro-4-oxoquinoline-3-carboxylic acid ethyl ester. Subsequently, an ethylation reaction is performed with bromoethane to obtain 1-ethyl-6-fluoro-7-chloro-1,4-dihydro-4-oxoquinoline-3-carboxylic acid ethyl ester, which is then hydrolysed and reacted with acetic anhydride and boric acid to form a boron chelate, followed by condensation with piperazine, and finally hydrolysis under alkaline conditions to yield the final product, which is norfloxacin.

III. Main instruments and reagents

i. Main instruments: analytical balance, beaker, constant pressure drip funnel, measuring cylinder, heating sleeve, glass rod, rotary evaporator, three-necked flask, reflux condensing tube, filter bottle, vacuum pump, Brinell funnel, oil pump, *etc.*

ii. Main reagents: diethyl ethoxy methylmalonate, fluorochloroaniline, potassium carbonate, bromoethane, toluene, acetone, dimethyl sulfoxide, dimethyl formamide, boric acid, acetic anhydride, zinc chloride, acetic acid, liquid paraffin, piperazine, *etc.*

IV. Experimental steps

i. Synthesis of 6-fluoro-7-chloro-1, 4-dihydro4-fluoro-quinolin-3-carboxylate ethyl ester

Add 4-chloroaniline and ethyl 2-ethoxy-2-methylpropanedioate into a three-necked flask equipped with a magnetic stirrer, a reflux condenser, and a thermometer. Stir repidly and heat to 120 ℃, maintaining the reaction for 2.5 h. After the reaction, cool to the room temperature. Convert the reflux apparatus to a distillation apparatus. Add 80 mL of paraffin and heat to 250 ℃. The ethanol will generate. Then recover the ethanol. After 30 minutes, cool the reaction to 60 ℃ for suction filtration. Wash the filter cake with toluene and acetone until the filter cake is greyish-white, followed by drying.

Note:

(i) The reaction has strict requirements for anhydrous conditions, so all instruments and reagents involved must be treated to be anhydrous.

(ii) The cyclic temperature of the reaction should be controlled at about 250 ℃. In order to avoid a temperature greater than 270 ℃, the reaction liquid phase is viscous at the beginning of the reaction.

In order to prevent the reaction system from being heated unevenly, the stirring speed should be accelerated appropriately.

(iii) This cyclic reaction is the Gould-Jacobs reaction, and the reaction conditions are not properly controlled, resulting in the formation of trans products.

ii. Synthesis of 1-ethyl-6-fluoro-7-chloro-1, 4-dihydro-4-quinolin-3-carboxylate

Add 6-fluoro-7-chloro-1, 4-dihydro-4-fluoro-quinoline-3-carboxylate ethyl ester, anhydrous potassium carbonate and dimethylformamide into a 250 mL three-necked flask equipped with a magnetic agitator, a reflux condensation tube, a thermometer and a constant pressure drip funnel. Under intense agitation, heat the temperature to 70-80℃, and add bromo-ethane within 60 min. Then further raise to 100-110°C and maintain for 6-8 h. After the reaction, remove the dimethylformamide under reduced pressure, and cool naturally to around 50℃. Then, add 200 mL of water, resulting in solid precipitation, which is suction filtered. Wash the filter cake multiple times with water, and dry to obtain the crude product, which is then recrystallised using ethanol.

Note:

(i) The reaction has strict requirements for anhydrous conditions, so the reagents used, dimethylformamide and anhydrous potassium carbonate must be treated to be anhydrous, as even a small amount of moisture can significantly affect the reaction.

(ii) The boiling point of the reaction reagent bromoethane is relatively low, so it is very volatile. In order to avoid loss, be careful to add, and pay attention to the air tightness of the reaction system.

(iii) Generally speaking, the filter cake should be washed thoroughly, and the solid formed into chunks should be crushed carefully. The excess potassium carbonate should be removed by washing with water several times.

(iv) Specific recrystallization steps of the final product: Take a certain amount of 1-ethyl-6-fluoro-7-chloro-1,4-dihydro-4-oxyquinoline-3-carboxylate crude product, add 4-5 times the amount of ethanol solvent. Under reflux state, add an appropriate amount of activated carbon, while hot filter. Then the filtrate is naturally cooled to about 10℃. There is some crystals precipitation, then filter. The filter cake is thoroughly washed and dried. The recrystallization product is obtained.

iii. Synthesis of boron chelates

Add 1 g of zinc chloride, 3.3 g of boric acid and 17 g of acetic anhydride into a three-neck round-bottom flask equipped with a magnetic agitator, a condensing tube, a thermometer and a constant pressure drip hopper. Stir vigorously and heat to 79℃ to initiate the reaction. Then heat the reaction system to 120℃ and slowly add acetic anhydride. After the addition reflux the mixture for 1.5 h and cool. Add 1-ethyl-6-fluoro-7-chloro-1, 4-dihydro-4-oxyquinoline-3-carboxylate ethyl ester. Reflux the reaction for 3 h, and natural cool to room temperature. Add appropriate water and filter. Wash with a small amount of cold ethanol until the color becomes gray. Dry to obtain the product.

Note:

(i) In the reaction, boric acid reacts with acetic anhydride to produce a triacetyl borate product.

(ii) In reaction, when washing the product with ethanol, it is necessary to ensure that the temperature of the product has been lowered, and then wash with cold ethanol to prevent product loss.

iv. Synthesis of norfloxacin

Add 10 g of chelate, 8 g of anhydrous piperazine and 30 g of dimethyl sulfoxide into a three necked flask equipped with a magnetic agitator, a reflux condensing tube and thermometera. Heat to 110℃, and maintain the reaction for 3 h. Then, lower the temperature of the reaction system to 90℃. Add 20 mL of 10% NaOH solution. After reflux reaction for 2 h, cool to room temperature. Add 50 mL of water. Adjust the pH to 7.1-7.2 with acetic acid. Pump the filtration and wash with water to obtain the crude products. Add the crude product and 100 mL of water into a 250 mL beaker. Adjust the pH to 4-5 with acetic acid, resulting in product precipitation. Filter the mixture and wash thoroughly. Dry to obtain the product, with is norfloxacin.

Note:

(i) Norfloxacin can be dissolved in alkali. In order to judge whether the reaction is complete, NaOH can be added to the reaction solution for reflux to observe whether the color of the solution is clear and transparent.

(ii) When filtering the coarse product, the filter cake should be fully washed to remove the acetate from it, so as to avoid bringing it into the refining step.

Ⅴ. Attention or thinking questions

i. Attention

(i) Try to write the reaction mechanism of the Gould-Jacobs reaction.

(ii) High temperature reaction is involved in synthesis. Please write down the safety precautions in the operation process of high temperature reaction.

ii. Thinking questions

(i) What are the by-products involved in the reaction process? Please list them.

(ii) If decompression removes dimethylformamide solvent, what should be noted?

实验四　来曲唑的合成

一、实验目的

1. 掌握来曲唑药物的合成反应机理。
2. 熟悉来曲唑药物的合成操作过程。
3. 了解来曲唑药物的理化性质及临床用途。

二、实验原理

（一）药物简介

来曲唑是新一代芳香化酶抑制剂，分子式为 $C_{17}H_{11}N_5$，为人工合成的苄三唑类衍生物。来曲唑通过抑制芳香化酶，使雌激素水平下降，从而消除雌激素对肿瘤生长的刺激作用，用于乳腺癌的内分泌治疗。

来曲唑在体内的活性比第一代芳香化酶抑制剂氨鲁米特强 150~250 倍。由于其选择性较高，不影响糖皮质激素、盐皮质激素和甲状腺功能，大剂量使用对肾上腺皮质类固醇类物质分泌无抑制作用，因此具有较高的治疗指数。

临床前研究表明，来曲唑对全身各系统及靶器官无潜在毒性，无诱变性及致癌作用，且不良反应较小，耐受性良好，与其他芳香化酶抑制剂和抗雌激素药物相比，抗肿瘤作用更强。适用于治疗抗雌激素治疗无效的晚期乳腺癌绝经后以及早期乳腺癌患者的治疗。

化学结构式为：

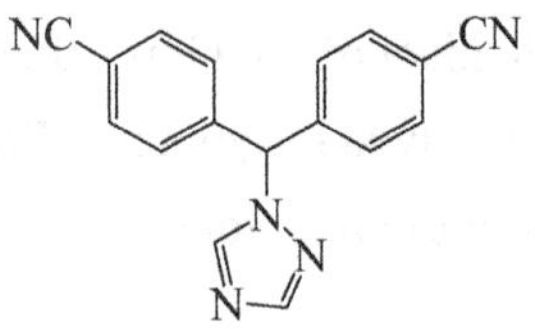

1-[双(4-氰基苯基)甲基]-1,2,4-三氮唑

（二）合成工艺路线

来曲唑的合成，主要是通过对氰基苄卤和 1H-1,2,4-三氮唑原料制得中间体 1-(4 氰基苄基)-1H-1,2,4-三氮唑后，与对卤代苯甲腈反应制备，其合成工艺路线具体如下：

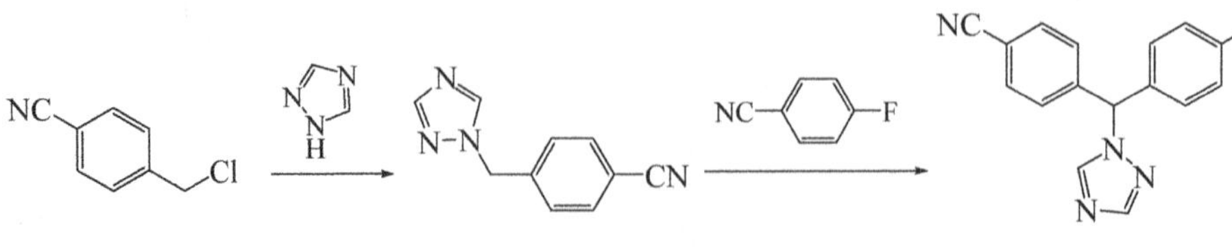

三、主要仪器和试剂

1. 主要仪器：三颈烧瓶、磁力搅拌器、温度计、分析天平、恒压滴液漏斗、冷凝管、干

燥管、分液漏斗、烧杯、橡胶管、锥形瓶、抽滤瓶、布氏漏斗、滤纸、真空泵等。

2. 主要试剂：对氰基苄氯、1,2,4-三氮唑、无水碳酸钾、碘化钾、乙腈、200-300 目硅胶、乙酸乙酯、二氯甲烷、二甲基甲酰胺、叔丁醇钾、对氟苯腈、饱和氯化铵、饱和氯化钠、无水硫酸镁、95%乙醇等。

四、实验步骤

（一）1-(4-氰基苄基)-1H-1,2,4-三氮唑合成

在配有磁力搅拌器、温度计、冷凝管、干燥管以及恒压滴液漏斗的 100 mL 三颈烧瓶中，加入对氰基苄氯 15.2 g、乙腈 40 mL，快速搅拌，完全溶解后，继续加入 1,2,4-三氮唑 10.4 g、无水碳酸钾 13.8 g 以及碘化钾 0.8 g，升温至 70℃，保温反应 5 h，反应结束后，自然冷却至室温，有固体产生，抽滤，用少许二氯甲烷洗涤滤饼，合并滤液，蒸去二氯甲烷，剩余物浓缩，用硅胶柱纯化，得到白色结晶，得 1-(4-氰基苄基)-1 H-1,2,4-三氮唑。

（二）来曲唑的合成

在配有磁力搅拌器、温度计、冷凝管以及恒压滴液漏斗的 250 mL 三颈烧瓶中，加入叔丁醇钾 4.8 g、二甲基甲酰胺 10 mL，用冰水浴冷却至 0℃，快速搅拌，随后滴加 1-(4-氰基苄基)-1 H-1,2,4-三氮唑 3.7 g 与二甲基甲酰胺 10 mL 组成的溶液，继续搅拌 60 min 后，继续滴加对氟苯腈 2.7 g 与二甲基甲酰胺 5 mL 组成的溶液，继续反应 60 min。反应结束后，向反应混合物中，加入适量饱和氯化铵溶液，若有固体，可加水直至固体溶解，至溶液澄清。用乙酸乙酯充分萃取，合并乙酸乙酯萃取液，用水充分洗涤，再用饱和氯化钠溶液洗 1 次。随后，有机相用无水硫酸镁，进行干燥，过夜，减压蒸馏，有固体析出。可以使用 95%乙醇重结晶，得到来曲唑纯品。

五、注意事项与思考题

1. 试写出本反应的反应机理。
2. 本反应中用到柱层析，熟悉柱层析的相关操作流程。

Experiment 4 Synthesis of letrozole

I. Purpose of the experiment

i. To master the synthetic reaction mechanism of letrozole.

ii. To familiar with the operation process of letrozole synthesis.

iii. To understand the physicochemical properties and clinical uses of letrozole.

II. Experimental principle

i. Drug introduction

Letrozole is a new generation aromatase inhibitor with the molecular formula $C_{17}H_{11}N_5$, which is a synthetic benzyl triazole derivative. Letrozole is used for endocrine treatment of breast cancer by inhibiting aromatases and reducing estrogen levels, thus eliminating the stimulating effect of estrogen on tumor growth.

Letrozole is 150 to 250 times more active in body than the first generation aromatase inhibitor, aminoglutethimide. Due to its high selectivity, it does not affect glucocorticoid, corticosteroids and thyroid function, and has no inhibitory effect on the secretion of corticosteroids in large doses, so it has a high therapeutic index.

Preclinical studies have shown that letrozole has no potential toxicity, no mutagenic or carcinogenic effect on all systems and target organs of the body, and has little toxic and side effects, good tolerance, and stronger anti-tumor effect compared with other aromatase inhibitors and anti-estrogen drugs. It is suitable for the treatment of postmenopausal patients with advanced breast cancer who have failed anti-estrogen therapy and the treatment of early breast cancer.

The chemical structure formula is:

1-[bis (4-cyanophenyl) methyl]-1, 2, 4-triazole

ii. Synthetic process route

The synthesis of letrozole mainly involves the preparation of intermediate 1-(4-cyanophenyl) -1h-1, 2, 4-triazole from p-cyanophenyl halide and 1h-1, 2, 4-triazole, then the reaction of intermediate 1-(4-cyanophenyl) -1h-1, 2, 4-triazole with *p*-halo benzonitrile. The synthesis process is as follows:

III. Main instruments and reagents

i. Main instruments: three-necked flask, magnetic stirrer, thermometer, analytical balance, constant pressure drip funnel, condensing tube, drying tube, separation funnel, beaker, rubber tube, conical bottle, suction bottle, Brinell funnel, filter paper, vacuum pump, *etc*.

ii. Main reagents: *p*-benzyl chloride, 1, 2, 4-triazole, anhydrous potassium carbonate, potassium iodide, acetonitrile, 200-300 mesh silica gel, ethyl acetate, methylene chloride, dimethylformamide, potassium tert-butanol, *p*-fluorobenzonitrile, ammonium chloride, sodium chloride, anhydrous magnesium sulfate, 95% ethanol, *etc*.

IV. Experimental steps

i. 1-(4-cyano-benzyl) -1h-1, 2, 4-triazole synthesis

Add 15.2g of *p*-cyanophenyl chloride and 40 mL of acetonitrile into a 100 mL three-neck flask equipped with a magnetic stirrer, a thermometer, a condensing tube, a drying tube and a constant pressure drip funnel. Stir quickly until completely dissolved, then add 10.4 g of 1, 2, 4-triazole, 13.8 g of anhydrous potassium carbonate and 0.8 g of potassium iodide. Heat to 70℃ and maintain for 5 h. After the reaction, cool naturally to room temperature. A solid will form. Filter by suction. Wash the cake with a little dichloromethane. Combin the filtrate. Evaporate the dichloromethane. Concentrate the reside. Purfy using a silica gel column to obtain white crystal yielding 1-(4-cyano-benzyl)-1 h-1, 2, 4-triazole.

ii. Synthesis of letrozole

Add 15.2 g of 4-cyanobenzyl chloride and 40 mL of acetonitrile into a 250 mL three-neck flask equipped with a magnetic stirrer, thermometer, condenser, and constant pressure dropping funnel. Cool to 0°C using an ice-water bath. Then add 3.7 g of 1-(4-cyanobenzyl)-1H-1,2,4-triazole in 10 mL of dimethylformamide, and continue stirring for 60 min. Then, continue to dropwise add a solution of 2.7 g of para-fluorobenzonitrile in 5 mL of dimethylformamide, and continue the reaction for another 60 minutes. After the reaction, add an appropriate amount of saturated ammonium chloride solution to the reaction mixture. If there is solid, add water until the solid dissolves, and the solution becomes clear. Extract thoroughly with ethyl acetate, combine the ethyl acetate extracts, wash with water, and then wash once with saturated sodium chloride solution. Subsequently, dry the organic phase with anhydrous magnesium sulfate overnight, and perform reduced pressure distillation to precipitate a solid. Recrystallization can be performed using 95% ethanol to obtain pure letrozole.

V. Attention or thinking questions

i. Try to write the reaction mechanism of this reaction.

ii. Used column chromatography in this reaction, familiar with the relevant operation process of column chromatography.

实验五　贝诺酯的合成

一、实验目的

1. 掌握贝诺酯药物合成中涉及的酯化反应方法。
2. 熟悉贝诺酯乙醇重结晶的方法。
3. 了解本反应中处理有害气体的方法。
4. 了解贝诺酯的理化性质及临床用途。

二、实验原理

(一) 药物简介

贝诺酯，又名扑炎痛，化学式为 $C_{17}H_{15}NO_5$，为阿司匹林与对乙酰氨基酚（扑热息痛）的酯化产物，是新型的消炎、解热、镇痛、治疗风湿病的药物。其作用机制基本与阿司匹林及对乙酰氨基酚相同。疗效与阿司匹林相似，不良反应比阿司匹林少。特点是较少引起胃肠道出血，患者易于耐受，作用时间比阿司匹林或对乙酰氨基酚长。

该药品为白色结晶性粉末，无味，不溶于水，易溶于热醇中，熔点 175～176℃。用于风湿性关节炎及其他发热而引起的中等疼痛的治疗。

化学结构式为：

4-乙酰氨基苯基乙酰水杨酸酯

(二) 合成工艺路线

贝诺酯的合成，主要是通过对乙酰氨基酚与乙酰水杨酸制得，阿司匹林与二氯亚砜在少量吡啶催化下，进行酸羟基的卤置换反应，生成 2-乙酰氧基苯甲酸。对乙酰氨基酚在氢氧化钠作用下，生成钠盐，再与 2-乙酰氧基苯甲酰氯进行酰基化反应，生成 2-乙酰氧基苯甲酸-4-乙酰氨基苯酯。其合成工艺路线具体如下：

三、主要仪器和试剂

1. 主要仪器：温度计、沸石、恒压滴液漏斗、分析天平、烧杯、量筒、加热套（油浴锅）、玻璃棒、旋转蒸发仪、三颈烧瓶、回流冷凝管、抽滤瓶、布氏漏斗、真空泵等。

2. 主要试剂：阿司匹林、氯化亚砜、吡啶、对乙酰氨基酚、氢氧化钠、无水乙醇等。

四、实验步骤

在配有磁力搅拌器、回流冷凝管、温度计的三颈烧瓶中，加入几粒沸石、阿司匹林 9 g、氯化亚砜溶剂 5 mL、碱性催化剂吡啶 1～2 滴，放置油浴上缓慢升温，在温度为 75℃时，剧烈搅拌，直至无气体逸出，整个过程大约 3 h。反应结束后，将回流装置调整成减压蒸馏装置，用油泵减压除去过量的氯化亚砜，随后冷却，得到乙酰水杨酰氯，加入丙酮 6 mL 干燥，待用。

在配有磁力搅拌器、恒压滴液漏斗、温度计的三颈烧瓶中，加入对乙酰氨基酚 8.6 g、水 50 mL，剧烈搅拌下，缓慢加入 18 mL 氢氧化钠水溶液（其中氢氧化钠为 3.3 g），反应体系的温度小于 15℃。随后，缓慢加入上步制得的乙酰水杨酰氯无水丙酮液，调节 pH 至 9.0～10.0，在室温下搅拌反应 2 h，反应结束后，减压抽滤，用水冲洗至中性，干燥，得其粗品。

在烧瓶中加入上步得到的贝诺酯粗品，分批次加入无水乙醇，一般乙醇的加入量为贝诺酯粗品的量的 8 倍，以在乙醇回流状态下刚好溶解为宜。回流后，将其冷却至室温，使得贝诺酯全部析出，减压抽滤，干燥，得到贝诺酯纯品。

五、注意事项与思考题

（一）注意事项

1. 本反应对无水的要求特别严格，反应中所涉及的仪器等均需做无水处理。对反应中生成的氯化氢和二氧化硫气体，具有刺激性，会污染空气，可以通过碱液吸收处理。
2. 为了便于搅拌，观察内温，使反应更趋完全，可适当增加氯化亚砜用量至 6～7 mL。
3. 反应中所用的吡啶主要是催化作用。

（二）思考题

1. 除了用二氯亚砜制备酰氯，还可以通过哪些方法制备酰氯？
2. 在由羧酸和氯化亚砜反应制备酰氯时，有何注意事项？
3. 试设计贝诺酯的其他合成路线？

Experiment 5 Synthesis of benorilate

I. Purpose of the experiment

i. To master esterification reaction methods involved in the synthesis of benorilate.

ii. To familiar with the recrystallization method of benorilate ethanol.

iii. To understand how to deal with harmful gases in this reaction.

iv. To understand the physicochemical properties and clinical uses of benorilate.

II. Experimental principle

i. Drug introduction

Benorilate, also known as propyl pain, has the chemical formula $C_{17}H_{15}NO_5$. It is an esterification product of aspirin and paracetamol. It is a new type of anti-inflammatory, antipyretic, analgesic, rheumatic drugs. Its mechanism of action is basically the same as aspirin and acetaminophen. The efficacy is similar to aspirin, with fewer adverse reactions than aspirin. It is less likely to cause gastrointestinal bleeding. It is easily tolerated by patients and lasts longer than aspirin or acetaminophen.

The drug is a white crystalline powder, tasteless. It is insoluble in water and easily soluble in hot alcohol. Its melting point is 175-176 ℃. It is used for the treatment of rheumatoid arthritis and other moderate pain caused by fever.

The chemical structure formula is:

4-acetamidophenyl-O-acetylsalicylate

ii. Synthetic process route

The synthesis of benorilate is mainly through the preparation of acetaminophen and acetyl salicylic acid. Under the catalysis of a small amount of pyridine, aspirin and thionyl chloride generates the halogenation of acid hydroxyl to produce 2-acetoxybenzoic acid. Under the action of sodium hydroxide, acetaminophen generates sodium salt, and then acylates with 2-acetoxy-benzoyl chloride to produce 2-acetoxy-benzoyl 4-acetaminophenate. The synthesis process is as follows:

$$\underset{\substack{\text{COCl}\\\text{OCOCH}_3}}{\bigcirc} + \text{NaO}-\bigcirc-\text{NHCOCH}_3 \longrightarrow \underset{\substack{\text{COO}\\\text{OCOCH}_3}}{\bigcirc}-\bigcirc-\text{NHCOCH}_3$$

III. Main instruments and reagents

i. Main instruments: thermometer, zeolite, constant pressure drip funnel, analysis balance, beaker, measuring cylinder, heating sleeve (oil bath), glass rod, rotary evaporator, three-necked flask, reflux condensing tube, filter bottle, Brinell funnel, vacuum pump, *etc.*

ii. Main reagents: aspirin, sulfoxide chloride, pyridine, paracetamol, sodium hydroxide, absolute alcohol, *etc.*

IV. Experimental steps

Add several zeolite, 9 g of aspirin, 5 mL of sulfone chloride solvent and 1-2 drops of alkaline catalyst pyridine into a three-necked flask equipped with a magnetic agitator, a reflux condensing tube and a thermometer. Gradually heat in an oil bath. Stirring vigorously at a temperature of 75°C until no gas is evolved, which takes about 3 h. After the reaction, adjust the reflux apparatus to a reduced pressure distillation setup, and use an oil pump to remove the excess thionyl chloride under reduced pressure, then cool to obtain acetylsalicylic acid chloride. Add 6 mL of acetone to dry for later use.

Add 8.6 g of paracetamol and 50mL of water into a three-neck round-bottom flask equipped with a magnetic stirrer, a constant-pressure drip hopper and a thermometer. Under vigorous stirring, slowly add 18 mL of sodium hydroxide aqueous solution (with 3.3 g of sodium hydroxide). Keep the temperature of the reaction system below 15°C. Then, slowly add the anhydrous acetone solution of acetylsalicylic acid chloride prepared in the previous step. Adjust the pH to 9.0-10.0. Stir at room temperature for 2 h. After the reaction, perform reduced pressure filtration, wash with water until neutral. Dry to obtain the crude product.

Add the crude benorilate obtained in the previous step into the flask. Then add anhydrous ethanol in batches, generally using 8 times the amount of ethanol that of the crude benorilate to ensure it to dissolve just under the condition of ethanol reflux. After refluxing, cool to room temperature to allow all benorilate to precipitate, perform reduced pressure filtration, and dry to obtain pure benorilate.

V. Attention or thinking questions

i. Attention

(i) This reaction is very strict to the anhydrous requirements. The reaction involved in the instrument, *etc.*, need to do anhydrous treatment. The hydrogen chloride and sulfur dioxide gases generated in the reaction are irritating and will pollute the air. They can be absorbed and treated by lye.

(ii) In order to facilitate agitation, observe the internal temperature and make the reaction more complete, the amount of sulfoxide chloride can be appropriately increased to 6~7 mL.

(iii) The pyridine used in the reaction is mainly catalytic.

ii. Thinking questions

(i) In addition to the preparation of acyl chloride by thionyl chloride, what other methods can be used to prepare acyl chloride?

(ii) What should be noted when preparing acyl chloride from carboxylic acid and sulfoxide chloride?

(iii) Try to design other synthesis routes of benorilate?

实验六　醋酸胍那苄的合成

一、实验目的

1. 掌握醋酸胍那苄药物的合成反应机理。
2. 熟悉醋酸胍那苄药物的合成实验操作过程。
3. 了解醋酸胍那苄的理化性质及临床用途。

二、实验原理

（一）药物简介

醋酸胍那苄，分子式为 $C_{10}H_{12}Cl_2N_4O_2$，1982 年于美国首次上市，是一种中枢作用的抗高血压药物。本药物为中枢性 α-2 受体激动药，具有类似胍乙啶的抑制去甲肾上腺素释放的外周性作用，产生良好的降压作用，可使总外周阻力下降，对心功能无显著影响，不改变心排出量、心输出量及肾小球滤过率（GFR）。

化学结构式为：

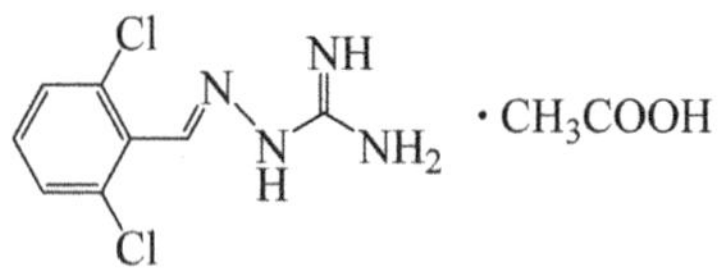

[(2,6-二氯苯亚甲基)氨基] 胍醋酸盐

（二）合成工艺路线

醋酸胍那苄的合成，主要通过二氯苯甲醛等制得，其合成工艺路线具体如下：

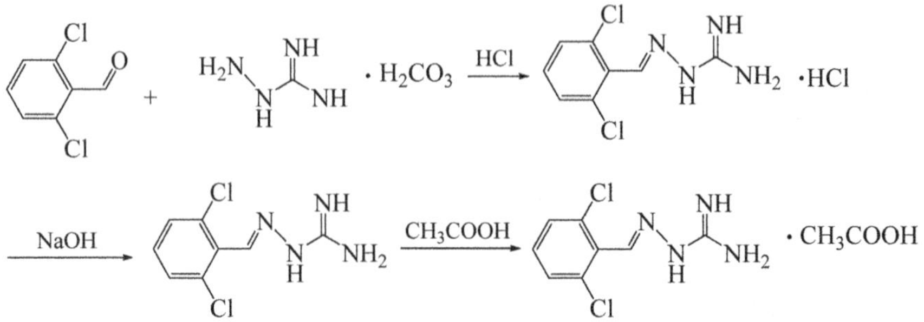

三、主要仪器和试剂

1. 主要仪器：三颈烧瓶、磁力搅拌器、分析天平、温度计、恒压滴液漏斗、冷凝管、分液漏斗、烧杯、橡胶管、锥形瓶、抽滤瓶、布氏漏斗、滤纸、真空泵等。
2. 主要试剂：2,6-二氯苯甲醛、氨基胍碳酸盐、浓盐酸、正丁醇、10% NaOH、丙酮、乙腈等。

四、实验步骤

（一）［（2,6-二氯苯亚甲基）氨基］胍盐酸盐的合成

在配有磁力搅拌器、温度计、冷凝管、干燥管以及恒压滴液漏斗的 500 mL 三颈烧瓶中，加入 2,6-二氯苯甲醛 9.3 g、氨基胍碳酸盐 7.8 g、浓盐酸 50 mL 以及正丁醇混合物 250 mL，升温至 120℃，回流反应，反应中生成的水共沸 4 h 除去。反应结束后，将反应体系冷却，有固体生成，抽滤，得产物，即为［（2,6-二氯苯亚甲基）氨基］胍盐酸盐白色结晶。

（二）［（2,6-二氯苯亚甲基）氨基］胍的合成

在配有磁力搅拌器、温度计、冷凝管以及恒压滴液漏斗的 500 mL 三颈烧瓶中，加入［（2,6-二氯苯亚甲基）氨基］胍盐酸盐 4.0 g、水 60 mL 以及 10% NaOH 溶液 13 mL。快速搅拌，有固体产生，抽滤，滤饼用水、丙酮洗涤，过滤，干燥，得固体。可以用乙腈重结晶，即为［（2,6-二氯苯亚甲基）氨基］胍产物的纯品。

（三）醋酸胍那苄的合成

上步制得的［（2,6-二氯苯亚甲基）氨基］胍与乙酸反应，可制得醋酸胍那苄。

五、注意事项与思考题

1. 试写出本反应的反应机理。
2. 反应中有氨产生，有较强刺激性，应在通风橱中操作。

Experiment 6　Synthesis of Guanabenz Acetate

Ⅰ. Purpose of the experiment

i. To master the synthetic reaction mechanism of Guanabenz Acetate.

ii. To familiar with the experimental operation process of Guanabenz Acetate synthesis.

iii. To understand the physicochemical properties and clinical applications of Guanabenz Acetate.

Ⅱ. Experimental principle

i. Drug introduction

Guanabenz Acetate, with the molecular formula $C_{10}H_{12}Cl_2N_4O_2$, was first marketed in the United States in 1982. It is a central acting antihypertensive drug. This drug is a central α-2 receptor agonist, which has a peripheral effect similar to guanethidine in inhibiting norepinephrine release. It produces a good antihypertensive effect and decreases total peripheral resistance. It has no significant effect on cardiac function and does not change cardiac output, cardiac output and GFR.

The chemical structure formula is:

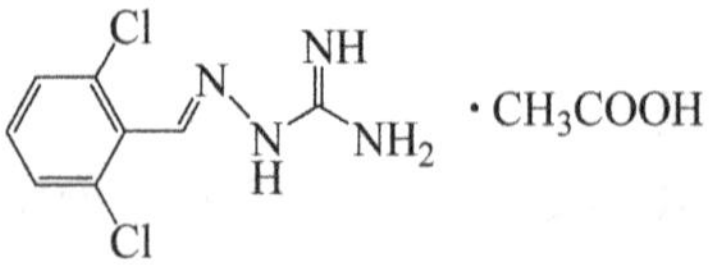

[(2, 6-dichlorophenylene) amino] guanidine acetate

ii. Synthetic process route

The synthesis of guanidine acetate is mainly prepared by dichlorobenzaldehyde, *etc*. The synthesis process is as follows:

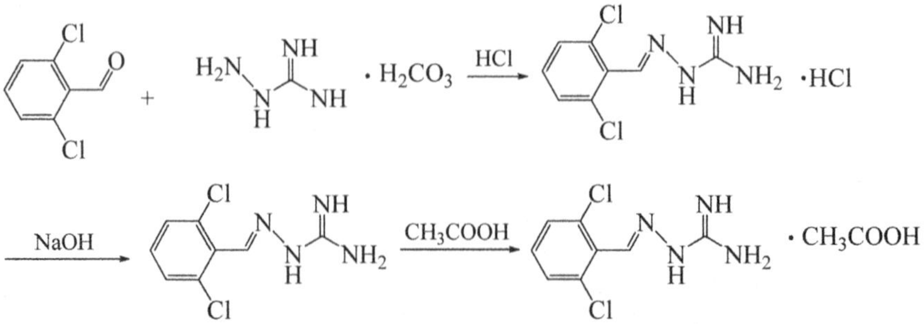

Ⅲ. Main instruments and reagents

i. Main instruments: three-neck flask, magnetic stirrer, analytical balance, thermometer, constant pressure drip funnel, condensing tube, separation funnel, beaker, rubber tube, conical bottle, suction bottle, Brinell funnel, filter paper, vacuum pump, *etc.*

ii. Main reagents: 2, 6-dichlorobenzaldehyde, aminoguanidine carbonate, concentrated hydrochloric acid, *n*-butanol, 10%NaOH, acetone, acetonitrile, *etc.*

Ⅳ. Experimental steps

i. Synthesis of [(2, 6-dichlorobenzene methylene) amino] guanidine hydrochloride

Add 9.3 g of 2,6-dichlorobenaldehyde, 7.8 g of aminoguanidine carbonate, 50 mL of concentrated hydrochloric acid and 250 mL of *n*-butanol into a 500 mL three-necked flask equipped with a magnetic agitator, a thermometer, a condensing tube, a drying tube and a constant pressure drip funnel. Heat to 120 ℃ and reflux the reaction. Remove the water generated during the reaction by azeotrope for 4 h. After the reaction, cool the reaction system, and the solid will form. Filter under vacuum to obtain the product, which is the white crystal of [(2, 6-dichlorobenzene methylene) amino] guanidine hydrochloride.

ii. Synthesis of [(2, 6-dichlorobenzene methylene) amino] guanidine

Add 4.0 g of [(2, 6-dichloro phenylene) amino] guanidine hydrochloride, 60 mL of water, and 13 mL of 10% NaOH solution into a 500 mL three-necked flask equipped with a magnetic stirrer, a thermometer, a condensing tube, and a constant-pressure drip funnel. Stir rapidly, and the solid will form. Filter under vacuum, wash the filter cake with water and acetone. After that, filter and dry to obtain the solid. Recrystallisation can be performed using acetonitrile to yield the pure product of (2, 6-dichlorobenzene methylene) amino] guanidine.

iii. Synthesis of guanidine acetate

The reaction of [(2, 6-dichloro phenylene) amino] guanidine with acetic acid can produce guanidine acetate.

Ⅴ. Attention or thinking questions

i. Try to write the reaction mechanism of this reaction.

ii. Ammonia is produced in the reaction, and there is strong irritation. It should be operated on the fume hood.

实验七 阿司匹林的合成

一、实验目的

1. 掌握阿司匹林的性状、特点和化学性质。
2. 熟悉和掌握酯化反应的原理和实验操作。
3. 熟悉、巩固重结晶的原理和实验方法。
4. 了解阿司匹林中杂质的来源和鉴别。

二、实验原理

（一）药物简介

阿司匹林又名乙酰水杨酸，分子式为 $C_9H_8O_4$，是一种白色结晶或结晶性粉末，无臭或微带醋酸臭，微溶于水，易溶于乙醇，可溶于乙醚、氯仿，水溶液呈酸性。该药品为水杨酸的衍生物，经近百年的临床应用，证明其对缓解轻度或中度疼痛，如牙痛、头痛、神经痛、肌肉酸痛及痛经效果较好，亦可用于感冒、流感等发热疾病的退热，治疗风湿病等。

近年来发现阿司匹林对血小板聚集有抑制作用，能阻止血栓形成，临床上用于预防短暂脑缺血发作、心肌梗死、人工心脏瓣膜和静脉瘘或其他手术后血栓的形成。

化学结构式为：

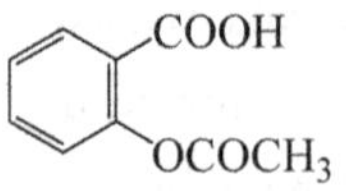

2-（乙酰氧基）-苯甲酸

（二）合成工艺路线

阿司匹林的合成，以邻羟基苯甲酸为原料，在乙酸酐的作用下制得，其合成工艺路线如下：

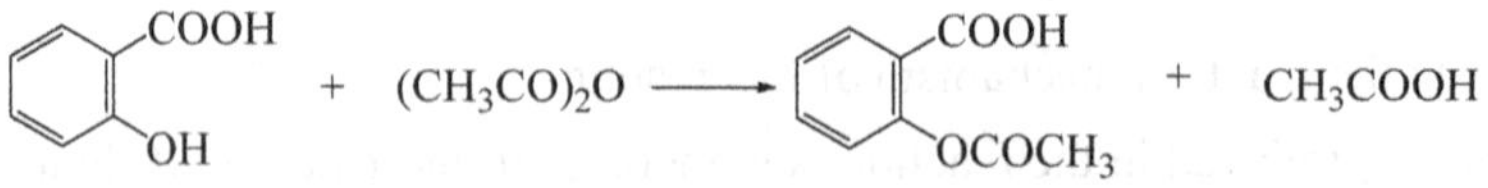

在本反应过程中，阿司匹林能够自身缩合，产生一种聚合物。利用阿司匹林和碱反应，生成水溶性钠盐的性质，从而与聚合物分离。

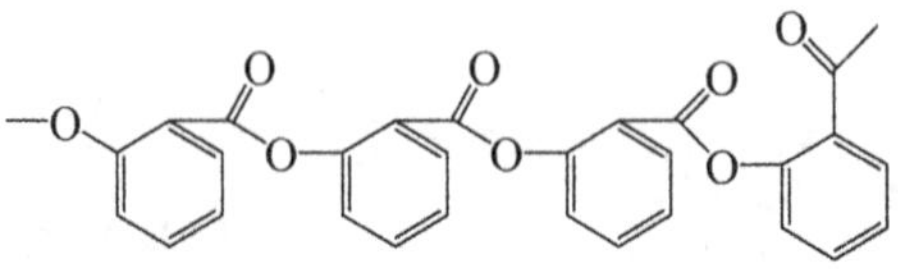

在阿司匹林产品中的一个主要副产物是水杨酸，其来源可能是酰化反应不完全的原料，

也可能是阿司匹林的水解产物。水杨酸可以在最后的重结晶中加以分离。

三、主要仪器和试剂

1. 主要仪器：温度计、磁力搅拌器、锥形瓶、分析天平、烧杯、量筒、加热套、玻璃棒、三颈烧瓶、抽滤瓶、布氏漏斗、真空泵等。

2. 主要试剂：水杨酸、乙酸酐、乙酸乙酯、浓硫酸、浓盐酸、饱和碳酸氢钠等。

四、实验步骤

在 500 mL 锥形瓶中，加入水杨酸 10.0 g、乙酸酐 25.0 mL，然后用胶头滴管滴入浓硫酸，同时缓慢旋摇锥形瓶，使水杨酸溶解。将锥形瓶缓慢加热至 85～95℃，保温维持 10 min。然后停止加热，使其缓慢冷却至室温。在冷却过程中，有阿司匹林固体缓慢从溶液中析出。在冷却到室温时，有较多晶体析出，向其中加入水 250 mL，并将该体系放置冰浴中冷却。待充分冷却后，有大量固体析出，抽滤，得到固体，用冷水洗涤，滤饼压紧抽干，得到粗产物。

将上步得到粗品放置在 250 mL 烧杯中，并向其中加入饱和碳酸氢钠水溶液 125 mL。剧烈搅拌，直至无气泡产生为止。有固体产生，真空抽滤，除去不溶物，同时用少量水洗涤。在另一支 250 mL 烧杯中，加入浓盐酸 17.5 mL 及水 50 mL，将前面得到的滤液，缓慢、分多次倒入该烧杯中，同时不停搅拌。阿司匹林固体将从溶液中析出。将其放置冰浴中冷却，抽滤，滤饼用冰水洗涤，抽滤，同时压干固体，得阿司匹林粗品。

将所得的阿司匹林粗品加入 25 mL 锥形瓶中，加入少量热乙酸乙酯（小于 15 mL），缓慢升高温度，直至固体溶解，接着将体系冷却至室温，亦可以用冰浴冷却，阿司匹林渐渐析出，抽滤，得到精制的阿司匹林。

注：

1. 在加热的过程中，如果使用水浴加热，注意勿让水蒸气进入锥形瓶中，以防止酸酐和生成的阿司匹林水解。

2. 本反应在冷却的过程中，如果阿司匹林没有析出，或者析出较为缓慢，可以用玻璃棒轻轻摩擦锥形瓶的内壁，促进晶体的析出。

3. 在加水的处理过程中，要特别注意，必需等到析晶充分析出后才能加入水。加水的过程中，要注意加入速度，要缓慢加入，此步有放热现象产生，亦有可能是溶液体系沸腾，进而产生一定量的乙酸蒸气，具有刺激性。

4. 在将碳酸氢钠水溶液加到阿司匹林中时，此时，会产生大量的气泡，注意加入的速度，要分批加入，边加边搅拌，以防气泡产生过多，引起溶液外溢。

5. 在将滤液加入盐酸之后，发现仍然没有固体析出，需要注意该溶液是否呈现酸性，如果酸度不够，需要进一步的补加盐酸，直至溶液的 pH 在 2 左右，将会有固体晶体析出。

6. 阿司匹林从乙酸乙酯中析出步骤，若发现没有固体析出，可能是乙酸乙酯的量加多了，需要将乙酸乙酯挥发除去一部分，重新冷却，程序操作，便有固体产生。

7. 阿司匹林纯度的检查：在干净的试管中，加入一定量的阿司匹林，同时加入乙醇 1 mL，让其溶解，在试管中滴加少量 10% $FeCl_3$ 溶液，纯度较高的阿司匹林的颜色应该是无色的，若出现紫色或者蓝紫色出现，说明阿司匹林不纯。

五、注意事项与思考题

1. 本反应阿司匹林的合成过程中,加入了一定的浓硫酸,有何作用?能否用其他酸进行替代?

2. 反应过程中,产生了哪些副产物?

3. 在阿司匹林的纯度检查时,为什么说若出现紫色或者蓝紫色出现,说明阿司匹林不纯?解释其原因。

4. 试写出本反应的反应机理。

Experiment 7   Synthesis of aspirin

I. Purpose of the experiment

i. To master the characters, characteristics and chemical properties of aspirin.

ii. To familiar with and master the principle and experimental operation of esterification reaction.

iii. To familiar with and consolidate the principles and experimental methods of recrystallization.

iv. To understand the sources and identification of impurities in aspirin.

II. Experimental principle

i. Drug introduction

Aspirin, also known as acetylsalicylic acid, has a molecular formula $C_9H_8O_4$. It is a white crystalline or crystalline powder. It is odorless or slightly odorless acetic acid. It is slightly soluble in water, easily soluble in ethanol, soluble in ethers and chloroform. And its aqueous solution is acidic. The drug is a derivative of salicylic acid. After nearly 100 years of clinical application, it has proven its effectiveness in alleviating mild to moderate pain, such as toothache, headache, neuralgia, muscle soreness and dysmenorrhea. It can also be used for reducing fever of febrile diseases such as cold and flu, and treating rheumatic pain.

In recent years, it has been found that aspirin has an inhibitory effect on platelet aggregation, which can prevent thrombus formation. It is clinically used to prevent thrombus formation after transient ischemic attack, myocardial infarction, artificial heart valve, venous fistula or other operations.

The chemical structure formula is:

$$\underset{\text{acetylsalicylic acid}}{\begin{array}{c}\text{COOH}\\ \text{C}_6\text{H}_4\\ \text{OCOCH}_3\end{array}}$$

ii. Synthetic process route

The synthesis of aspirin is mainly made by *o*-hydroxybenzoic acid under the action of acetic anhydride. The synthesis process is as follows:

$$\text{C}_6\text{H}_4(\text{OH})(\text{COOH}) + (\text{CH}_3\text{CO})_2\text{O} \longrightarrow \text{C}_6\text{H}_4(\text{OCOCH}_3)(\text{COOH}) + \text{CH}_3\text{COOH}$$

In this process, aspirin is able to condense itself to produce a polymer. Using the properties of a reaction between aspirin and a base to form a water-soluble sodium salt, thus separating it from a polymer.

The other major by-product in aspirin products is salicylic acid, which can be derived either from incomplete acylation of the raw material or from the hydrolyzed product of aspirin. Salicylic acid can be separated in the final recrystallization.

III. Main instruments and reagents

i. Main instruments: thermometer, magnetic stirrer, conical flask, analytical balance, beaker, measuring cylinder, heating sleeve, glass rod, three-necked round bottom flask, filter flask, Brinell funnel, vacuum pump, *etc*.

ii. Main reagents: salicylic acid, acetic anhydride, ethyl acetate, concentrated sulfuric acid, concentrated hydrochloric acid, sodium bicarbonate, *etc*.

IV. Experimental steps

Add 10.0 g of salicylic acid and 25.0 mL of acetic anhydride into a 500 mL conical flask. Then drop concentrated sulfuric acid with a glue-head dropper. Slowly swirling the flask to dissolve the salicylic acid. Slowly heat to 85-95 ℃ and maintain for 10 min. Then stop heating and allow it to cool slowly to room temperature. During the cooling process, aspirin solid will slowly precipitate from the solution. During cooling to room temperature, a significant amount of crystals will form. Add 250 mL of water to the mixture and place it in an ice bath to cool. To be fully cooled, a large number of solids will precipitate. Filter to obtain the solid. Wash the filter cake with cold water and press. Dry to obtain the crude product.

Place the coarse product into a 250 mL beaker and add 125 mL of saturated sodium bicarbonate water solution to it. Stir vigorously until no bubbles form. A solids will form, and then vacuum filter to remove insoluble matter, while washing with a little water. In another 250 mL beaker, add 17.5 mL of concentrated hydrochloric acid and 50 mL of water. Pour the filtrate into the beaker slowly several times. Stir constantly. The aspirin solids will precipitate from the solution. Place it in an ice bath to cool, and then filter. Wash the filter cake with ice water, and filter again while pressing the solid dry to obtain crude aspirin.

Add the obtained crude aspirin into a 25 mL conical flask. Then add a small amount of hot ethyl acetate (less than 15 mL). Slowly raise the temperature until the solid dissolves. Then cool the system to room temperature or in an ice bath, allowing aspirin to gradually separated out. Filter to obtain refined aspirin.

Note:

i. During the heating process, if using a water bath, be careful not to let water vapour enter the conical flask to prevent the anhydride and the generated aspirin from hydrolysing.

ii. During the cooling process, if aspirin does not precipitate or precipitates slowly, the glass rod can be used to gently rub the inner wall of the conical bottle to promote the crystal precipitation.

iii. In the process of adding water, it is necessary to wait until the crystallization is sufficient before adding water. In the process of adding water, we should pay attention to the speed of adding it, while it should be done slowly. This step has an exothermic phenomenon, and it may also be boiling of the solution system. Then produce a certain amount of acetic acid vapor, which is stimulating.

iv. When adding sodium bicarbonate aqueous solution to aspirin, a large number of bubbles will be produced at this time. Pay attention to the speed of adding.

v. After adding hydrochloric acid to the filtrate, if no solid precipitates are observed, check whether the solution is acidic. If the acidity is insufficient, further hydrochloric acid should be added until the acidity of the solution is around pH 2, at which point solid crystals will precipitate.

vi. In the step of precipitating aspirin from ethyl acetate, if no solid is observed, it may be due to an excess of ethyl acetate; you need to evaporate some of the ethyl acetate, cool it again, and follow the procedure to obtain solid.

vii. Check the purity of aspirin: In a clean test tube, add a certain amount of aspirin. At the same time, add 1 mL of ethanol to dissolve it. Add a small amount of 10% $FeCl_3$ solution into the test tube. The color of high purity aspirin should be colorless. If purple or blue purple appears, it indicates that aspirin is not pure.

Ⅴ. Attention or thinking questions

i. In the synthesis process of aspirin, what is the effect of adding concentrated sulfuric acid? Can other acids be used instead?

ii. What by-products are produced during the reaction?

iii. In the purity check of aspirin, why is it said that if purple or blue purple appears, it indicates that aspirin is not pure?

iv. Try to write the reaction mechanism of this reaction.

实验八　盐酸萘替芬的合成

一、实验目的

1. 掌握盐酸萘替芬药物的合成反应机理。
2. 熟悉盐酸萘替芬药物的合成实验操作流程。
3. 了解盐酸萘替芬药物的理化性质及临床用途。

二、实验原理

（一）药物简介

盐酸萘替芬，化学式为 $C_{21}H_{22}ClN$，20 世纪 80 代末在美国上市。本药物为白色或类白色结晶性粉末，无臭。在甲醇、三氯甲烷中易溶，在水中几乎不溶，剂型为乳膏和凝胶，为丙烯胺类局部抗真菌药。其作用机制为抑制真菌角鲨烯环氧化酶，干扰真菌细胞壁的麦角固醇的生物合成，影响真菌的脂质代谢，使真菌细胞损伤或死亡而起到杀菌和抑菌作用。

药品类别为皮肤科用药，适用于敏感真菌所致的皮肤真菌病如体股癣、手足癣、头癣、甲癣、花斑癣、浅表念珠菌病。

化学结构式为：

（E)-N-甲基-N-(3-苯基-2-丙烯基)-1-萘甲胺盐酸盐

（二）合成工艺路线

盐酸萘替芬的合成，主要通过萘、多聚甲醛、冰醋酸等原料制得，其合成工艺路线如下：

三、主要仪器和试剂

1. 主要仪器：三颈烧瓶、磁力搅拌器、分析天平、温度计、恒压滴液漏斗、冷凝管、分液漏斗、烧杯、橡胶管、锥形瓶、抽滤瓶、布氏漏斗、滤纸、真空泵等。

2. 主要试剂：萘、多聚甲醛、冰醋酸、85%磷酸、浓盐酸、10%碳酸钾、乙醚、甲胺、无水乙醇、PEG 400、二氯甲烷、10%氢氧化钠、无水硫酸钠、苯乙烯、乙腈、PEG 600、异丙醇等。

四、实验步骤

（一）1-氯甲基萘的合成

在配有磁力搅拌器、温度计、冷凝管、干燥管以及恒压滴液漏斗的 250 mL 三颈烧瓶中，加入萘 128 g、多聚甲醛 55 g、冰醋酸 130 mL、85%磷酸 85 mL 以及浓盐酸 21 mL，快速搅拌，升温至 80~85℃，保温反应 6 h。反应结束后，冷却至室温，倒至分液漏斗中，分出有机层，分别加入冰水、冷的 10%碳酸钾溶液充分洗涤，向有机层加乙醚，同时用无水硫酸钠进行干燥，过夜，抽滤，回收溶剂，减压蒸馏，得到产物。

（二）N-甲基-1-萘甲基胺的合成

在配有磁力搅拌器、温度计、冷凝管、干燥管以及恒压滴液漏斗的 500 mL 三颈烧瓶中，加入 30%甲胺 41.5 g、无水乙醇 100 mL、碳酸钾 13.8 g、PEG 400 5.0 g，反应体系用冰浴冷却，快速搅拌，缓慢滴加 1-氯甲基萘 60 g 以及无水乙醇 200 mL 组成的溶液。滴加结束后，搅拌反应 3 h。减压回收溶剂，剩余物加入二氯甲烷 175 mL。分别用 10%氢氧化钠溶液 300 mL、水充分洗涤，随后，用无水硫酸钠充分干燥，抽滤，减压蒸馏，得产物，即为 N-甲基-1-萘甲基胺。

（三）3-氯-1-苯丙烯的合成

在配有磁力搅拌器、温度计、冷凝管、干燥管以及恒压滴液漏斗的 100 mL 三颈烧瓶中，加入浓盐酸 30 mL、多聚甲醛 25 g，快速搅拌，随后加入苯乙烯 52 g，升温至回流，保温反应 3 h，反应结束后，静置分层，倒去下层，上层用冰水充分洗涤至中性，用无水硫酸钠干燥，抽滤、干燥，得固体，室温为液体，蒸馏，得产物，即为 3-氯-1-苯丙烯。

（四）N-甲基-N-（3-苯基-2-丙烯基）-1-萘甲胺盐酸盐的合成

在配有磁力搅拌器、温度计、冷凝管、干燥管以及恒压滴液漏斗的 100 mL 三颈烧瓶中，加入 3-氯-1-苯丙烯 10 g、N-甲基-1-萘甲基胺 12 g、乙腈 100 mL、碳酸钾 10.5 g 以及 PEG 600 6.0 g，升温至回流，快速搅拌，保温反应 3 h，反应结束后，自然冷却至室温，随后加水 75 mL、乙醚 50 mL，快速搅拌 30 min，静置 2 h，随后分出有机层。水层用乙醚充分萃取，合并有机层，用水充分洗涤。有机层中加入冰水 15 mL、浓盐酸 15 mL，充分搅拌 1 h。静置后，分出油状物，用玻璃棒摩擦油状物使其固化，过滤，水洗，干燥，得到产物，即为盐酸萘替芬。可以用异丙醇-乙醚酸，进行重结晶，得纯品。

五、注意事项与思考题

（一）注意事项

1. 试写出本反应的反应机理。
2. 熟悉重结晶实验操作过程。

（二）思考题

PEG 400，PEG 600 的作用是什么？

Experiment 8 Synthesis of Naftifine hydrochloride

I. Purpose of the experiment

i. To master the synthetic reaction mechanism of Naftifine hydrochloride.

ii. To familiar with the experimental procedures for the synthesis of Naftifine hydrochloride.

iii. To understand the physicochemical properties and clinical applications of Naftifine hydrochloride.

II. Experimental principle

i. Drug introduction

Naftifine hydrochloride, with the molecular formula $C_{21}H_{22}ClN$, was l in the United States in the late 1980s. This drug is a white or white-like crystalline powder, odorless. It is easily soluble in methanol and trichloromethane, almost insoluble in water. It is available in cream and gel form as a local antifungal agent of the a crylamide class. Its mechanism of action involves inhibiting fungal squalene cycloxygenase, interfering with the biosynthesis of ergosterol in the fungal cell wall, affecting the fungal lipid metabolism of fungi, leading to damage or death.

The drug category dermatological, which is suitable for skin fungal diseases caused by sensitive fungi, such as tinea corporis, tinea pedis, tinea capitis, tinea unguium, tinea versicolor and superficial candidiasis.

The chemical structure formula is:

(E)-N-cirmamyl-N-mcthyl (1-naphthylmethyl) amine hydrochloride

ii. Synthetic process route

Naftifine hydrochloride is synthesized mainly from naphthol, paraformaldehyde, glacial acetic acid and other raw materials. The synthesis process is as follows:

III. Main instruments and reagents

i. Main instruments: three necked flask, magnetic stirrer, analytical balance, thermometer, constant pressure drip funnel, condensing tube, separation funnel, beaker, rubber tube, conical bottle, suction bottle, Brinell funnel, filter paper, vacuum pump, *etc*.

ii. Main reagents: naphthalene, polyformaldehyde, glacial acetic acid, phosphoric acid, concentrated hydrochloric acid, potassium carbonate, ether, methylamine, anhydrous ethanol, PEG 400, methylene chloride, sodium hydroxide, anhydrous sodium sulfate, styrene, acetonitrile, PEG600, isopropyl alcohol, *etc*.

IV. Experimental steps

i. Synthesis of 1-chloromethyl naphthalene

Add 128 g of naphthalene, 55 g of paraformaldehyde, 130 mL of glacial acetic acid, 85 mL of 85% phosphoric acid and 21 mL of concentrated hydrochloric acid into a 250 mL three-necked flask equipped with a magnetic stirrer, a thermometer, a condense tube, a drying tube and a constant pressure drip funnel. Stir rapidly and heat to 80-85 ℃, maintaining for 6 h. After the reaction, cool to room temperature. Pour into the liquid separation funnel. Separate the organic layer. Wash thoroughly with ice water and cold 10% potassium carbonate solution. Add ether to the organic layer. At the same time, dry with anhydrous sodium sulfate overnight. Then filter and recovery the solvent. Perform reduced pressure distillation to obtain the products.

ii. Synthesis of N-methyl-1-naphthylmethylamine

Add 41.5 g of 30% methylamine, 100mL of ethanol, 13.8 g of potassium carbonate and 5.0 g of PEG 400 into a 500 mL three necked flask equipped with a magnetic agitator, a thermometer, a condensing tube, a drying tube and a constant pressure drip funnel. Cool the reaction system with an ice bath and stir rapidly. Slowly add 60 g of 1-chloromethyl naphthalene and 200 mL of anhydrous ethanol solution. After the addition, stir the reaction for 3 h. Recovery the solvent under reduced pressure. Add 175 mL of methylene chloride to the remaining material. Wash thoroughly with 300mL of 10% sodium hydroxide solution and water. And then fully dry with anhydrous sodium sulfate. Pump and filter. Perform reduced pressure to obtain the product, which is N-methyl-1-naphthylmethylamine.

iii. Synthesis of 3-chloro-1-phenylpropene

Add 30 mL of concentrated hydrochloric acid and 25 g of paraformaldehyde into a 100 mL three necked flask equipped with a magnetic agitator, a thermometer, a condensing tube, a drying tube and a constant pressure drip funnel. Stir quickly. Then add 52 g of styrene. Heat up to reflux, and maintain the reaction for 3 h. After the reaction, let it stand to separate, and pour off the lower layer. Wash the upper layer thoroughly with ice water until neutral. Dry with anhydrous sodium sulfate and filter. Dry to obtain a solid, which is liquid at room temperature. Distill to obtain the product, which is 3-chloro-1-propene.

iv. Synthesis of N-methyl-N -(3-phenyl-2-allyl) -1-naphthylmethylamine hydrochloride

Add 10 g of 3-chloro-1-styrene, 12 g of n-methyl-1-naphthylmethylamine, l00 mL of acetonitrile, 10.5 g of potassium carbonate and 6.0 g of PEG 600 into a 100 mL three necked flask equipped with a magnetic stirrer, a thermometer, a condensing tube, a drying tube and a constant pressure drip funnel. Heat to reflux and stir rapidly, maintaining the reaction for 3 h. After the reaction, cool to room temperature naturally. Then add 75 mL of water and 50 mL of ether. Stir rapidly for 30 min, and leave for 2 h. Separate the organic layer. Extract the water layer with ethyl ether. Combine the organic layer, and wash thoroughly with water. Add 15 mL of ice water and 15 mL of concentrated hydrochloric acid into the organic layer and stir thoroughly for 1 h. After standing, separate the oil substance, and rub it with a glass rod to cure. Filter and wash with water. Dry to obtain the product, which is Naftifine hydrochloride. Recrystallisation can be performed using isopropyl alcohol-ether acid to obtain the pure product.

V. Attention or thinking questions

i. Attention

(i) Try to write the reaction mechanism of this reaction.
(ii) Familiar with the operation process of the recrystallization experiment.

ii. Thinking questions

What are the functions of PEG 400 and PEG 600?

实验九　依达拉奉的合成

一、实验目的

1. 掌握依达拉奉合成实验的操作过程。
2. 熟悉吡唑环的合成的反应机理。
3. 了解无水苯肼的性质以及在使用过程中的注意事项。
4. 了解依达拉奉的理化性质及临床用途。

二、实验原理

（一）药物简介

依达拉奉又名依达拉丰，分子式为 $C_{10}H_{10}N_2O$，是一种白色结晶性粉末。依达拉奉是一种脑保护剂（自由基清除剂）。溶于水，微溶于醇和苯，不溶于醚、石油醚及冷水。主要作为合成医药品安替比林、氨基比林的原料，也用于染料及彩色胶片染料、农药及有机合成工业中。

临床研究表明，N-乙酰门冬氨酸是特异性的存活神经细胞的标志，脑梗死发病初期含量急剧减少。脑梗死急性期患者给予依达拉奉，可抑制梗死周围局部脑血流量的减少，使发病后第 28 天脑中 N-乙酰门冬氨酸含量，较甘油对照组明显升高，静脉给予依达拉奉可阻止脑水肿和脑梗死的进展，并缓解所伴随的神经症状，抑制迟发性神经元死亡，依达拉奉可清除自由基，抑制脂质过氧化，从而抑制脑细胞、血管内皮细胞、神经细胞的氧化损伤。

化学结构式为：

3-甲基-1-苯基-2-吡唑啉-5-酮

（二）合成工艺路线

依达拉奉的合成，以苯肼与乙酰乙酸乙酯等原料，经成环反应制得，其合成工艺路线如下：

$$\text{C}_6\text{H}_5\text{—NHNH}_2 \xrightarrow[\text{(2) HCl　(3) 10\%NaOH}]{\text{(1) CH}_3\text{COCH}_2\text{COOC}_2\text{H}_5/\text{C}_2\text{H}_5\text{OH}}$$ 3-甲基-1-苯基-2-吡唑啉-5-酮

三、主要仪器和试剂

1. 主要仪器：温度计、恒压滴液漏斗、三颈烧瓶、磁力搅拌器、分析天平、烧杯、量筒、加热套、玻璃棒、抽滤瓶、布氏漏斗、真空泵等。

2. 主要试剂：苯肼、乙酰乙酸乙酯、无水乙醇、浓盐酸、10%氢氧化钠等。

四、实验步骤

在配有回流冷凝管、温度计、恒压滴液漏斗以及磁力搅拌器的三颈烧瓶中，加入乙酰乙酸乙酯 13 g 以及 70%乙醇 5 mL，搅拌，升温至 45℃时，缓慢滴加苯肼 10.8 g 与无水乙醇 3 mL 配成的溶液，大约 30 min 内滴加完毕，保温反应 30 min 后，将反应体系冷却至 20℃，滴加浓盐酸 1 mL，升高温度至 45℃，继续反应 2 h，滴加 10% NaOH 调节 pH 至 7，加水 20 mL，降至室温，继续搅拌 1 h，减压抽滤，滤饼用冷的无水乙醇洗涤三次，干燥，得淡黄色结晶粗品。

精制时，可以用混合溶剂乙酸乙酯/无水乙醇（2∶1，30 mL）进行重结晶，最终得到白色结晶。

注：

1. 一般来说，游离的苯肼不稳定，接触空气会冒烟并很快变质，操作时，要快速，同时，在滴加过程中，最好使用氮气球或者用流通氮气进行保护。

2. 精制时，重结晶可以加活性炭进行脱色，第一次重结晶所得产品时，如果颜色较深，再进行重结晶一次，可得到白色晶体。

五、注意事项与思考题

（一）注意事项

试写出本反应的反应机理。

（二）思考题

1. 导致苯肼在空气中不稳定的因素是什么？
2. 本反应中浓盐酸作用是什么？

Experiment 9 Synthesis of Edaravone

I. Purpose of the experiment

i. To master the operation process of the Edaravone synthesis experiment.

ii. To familiar with the synthetic reaction mechanism of the pyrazole ring.

iii. To understand the properties of anhydrous phenylhyzrazine and matters needing attention in the process of use.

iv. To understand the physicochemical properties and clinical use of Edaravone.

II. Experimental principle

i. Drug introduction

Edaravone has a molecular formula of $C_{10}H_{10}N_2O$. It is a white crystalline powder. It acts as a brain protectant (free radical scavenger). It is soluble in water, slightly soluble in alcohol and benzene, insoluble in ether, petroleum ether and cold water. It is mainly used as raw material for the synthesis of medical drugs antipyrine, aminopyrine, and also used in dyes and color film dyes, pesticides and organic synthesis industry.

Clinical studies have shown that n-acetylaspartate is a specific marker for surviving neurons, with its levels sharply decreasing in the early stages of cerebral infarction. Administering edaravone to patients in the acute phase of cerebral infarction can inhibit the reduction of local cerebral blood flow around the infarct, resulting in a significantly increased level of n-acetylaspartate in the brain on day 28 post-onset compared to the glycerol control group. Intravenous administration of edaravone can prevent the progression of cerebral oedema and cerebral infarction, alleviate accompanying neurological symptoms, and inhibit delayed neuronal death. Edaravone can scavenge free radicals and inhibit lipid peroxidation, thereby reducing oxidative damage to brain cells, vascular endothelial cells, and neurons.

The chemical structure formula is:

3-methyl-1-phenyl-2-pyrazolin-5-one

ii. Synthetic process route

The synthesis of Edaravone is mainly made by ring formation of phenyl hydrazine and ethyl acetoacetate and other raw materials. The synthesis process is as follows:

III. Main instruments and reagents

i. Main instruments: thermometer, constant pressure drip funnel, three-neck flask, magnetic stirrer, analytical balance, beaker, measuring cylinder, heating sleeve, glass rod, filter bottle, Brinell funnel, vacuum pump, *etc.*

ii. Main reagents: phenyl hydrazine, ethyl acetoacetate, ethanol, concentrated hydrochloric acid, 10% sodium hydroxide, *etc.*

IV. Experimental steps

Add 13 g of ethyl acetoacetate and 5 mL of 70% ethanol into a three-neck flask equipped with a reflux condensing tube, a thermometer, a constant-pressure drip funnel and a magnetic stirrer, and stir. When the temperature rises to 45 ℃, slowly add 10.8 g of phenyl hydrazine and 3 mL of anhydrous ethanol solution in about 30 minutes. After maintaining the reaction for 30 min, cool the reaction system to 20 ℃. Add 1 mL of concentrated hydrochloric acid. Raise the temperature to 45 ℃ and maintain the reaction for 2 h. Adjust the pH to 7 with 10% NaOH. Add 20 mL of water. Cool to room temperature and stir for 1 h. Wash the filter cake three times with cold anhydrous ethanol. Dry to obtain a coarse yellowish crystal.

During refining, recrystallization can be carried out with mixed solvent of ethyl acetate/anhydrous ethanol (2:1, 30 mL) to ultimately obtain white crystals.

Note:

i. Generally, free phenylhydrazine is unstable, smoking and deteriorating quickly upon exposure to air. Operations should be performed quickly, and during the addition process, it is best to use a nitrogen balloon or protect with flowing nitrogen.

ii. During purification, activated carbon can be added during recrystallisation for decolorisation. If the product obtained from the first recrystallisation is too dark, a second recrystallisation can yield white crystals.

V. Attention or thinking questions

i. Attention

Try to write the reaction mechanism of this reaction.

ii. Thinking questions

(i) What causes phenyl hydrazine to become unstable in the air?

(ii) What is the action of concentrated hydrochloric acid in this reaction?

实验十　盐酸苯海索的合成

一、实验目的

1. 掌握盐酸苯海索药物的合成反应机理。
2. 熟悉盐酸苯海索药物的合成实验操作过程。
3. 了解盐酸苯海索药物的理化性质及临床用途。

二、实验原理

（一）药物简介

盐酸苯海索又名安坦，化学式为 $C_{20}H_{32}ClNO$，为中枢抗胆碱抗帕金森病药，选择性阻断纹状体的胆碱能神经通路，而对外周作用较小，从而有利于恢复帕金森病患者脑内多巴胺和乙酰胆碱的平衡，改善帕金森病症状。

化学结构式为：

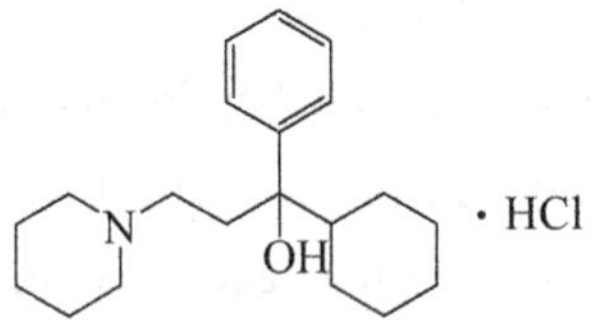

α-环己基-α-苯基-1-哌啶丙醇盐酸盐

（二）合成工艺路线

盐酸苯海索的合成，以苯乙酮、甲醛、哌啶盐酸盐为原料，进行 Mannich 反应制得 β-哌啶基苯丙酮盐酸盐中间体，再与氯代环己烷、金属镁作用制备的 Grignard 试剂反应制得，其合成工艺路线具体如下：

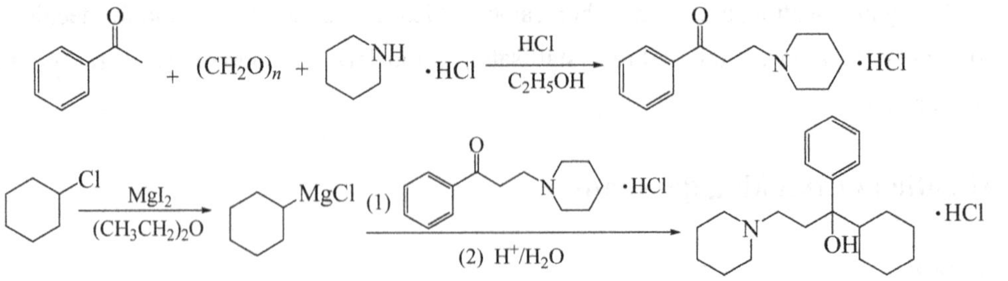

三、主要仪器和试剂

1. 主要仪器：三颈烧瓶、磁力搅拌器、温度计、分析天平、恒压滴液漏斗、冷凝管、干燥管、分液漏斗、烧杯、橡胶管、锥形瓶、抽滤瓶、布氏漏斗、滤纸、真空泵等。
2. 主要试剂：哌啶、95%乙醇、浓盐酸、苯乙酮、多聚甲醛、氯代环己烷、镁屑、碘、无水乙醚、活性炭等。

四、实验步骤

(一) 哌啶盐酸盐的合成

在配有磁力搅拌器、温度计、尾气吸收装置、冷凝管、干燥管以及恒压滴液漏斗的 250 mL 三颈烧瓶中，加入哌啶 17.4 mL、95%乙醇 30 mL，快速搅拌下，缓慢滴加浓盐酸约 16 mL，同时调节 pH 至 2，然后减压除去乙醇和水，反应体系呈糊状，停止蒸馏，自然冷却至室温，减压抽滤，得滤饼，干燥，得到白色结晶，即为哌啶盐酸盐。

(二) β-哌啶苯丙酮盐酸盐的合成

在配有磁力搅拌器、温度计、尾气吸收装置、冷凝管、干燥管以及恒压滴液漏斗的 250 mL 三颈烧瓶中，分别加入苯乙酮 8.8 mL、95%乙醇 18 mL、哌啶盐酸盐 9.1 g、多聚甲醛 3.8 g 以及浓盐酸 0.25 mL，快速搅拌，升温至 80~85℃回流，搅拌反应 4 h，反应结束后，用冰水浴，抽滤，用 95%乙醇洗涤滤饼 3 次（每次约 5 mL），直至洗出液呈中性。抽滤，干燥滤饼，得白色鳞片状结晶，即为 β-哌啶苯丙酮盐酸盐。

(三) 盐酸苯海索的合成

在配有磁力搅拌器、温度计、冷凝管、干燥管以及恒压滴液漏斗的 250 mL 三颈烧瓶中，分别加入镁屑 2.1 g、无水乙醚 15 mL 以及碘颗粒 1 粒，随后，将氯代环己烷 11.2 g 以及无水乙醚 15 mL 组成的混合液，缓慢滴入反应体系中，缓慢升温至微沸，碘的颜色渐渐退去，反应物呈乳灰色混浊状，表示反应开始。滴加结束后，继续搅拌回流反应 30 min，镁屑全部消失，停止反应。将反应体系置于冰水浴中。

在快速搅拌下，分批加入上步制得的 β-哌啶苯丙酮盐酸盐，继续搅拌回流反应 2 h。反应结束后，冰水浴冷却到小于 15℃，随后将反应液缓慢倒至盛有稀盐酸（浓盐酸 11 mL 和水 33 mL）的锥形瓶中，继续用冰水浴，将其冷却至小于 5℃，有固体析出，减压抽滤，用水充分洗涤滤饼至中性，得粗产物。粗产物用以适量 95%乙醇加热溶解，加活性炭脱色，趁热抽滤，滤液用冰水浴冷却至 10℃以下，有晶体析出，过滤，用少许乙醇洗涤，抽滤，干燥，得白色固体，即为盐酸苯海索纯品。

五、注意事项与思考题

(一) 注意事项

1. 试写出本反应的反应机理。
2. 反应过程中，多聚甲醛逐渐溶解，反应结束后，反应液中若有较多的多聚甲醛颗粒存在时，需增加反应时间。
3. 本反应为格氏反应，对无水条件要求较为严格，反应前需对试剂以及仪器做无水处理。

(二) 思考题

本反应中哌啶苯丙酮盐酸盐合成盐酸苯海索时，有哪些注意事项？

Experiment 10   Synthesis of trihexyphenidyl hydrochloride

I. Purpose of the experiment

i. To master the synthetic reaction mechanism of trihexyphenidyl hydrochloride.

ii. To familiar with the experimental operation process of synthesis of trihexyphenidyl hydrochloride.

iii. To understand the physicochemical properties and clinical applications of trihexyphenidyl hydrochloride.

II. Experimental principle

i. Drug introduction

Trihexyphenidyl hydrochloride, also known as Antan, has the molecular formula of $C_{20}H_{32}ClNO$. It is a central anti-cholinergic drug for anti-Parkinson's disease. It acts by selectively blocking cholinergic neural pathways in the striatum, while having little peripheral effects, which is beneficial to restoring the balance of dopamine and acetylcholine in the brain of patients with Parkinson's disease and improving the symptoms of the disease.

The chemical structure formula is:

α-cyclohexyl-α-phenyl-1-piperidine propyl hydrochloride

ii. Synthetic process route

The synthesis of trihexyphenidyl hydrochloride is mainly made by Mannich reaction of acetophenone with formaldehyde and piperidine hydrochloride to produce β-piperidine phenylacetone hydrochloride intermediates, and then react with Grignard reagent prepared by chlorocyclohexane and metal magnesium. The synthesis process is as follows:

III. Main instruments and reagents

i. Main instruments: three-neck flask, magnetic stirrer, thermometer, analytical balance, constant pressure drip funnel, condensing tube, separation funnel, beaker, rubber tube, drying tube, conical bottle, suction bottle, Brinell funnel, filter paper, vacuum pump, *etc.*

ii. Main reagents: piperidine, 95% ethanol, concentrated hydrochloric acid, acetophenone, polyformaldehyde, chlorocyclohexane, magnesium dust, iodine, anhydrous ether, activated carbon, *etc.*

IV. Experimental steps

i. Synthesis of piperidine hydrochloride

Add 17.4 mL of piperidine and 30 mL of 95% ethanol into a 250 mL three-neck flask equipped with a magnetic agitator, a thermometer, a tail gas absorption device, a condensing tube, a drying tube and a constant pressure drip funnel. After rapidly stirring, slowly add about 16 mL of concentrated hydrochloric acid to adjust the pH to 2. Then remove ethanol and water under pressure. When the reaction system is pasty, stop the distillation, and cool to room temperature naturally. Extract the filter cake under pressure to obtain the white crystal, which is piperidine hydrochloride.

ii. Synthesis of β-piperidine phenylacetone hydrochloride

Add 8.8 mL of acetophenone, 18 mL of 95% ethanol, 9.5 g of piperidine hydrochloride, 3.8 g of paraformaldehyde and 0.25 mL of concentrated hydrochloric acid into a 250 mL three-neck flask equipped with a magnetic stirrer, a thermometer, a tail gas absorption device, a condensation tube, a drying tube and constant pressure drip funnel. Stir rapidly and heat to reflux at 80-85℃ for 4 h. After the reaction, cool in an ice water bath and filter. Wash the filter cake with 95% ethanol 3 times (about 5 mL each time) until the liquid is neutral. After filtration, dry the cake to obtain white scaly crystals, which are β-piperidine phenylacetone hydrochloride.

iii. Synthesis of trihexyphenidyl hydrochloride

Add 2.1 g of magnesium, 15 mL of anhydrous ether and 1 iodine particle into a 250 mL three-neck flask equipped with a magnetic agitator, a thermometer, a condensing tube, a drying tube and a constant-pressure drip funnel. Then, slowly add the mixture consists of 11.2 g chlorocyclohexane and 15 mL of anhydrous ether into the reaction system. Gradually raise to a gentle boil. The color of iodine gradually recedes. The reactant become milky gray turbidity, indicating the beginning of the reaction. After the addition, continue stirring the reflux reaction for 30 minutes until all magnesium turnings disappear, at which point the reaction is stopped. Place the reaction system in an ice bath.

Add β-piperidine phenylacetone hydrochloride prepared in the previous step in batches under rapidly stirring. Continue stirring to reflux for 2 h. After the reaction, cool in an ice bath to below 15℃. Then slowly pour the reaction liquid into a conical bottle containing diluted hydrochloric acid (11 mL of concentrated hydrochloric acid and 33 mL of water). Continue cooling in an ice bath to below 5℃, resulting in solid precipitation. Filter the mixture iunder reduced pressure, and thoroughly wash the filter cake with water until neutral, yielding the crude product. Dissolve the crude product in an appropriate amount of 95% ethanol, decolorised with activated carbon, and filter while hot. Cool in an ice bath to below 10°C, resulting in crystal precipitation. Filter and wash with a small amount of ethanol. Then filter again. Dry to yield a white solid, which is the pure hydrochloride of benzhydroxylamine.

V. Attention or thinking questions

i. Attention

(i) Try to write the reaction mechanism of this reaction.

(ii) During the reaction process, paraformaldehyde is gradually dissolved. After the reaction, if there are more paraformaldehyde particles in the reaction liquid, the reaction time should be increased.

(iii) This reaction is a the Grigg reaction, which has strict requirements for anhydrous condition. Before the reaction it must be treated to ensure reagents and instruments anhydrous.

ii. Thinking questions

What precautions should be taken when synthesising hydrochloride of benzhydroxylamine from piperidyl phenylacetone hydrochloride in this reaction?

实验十一 葡甲胺的合成

一、实验目的

1. 掌握加氢反应的操作过程。
2. 熟悉反应中高压反应釜的装置结构。
3. 了解还原胺化反应的操作过程。
4. 了解葡甲胺的理化性质及临床用途。

二、实验原理

（一）药物简介

葡甲胺又名 N-甲基-D-葡胺、N-甲基-D-葡萄糖胺、N-甲基-D-葡糖胺，其分子式为 $C_7H_{17}NO_5$，为白色结晶性粉末。熔点 128~129℃。易溶于水，微溶于乙醇，几乎不溶于氯仿，味微甜而带咸涩。

葡甲胺能够增强心肌收缩力，具有强心作用，还有扩张外周血管、减轻心脏负荷的作用，临床上常用于冠心病、高血压心脏病等引起的胸闷、气短、呼吸困难、双下肢水肿等心衰症状。并能够改善心肌代谢，降低冠状动脉阻力，抑制血小板活性，故也能用于治疗冠状动脉粥样硬化性心脏病的稳定型劳力性心绞痛和急性心肌梗死等。

葡甲胺还可用于治疗心肌炎、扩张型心肌病等疾病，能够缩短心律失常持续时间，调节窦房结细胞的功能，具有抗心律失常作用，临床上可用于病窦综合征，如窦房阻滞、慢-快综合征及室性早搏等心律失常的辅助治疗。

化学结构式为：

1-去氧-1-甲胺基山梨醇

（二）合成工艺路线

葡甲胺是由葡萄糖与甲胺在镍催化剂作用下，还原胺化制得。其合成工艺路线具体如下：

三、主要仪器和试剂

1. 主要仪器：温度计、分析天平、磁力搅拌器、锥形瓶、烧杯、量筒、加热套、回流冷凝管、玻璃棒、抽滤瓶、布氏漏斗、滤纸、真空泵等。
2. 主要试剂：氢氧化钠、铝镍合金、95%乙醇、乙二胺四乙酸、甲胺、葡萄糖等。

四、实验步骤

（一）催化剂 Raney-Ni 的制备

在 1 L 烧杯中，加入氢氧化钠 50 g 以及蒸馏水 200 mL，用玻璃棒搅拌至溶解。在水浴上加热至 85℃，快速搅拌状态下，分批加入铝镍合金 50 g，整个过程在大约 1 h 内加完。然后在 100℃的温度下，搅拌 30 min，静置，使镍沉降，除去上层清液。固体用蒸馏水充分洗涤至中性，再次用 95%乙醇充分洗涤 3 次，每次 50 mL，用乙醇覆盖备用。

（二）甲胺醇溶液的合成

在 1 L 锥形瓶（吸收瓶）中，加入 95%乙醇 480 mL。在 1 L 蒸发瓶中放置甲胺水溶液 500 mL。

小心加热蒸发瓶，使甲胺缓慢蒸发，产生的甲胺气体通过回流冷凝管顶端，导入装有固体氢氧化钠的干燥塔，干燥后进入吸收瓶吸收。当蒸发瓶中甲胺水溶液温度上升到 90℃时，停止蒸馏，测定甲胺醇溶液的甲胺含量，应在 15%以上。若含量不足继续通甲胺，浓度过高，则加入乙醇稀释到 15%。

注：

甲胺的含量测定方法：精密吸取 1 mL 甲胺醇溶液于 100 mL 容量瓶中，加水至刻度，摇匀。吸取 20 mL 置于锥形瓶中，加 0.1 mol/L HCl 标准溶液 40 mL 及酚酞指示液数滴。用 0.1 mol/L NaOH 溶液滴定到显红色不褪为止。

（三）葡甲胺的制备

在 100 mL 高压反应釜中，加入葡萄糖 6 g、15%甲胺乙醇溶液 29 g 及 Raney-Ni 1.3 g，使用少量乙醇冲洗附着在釜壁上残留的 Raney-Ni。盖上釜盖，逐步对称的拧紧螺帽。按规定顺序排除釜内空气，通氢气使釜内压力达到 15 kg/cm^2，关闭进气阀，启动搅拌，待正常后开始加热，维持温度在 68℃。随时观察釜内压力变化，当压力降到 10 kg/cm^2 时，补充氢气到 15 kg/cm^2。如此反复通氢气至氢压不再变化为止，约需 6 h。停止搅拌，冷却至室温，打开排气阀排尽釜内残余空气，拧松螺帽，移开釜盖，吸出物料，过滤除去触媒。滤液冷却到 5℃ 以下，析出结晶，抽滤，得葡甲胺粗品。

粗品用 6～8 倍的蒸馏水溶解，加适量活性炭及 0.5 g 乙二胺四乙酸的水溶液，加热回流，抽滤，滤液在搅拌下，缓慢加入一定量的乙醇中。将其冷却到 5℃，有晶体析出，抽滤，干燥，得到精制葡甲胺产物。

注：

1. 本反应用的葡萄糖，在投料之前，一般经过 50～55℃干燥 24 h 后备用。
2. 高压反应釜中排除空气的操作步骤：拧开进气阀，通入氢气到 3 kg/cm^2，关闭进气阀，经检查无漏气现象后拧松排气阀,将气体放出,关闭排气阀后重复以上操作 2 次,使高压釜中的空气全部排除,最后通入氢气至所需压力($15\,kg/cm^2$),拧紧进气阀,关闭钢瓶阀门,进行氢化。

3. 反应后的 Raney-Ni 催化剂,仍有相当的活性,过滤时切勿滤干,以防催化剂燃烧。

五、注意事项与思考题

1. 请解释葡萄糖与甲胺在催化剂作用下,进行还原胺化的反应机理?
2. 在加压还原胺化反应中甲胺需要过量,为什么?

Experiment 11 Synthesis of meglumine

I. Purpose of the experiment

i. To master the operation process of hydrogenation reaction.

ii. To familiar with the device structure of high-pressure reactor in reaction.

iii. To understand the operation process of reduction amination reaction.

iv. To understand the physicochemical properties and clinical uses of meglumine.

II. Experimental principle

i. Drug introduction

Meglumine, also known as N-methyl-D-meglumine, *n*-methyl-D-glucosamine, n-methyl-D-glucosamine, has the molecular formula of $C_7H_{17}NO_5$. It is white crystalline powder. Its melting point is 128~129℃. It is soluble in water, slightly soluble in ethanol, almost insoluble in chloroform. It tastes slightly sweet and salty.

Meglumine can increase the contractility of the myocardium. It has the effect of strengthening the heart. And it expands the peripheral blood tube and reduces the heart load. Clinically, it is commonly used in coronary heart disease and hypertensive heart disease to cause chest tightness, shortness of breath, dyspnea, edema of both lower limbs and other heart failure symptoms. And it can improve myocardial metabolism, reduce coronary artery resistance, inhibit platelet activity. So it can be used in the treatment of coronary atherosclerotic heart disease stable exertive angina pectoris and acute myocardial infarction.

Meglumine can also be used for myocarditis, dilated cardiomyopathy and other diseases. In addition, it can shorten the duration of arrhythmia, regulate the function of sinoatrial node cells. And it has the effect of antiarrhythmic, which can be used in clinical adjuvant therapy for sick sinus syndrome such as sinus block, slow-fast syndrome and ventricular premature beat and other arrhythmias.

The chemical structure formula is:

$$\begin{array}{c} CH_2NHCH_3 \\ H-\!\!\!\!-OH \\ HO-\!\!\!\!-H \\ H-\!\!\!\!-OH \\ H-\!\!\!\!-OH \\ CH_2OH \end{array}$$

1-Deoxy-1-methylaminosorbitol

ii. Synthetic process route

Meglumine is mainly produced by reducing amination of glucose and methylamine under the action of the nickel catalyst. The synthesis process is as follows:

$$\begin{array}{c} CHO \\ H-OH \\ HO-H \\ H-OH \\ H-OH \\ CH_2OH \end{array} + CH_3NH_2 \xrightarrow[C_2H_5OH]{Ra\text{-}Ni,\ 15Kg/cm^2} \begin{array}{c} CH_2NHCH_3 \\ H-OH \\ HO-H \\ H-OH \\ H-OH \\ CH_2OH \end{array}$$

III. Main instruments and reagents

i. Main instruments: thermometer, analytical balance, magnetic stirrer, conical flask, beaker, condenser tube, measuring cylinder, heating sleeve, glass rod, suction bottle, Brinell funnel, filter paper, vacuum pump, *etc*.

ii. Main reagents: sodium hydroxide, aluminum nickel alloy, 95% ethanol, ethylenediamine tetraacetic acid, methylamine, glucose, *etc*.

IV. Experimental steps

i. Preparation of catalyst Raney-Ni

Add 50 g of sodium hydroxide and 200 mL of distilled water into a 1 L beaker. Stir with a glass rod until dissolved. Heat to 85 ℃ in a water bath and add 50 g of Al-Ni alloy in batches under rapidly stirring. Complete the whole process within about 1 hour. Then heat to 100 ℃, stirring for 30 min. Let it stand to allow nickel to settle, and remove the upper clear liquid. Wash the solid thoroughly with distilled water until neutral, then wash three times with 95% ethanol, 50 mL each time, and cover with ethanol for later use.

ii. Synthesis of methylamine alcohol solution

Add 480 mL of 95% ethanol into a 1 L conical bottle (absorption bottle). Place 500 mL of methylamine aqueous solution in a 1 L evaporative bottle.

The evaporation bottle is carefully heated to make the methylamine evaporate slowly. The methylamine gas produced is fed into a drying tower with solid sodium hydroxide through the top of the reflux condenser, and then absorbed into the absorption bottle after drying. When the temperature of methylamine aqueous solution in the evaporation bottle rises to 90 ℃, the distillation is stopped to determine the methylamine content of the methylamine alcohol solution, which should be above 15%. If the content is insufficient, continue to pass methylamine. If the concentration is too high, add ethanol dilution to 15%.

Note:

Method for the determination of methylamine content: Accurately absorb 1mL of methylamine alcohol solution into a 100 mL volumetric bottle. Add water to the scale, and shake well. Absorb 20 mL of the above solution into a conical flask, add 40 mL of 0.1 mol/L HCl standard solution and several drops of phenolphthalein indicator solution. Titrate with 0.1 mol/L NaOH solution until the stable red is achieved.

iii. Preparation of meglumine

Add 6 g of glucose, 29 g of 15% methylamine ethanol solution and 1.3 g of Raney-Ni into a 100 mL high-pressure reactor. Use a small amount of ethanol to wash the residual Raney-Ni attached to the reactor wall. Cover the kettle and tighten the nuts symmetrically step by step. Exhaust the air in the kettle in the prescribed order. Pass hydrogen to make the pressure in the kettle reach 15 kg/cm^2. Close the intake valve. Start stirring, and start heating when normal. Then maintain the temperature at 68℃. Observe the pressure change in the kettle at any time. When the pressure drops to 10 kg/cm^2, add hydrogen to 15 kg/cm^2. It takes about 6 hours to pass through the hydrogen repeatedly until the hydrogen pressure no longer changes. Stop stirring. Cool to room temperature. Open the exhaust valve to exhaust the residual air in the kettle. Loosen the nut. Remove the kettle cover. Extract the material. Filter and remove the catalyst. Cool the filtrate to below 5℃ to precipitate crystals, then perform suction filtration to obtain crude meglumine.

Dissolve the crude product with 6-8 times the amount of distilled water. Add an appropriate amount of activated carbon and 0.5 g of ethylenediamine tetraacetic acid aqueous solution. Heat to reflux, and perform suction filtration. Slowly add a certain amount of ethanol under stirring. Cool to 5℃, allowing crystals to precipitate, then perform suction filtration and dry to obtain the refined meglumine product.

Note:

i. The glucose used in this reaction should be generally dried for 24 hours at 50-55℃ before use.

ii. Steps to remove air from the high-pressure reactor: Open the gas inlet valve, introduce hydrogen gas to 3 kg/cm^2, close the gas inlet valve, check for leaks, then loosen the exhaust valve to release gas, close the exhaust valve, and repeat this operation twice to completely remove air from the high-pressure reactor. Finally, introduce hydrogen gas to the required pressure (15 kg/cm^2), tighten the gas inlet valve, close the cylinder valve, and proceed with hydrogenation.

iii. The Raney-Ni catalyst still has considerable activity after the reaction. Do not drain it during filtration to prevent the catalyst from burning.

V. Attention or thinking questions

i. Please explain the mechanism of reducing amination between glucose and methylamine under the action of catalyst?

ii. Why is an excess of methylamine required in pressurized reduction amination?

实验十二 利巴韦林的合成

一、实验目的

1. 掌握利巴韦林药物的合成反应机理。
2. 熟悉利巴韦林药物的合成实验操作过程。
3. 了解利巴韦林药物的理化性质及临床用途。

二、实验原理

（一）药物简介

利巴韦林又名病毒唑、三氮唑核苷等，化学名为 1-β-D-呋喃核糖基-1H-1,2,4-三氮唑-3-羧酰胺，分子式为 $C_8H_{12}N_4O_5$，为抗非逆转录病毒药。本药品为白色或类白色结晶性粉末，无臭。在水中易溶，在乙醇中微溶，在乙醚或二氯甲烷中不溶。

利巴韦林用于治疗拉沙热或流行性出血热（具肾脏综合征或肺炎表现者）（静脉滴注或口服）、慢性丙型肝炎的治疗，防治病毒性上呼吸道感染（滴鼻）。

本药物吸入用药，可导致肺功能退化、细菌性肺炎、气胸和心血管反应（血压下降以及心脏停搏）等，罕见贫血和网织红细胞过多的报道。也有结膜炎、皮疹发生。静脉或口服给药后主要的不良反应有溶血性贫血、血红蛋白减低及贫血、乏力等，停药后可消失。较少见的不良反应有疲倦、头痛、失眠等，多见于应用大剂量者，并可有食欲减退、恶心等。静脉推注可引起寒战，与干扰素合用可引起严重抑郁、自杀意念、溶血性贫血、骨髓抑制、自身免疫和感染疾患、肺功能紊乱（呼吸困难、肺浸润、局限性肺炎、肺炎等）、胰腺炎、糖尿病等不良反应。

化学结构式为：

1-β-D-呋喃核糖基-1H-1,2,4-三氮唑-3-羧酰胺

（二）合成工艺路线

利巴韦林的合成，以四乙酰核糖和 1,2,4-三氮唑-3-羧酸甲酯为原料，在催化剂双（对硝基苯基）磷酸酯作用下，缩合后，再用氨-甲醇溶液氨解，制得利巴韦林。其合成工艺路线具体如下：

三、主要仪器和试剂

1. 主要仪器：三颈烧瓶、磁力搅拌器、温度计、分析天平、恒压滴液漏斗、冷凝管、干燥管、分液漏斗、烧杯、橡胶管、锥形瓶、抽滤瓶、布氏漏斗、滤纸、真空泵等。
2. 主要试剂：四乙酰核糖、1,2,4-三氮唑-3-羧酸甲酯、无水甲醇、乙醇、氨气、乙腈、苯、氯仿、吡啶、乙酸乙酯、三氯氧磷、对硝基苯酚、双（对硝基苯基）磷酸酯、氢氧化钠、盐酸、刚果红试纸等。

四、实验步骤

（一）双（对硝基苯基）磷酸酯的合成

在配有磁力搅拌器、温度计、冷凝管、干燥管以及恒压滴液漏斗的 100 mL 三颈烧瓶中，加入乙腈 15 mL、苯 120 mL，随后加入对硝基苯酚 13 g、三氯氧磷 7.3 g。冰水浴冷却至 0 ℃，快速搅拌，缓慢滴加吡啶 13 mL，30 min 内滴加结束，然后，自然升温至室温，继续搅拌反应 6 h。有固体产生，抽滤，除去白色沉淀，滤液，减压回收溶剂，反应体系中的残留物，溶解在 30 mL 氯仿中，同时，在剧烈搅拌下，向其体系中加入 15%氢氧化钠溶液 40 mL，发现有黄色沉淀生成，冷却体系，过滤，滤饼用水充分洗涤，干燥，得粗产物。

将粗产物溶解在 100 mL 热水中，抽滤，用以除去少许的磷酸三苯酯。同时，滤液用 5 mol/L 盐酸调节酸碱度，至刚果红试纸为红色时停止。冷却，有固体生成，滤饼用水充分洗涤，干燥，得产物。产物可以用乙酸乙酯进行重结晶，得无色固体纯品。

（二）利巴韦林的合成

在配有磁力搅拌器、温度计、冷凝管、干燥管以及恒压滴液漏斗的 100 mL 三颈烧瓶中，加入四乙酰核糖 15.9 g、1,2,4-三氮唑-3-羧酸甲酯 5.8 g，快速搅拌混匀后，将反应体系升温至 170～180 ℃，剧烈搅拌，将混合物近完全溶解后，加入催化剂双（对硝基苯基）磷酸酯 0.15 g，用真空泵抽真空，在上述温度下减压反应 30 min。反应结束后，将反应体系冷却至 60～70 ℃，加入无水甲醇，充分溶解，冷却至室温，随后置于冰箱内降温，放置过夜，有白色固体析出。抽滤，将滤饼用冷无水甲醇充分洗涤 3 次，真空干燥，得无色固体 1-（2,3,5-三-O-乙酰基-β-D-呋喃核糖基）-1,2,4-三氮唑-3-羧酸甲酯。

将上述所得 1-（2,3,5-三-O-乙酰基-β-D-呋喃核糖基）-1,2,4-三氮唑-3-羧酸甲酯，悬浮于无水甲醇中，随后，冷却体系至-5 ℃，在快速搅拌下，通入干燥氨气，当固体全部溶解后，继续通干燥氨气 0.5 h，密闭反应体系，并于室温下静置 48 h。过滤，滤饼用无水甲醇充分洗涤，干燥，得粗产物。可以用 90%乙醇进行重结晶，真空干燥，得白色絮状晶体的利巴韦林纯品。

五、注意事项与思考题

1. 试写出本反应的反应机理。
2. 双（对硝基苯基）磷酸酯的制备中使用的乙腈和苯，有毒，可致癌，操作时需在通风橱中进行，同时避免吸入和接触到皮肤。
3. 缩合后将反应物冷却至 60～70 ℃时，即可加入无水甲醇，温度太高时加入甲醇可能造成甲醇的冲出，但是温度太低，产物容易变黏，加入甲醇后不易溶解。

Experiment 12 Synthesis of ribavirin

Ⅰ. Purpose of the experiment

i. To master the synthetic reaction mechanism of ribavirin.

ii. To familiar with the experimental operation process of ribavirin synthesis.

iii. To understand the physicochemical properties and clinical uses of ribavirin.

Ⅱ. Experimental principle

i. Drug introduction

Ribavirin, also known as ribavirin and ribavirin, has the chemical name 1-β-d-furanoribosyl-1h-1, 2, 4-triazole-3-carboxyamide, with the molecular formula is $C_8H_{12}N_4O_5$, which is an anti-retroviral drug. This drug is white or white -like a crystalline powder, odorless. It is soluble in water, slightly soluble in ethanol, insoluble in ethyl ether or methylene chloride.

Ribavirin is used for the treatment of Lassa fever or epidemic hemorrhagic fever (in persons with renal syndrome or pneumonia manifestations) (intravenous or oral), for the treatment of chronic hepatitis C, and for the prevention and treatment of viral upper respiratory infections (nasal drops).

When administered by inhalation, this drug has been reported to cause pulmonary degeneration, bacterial pneumonia, pneumothorax and cardiovascular reactions (decreased blood pressure and cardiac arrest), with rare reports of anemia and reticulocyte. Conjunctivitis and rash also occur. The main adverse reactions after intravenous or oral administration include hemolytic anemia, decreased hemoglobin, anemia, fatigue, *etc.*, which can disappear after drug withdrawal. Less common adverse reactions include tiredness, headache, insomnia, *etc.*, which are more common in patients applying large doses, loss of appetite, nausea, *etc*. Intravenous injections can cause chills. And combined with interferons can cause severe depression, suicidal ideation, hemolytic anemia, bone marrow suppression, autoimmune and infectious diseases, lung disorders (dyspnea, lung infiltration, localized pneumonia, pneumonia, *etc.*), pancreatitis, diabetes and other adverse reactions.

The chemical structure formula is:

1-β-d-furanoriboyl-1H-1, 2, 4-triazole-3-carboxyamide

ii. Synthetic process route

The synthesis of ribavirin is mainly through tetraacetyl ribose and 1, 2, 4-triazole-3-fusoate methyl ester as raw materials, under the action of catalyst double (*p*-nitrophenyl) phosphate, after condensation, and then ammonia-methanol solution ammonolysis, ribavirin. The synthesis process is as follows:

III. Main instruments and reagents

i. Main instruments: three-necked flask, magnetic stirrer, thermometer, analytical balance, constant pressure drip funnel, condensing tube, drying tube, separation funnel, beaker, rubber tube, conical bottle, suction bottle, Brinell funnel, filter paper, vacuum pump, *etc.*

ii. Main reagents: tetraacetyl ribose, 1, 2, 4-triazole-3-carboxylate methyl ester, anhydrous methanol, ethanol, ammonia, acetonitrile, benzene, chloroform, pyridine, ethyl acetate, phosphorus trichloride, *p*-nitrophenol, sodium hydroxide, hydrochloric acid, Congo red test paper, *etc.*

IV. Experimental steps

i. Synthesis of double (*p*-nitrophenyl) phosphate esters

Add 15 mL of acetonitrile and 120 mL of benzene into a 100 mL three-necked flask equipped with a magnetic agitator, a thermometer, a condensing tube, a drying tube and a constant pressure drip funnel. Then add 13 g of *p*-nitrophenol and 7.3 g of phosphorus trichloride. Cool in an ice water bath to 0℃. Stir quickly, and slowly add 13 mL of pyridine within 30 min. Naturally raise the temperature to room temperature, and continue stirring for 6 h. Dissolve the residue in the reaction system in 30 mL of chloroform. At the same time, add 40 mL of 15% sodium hydroxide solution to the system while stirring rigorously, resulting in the formation of a yellow precipitate. Cool and filter. Wash thoroughly the filter cake with water and dry to obtain crude products.

Dissolve the crude product in 100 mL of hot water and filter to remove a small amount of triphenyl phosphate. Adjusted the pH with 5 mol/L hydrochloric acid until Congo red test paper turns red. Upon cooling, a solid forms. Filter and thoroughly wash with water. Dry to obtain the product. The product can be recrystallised using ethyl acetate to yield a colourless solid pure product.

ii. Synthesis of ribavirin

Add 15.9 g of tetraacetyl ribose, 5.8 g of 1, 2, 4-triazole-3-carboxylate methyl ester into a 100 mL three necked flask equipped with a magnetic agitator, a thermometer, a condensation tube, a drying tube and a constant pressure drip hopper. After rapidly mixing, heat to 170-180℃ and stir vigorously until the mixture is nearly completely dissolved. Add 0.15 g of catalyst double (*p*-nitrophenyl) phosphate ester. Evacuate with a vacuum pump, maintaining reduced pressure for 30 min at the above temperature. After the reaction, cool the reaction system to 60-70℃. Add anhydrous methanol to fully dissolve, and cool to room temperature. Then place in the refrigerator to cool down overnight, resulting in the precipitation of a white solid. Filter and wash thoroughly with cold anhydrous methanol 3 times Dry by vacuum to obtain a colorless solid 1-(2, 3, 5-tri-o-acetyl-β-D-ribofuranose) -1, 2, 4-triazole-3-carboxylate methyl ester.

Suspend the obtained 1-(2, 3, 5-tri-o-acetyl-β-D-ribofuranose) -1, 2, 4-triazole-3-carboxylate methyl ester in anhydrous methanol. Then, cool to −5 ℃. Add dry ammonia gas under rapidly stirring. After all solids are dissolved, the dry ammonia gas is continued for 0.5 h in a closed reaction system. And let stand at room temperature for 48 h. Filter and wash thoroughly with anhydrous methanol. Dry to obtain crude product. Ribavirin can be recrystallized with 90% ethanol and vacuum dried to yield white flocculent crystals.

Ⅴ. Attention or thinking questions

i. Try to write the reaction mechanism of this reaction.

ii. The acetonitrile and benzene used in the preparation of double (*p*-nitrophenyl) phosphate ester are toxic and carcinogenic. The operation should be carried out in the fume hood, while avoiding inhalation and contact with the skin.

iii. After condensation, when the reactant is cooled to 60-70 ℃, anhydrous methanol can be added. When the temperature is too high, methanol may be flushed out, but when the temperature is too low, the product easily becomes sticky, and it is not easy to dissolve after methanol is added.

实验十三　L-抗坏血酸棕榈酸酯的合成

一、实验目的

1. 掌握 L-抗坏血酸棕榈酸酯合成反应机理。
2. 熟悉对温度敏感的反应的操作方法。
3. 了解 L-抗坏血酸棕榈酸酯的理化性质及用途。

二、实验原理

（一）药物简介

L-抗坏血酸棕榈酸酯，分子式为 $C_{22}H_{38}O_7$，为白色或黄白色粉末，有轻微柑橘香味，是一种无毒无害的多功能营养性抗氧保鲜剂。具有抗氧化及营养强化功能，用作抗氧增白剂，在油脂中抗氧效果非常明显，且耐高温，适用于医药、保健品、化妆品等。常用于含油食品、食用油、动植物油及高级化妆品中，也可用于各种婴幼儿食品及奶粉中。

抗坏血酸棕榈酸酯为棕榈酸与 L-抗坏血酸等天然成分酯化而成，是一种高效的氧清除剂和增效剂，被世界卫生组织食品添加剂委员会，评定为具有营养性、无毒、高效、使用安全的食品添加剂。

化学结构式为：

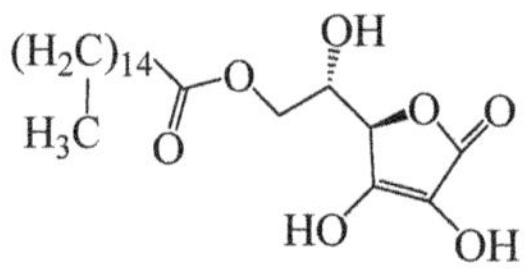

抗坏血酸棕榈酸酯

（二）合成工艺路线

L-抗坏血酸先与浓硫酸反应，生成硫酸酯，然后与长链脂肪酸发生酯交换反应得到棕榈酸酯。由于 L-抗坏血酸受热易氧化变性，因此反应不能加热。其合成工艺路线具体如下：

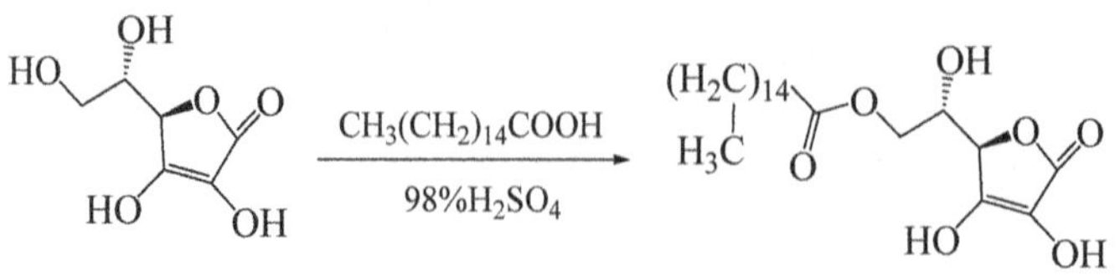

三、主要仪器和试剂

1. 主要仪器：三颈烧瓶、分析天平、磁力搅拌器、烧瓶、烧杯、量筒、玻璃棒、抽滤瓶、布氏漏斗、滤纸、真空泵等。
2. 主要试剂：L-抗坏血酸、浓硫酸、棕榈酸、正己烷、乙醚、石油醚等。

四、实验步骤

在装有搅拌器、温度计及磁力搅拌器的三颈烧瓶中,加入 19.5 mL 浓硫酸,剧烈搅拌下,在 20℃的条件下,缓慢加入 L-抗坏血酸 5.5 g,搅拌状态下,使 L-抗坏血酸全部溶解,同时,加入棕榈酸 10.7 g,快速搅拌,所有固体溶解,在室温下反应 48 h。反应结束后,剧烈搅拌下,将反应液倒入 50 g 碎冰中,抽滤,滤饼用冰水充分洗涤 3 次,压紧,再用正己烷充分洗涤 3 次,干燥,得粗品。

L-抗坏血酸棕榈酸酯粗品的精制,采用乙醚-石油醚混合溶剂,进行重结晶,得精制的 L-抗坏血酸棕榈酸酯。

五、注意事项与思考题

(一)注意事项

1. 本反应过程中,需在所有固体溶解后再加棕榈酸。
2. 本反应对温度较为敏感,需控制加入速度,使反应平稳进行反应,若温度上升较快,可用水或冰浴,控制反应体系温度小于 30℃。
3. 滤饼用正己烷洗涤,用于除去未反应的棕榈酸。

(二)思考题

L-抗坏血酸的酯化方法有哪些?

Experiment 13 Synthesis of L-ascorbyl palmitate

I. Purpose of the experiment

i. To master the synthesis mechanism of L-ascorbic palmitate.

ii. To familiar with the operation method of temperature-sensitive reaction.

iii. To understand the physicochemical properties and applications of L-ascorbic palmitate.

II. Experimental principle

i. Drug introduction

L-ascorbic palmitate, with the molecular formula of $C_{22}H_{38}O_7$, is a white or yellowish-white powder with a slight citrus flavor. It is a non-toxic and harmless multifunctional nutritional antioxidant preservative. It has antioxidant and nutrient strengthening functions. It can be used as an antioxidant whitening agent. It has a very obvious antioxidant effect in oil and oil, because it can resist high temperature. It is suitable for medicine, health products, cosmetics and so on. It is often used in oily food, edible oil, animal and plant oil and high-grade cosmetics. It can also be used in a variety of infant food and milk powder.

Ascorbic acid palmitate is esterified with natural ingredients such as L-ascorbic acid. It is a highly efficient oxygen scavenger and synergist. It has been assessed as a nutritive, non-toxic, highly efficient and safe food additive by the World Health Organization Committee on Food Additives.

The chemical structure formula is:

6-O-palmitoyl-L-ascorbic acid

ii. Synthetic process route

L-ascorbic acid first reacts with concentrated sulfuric acid to form sulfate, and then has a transesterification reaction with long chain fatty acids to get palmitate. Since L-ascorbic acid is easily oxidized and denatured by heat, the reaction cannot be heated. The synthesis process is as follows:

III. Main instruments and reagents

i. Main instruments: three-necked flask, analytical balance, magnetic stirrer, flask, beaker, measuring cylinder, glass rod, suction bottle, Brinell funnel, filter paper, vacuum pump, *etc.*

ii. Main reagents: ascorbic acid, concentrated sulfuric acid, palmitic acid, n-hexane, ether, petroleum ether, *etc.*

IV. Experimental steps

Add 19.5 mL of concentrated sulfuric acid into a three-necked flask equipped with a stirrer, a thermometer and a magnetic stirrer. Under vigorous stirring at 20℃, slowly add 5.5 g of L-ascorbic acid to dissolve all L-ascorbic acid. Meanwhile, add 10.7 g of palmitic acid and stir rapidly to dissolve all solids. It reacts at room temperature for 48 h. After the reaction, pour the reaction liquid into 50 g of crushed ice under vigorous stirring, and filter wash the filter cake thoroughly with ice water for 3 times. Press tightly, and then thoroughly wash with n-hexane for 3 times. Dry to obtain the crude product.

Perform the crude product of L-ascorbic acid palmitate using a mixed solvent of diethyl ether and petroleum ether for recrystallization, yielding the refined L-ascorbic acid palmitate.

V. Attention or thinking questions

i. Attention

(i) In this reaction process, palmitic acid should be added after all solids have dissolved.

(ii) This reaction is sensitive to temperature, so it is necessary to control the adding speed so that the reaction can be carried out smoothly. If the temperature rises quickly, water or ice bath can be used to control the temperature of the reaction system below 30℃.

(iii) The filter cake is washed with n-hexane in order to remove unreacted palmitic acid.

ii. Thinking questions

What methods are available for the esterification of L-ascorbic acid?

实验十四　奥沙普秦的合成

一、实验目的

1. 掌握奥沙普秦合成反应的机理。
2. 熟悉奥沙普秦的理化性质与临床用途。
3. 了解奥沙普秦合成过程中的实验操作过程。

二、实验原理

（一）药物简介

奥沙普秦又名 4,5-二苯基-2-噁唑丙酸、贝美格、恶丙嗪，分子式为 $C_{18}H_{15}NO_3$。本药物主要用于治疗风湿性关节炎、类风湿关节炎、骨关节炎、强直性脊椎炎、颈肩腕综合征、肩周炎、痛风及外伤和手术后的消炎镇痛。

化学结构式为：

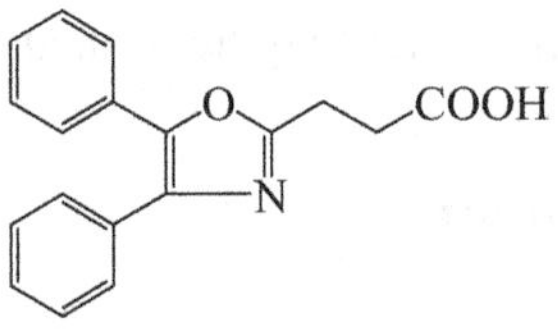

4,5-二苯基-2-噁唑丙酸

（二）合成工艺路线

奥沙普秦的合成，以二苯乙醇酮、丁二酸、乙酸酐等为原料，进行制备，其合成工艺路线具体如下：

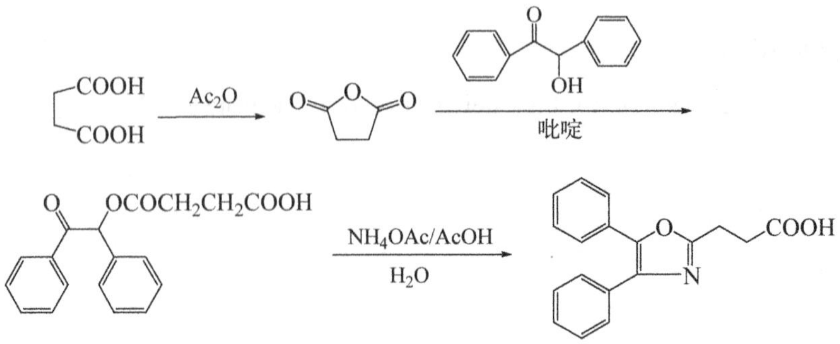

三、主要仪器和试剂

1. 主要仪器：三颈烧瓶、磁力搅拌器、分析天平、温度计、恒压滴液漏斗、冷凝管、干燥管、分液漏斗、烧杯、橡胶管、锥形瓶、抽滤瓶、布氏漏斗、滤纸、真空泵等。
2. 主要试剂：无水乙醚、丁二酸、二苯乙醇酮、乙酸铵、乙酸酐、吡啶、冰醋酸、甲醇、无水乙醇、氢氧化钠等。

四、实验步骤

（一）丁二酸酐的合成

在装有磁力搅拌器、温度计、回流冷凝管以及干燥管的三颈烧瓶中，加入丁二酸 71 g、乙酸酐 120 mL，升温至回流，保温反应 1 h。反应结束后，将反应液倒入干燥的烧杯中，静置 0.5 h。自然冷却至室温，冰箱冷冻，抽滤，干燥，得丁二酸酐粗产物，用乙醚洗涤，得白色丁二酸酐。

（二）奥沙普秦的合成

在装有磁力搅拌器、温度计、回流冷凝管以及干燥管的三颈烧瓶中，加入上步制得的丁二酸酐 40 g、二苯乙醇酮 62 g、吡啶 35 g，剧烈搅拌，升温至 90~95℃，保温反应 1 h 后。向反应体系中加入乙酸铵 45 g 以及冰醋酸 150 g，继续保温搅拌，反应 2.5 h。随后加 80 mL 加水，继续保温搅拌反应 1 h。反应结束后，冷却至室温，有晶体析出，抽滤、将滤饼水洗，干燥，得粗产物，用甲醇重结晶，得白色奥沙普秦纯品。

五、注意事项与思考题

1. 影响反应收率的因素有哪些？
2. 试写出奥沙普秦的合成反应原理。

Experiment 14 Synthesis of oxaprozin

Ⅰ. Purpose of the experiment

i. To master the mechanism of oxaprozin synthesis reaction.

ii. To familiar with the physicochemical properties and clinical uses of oxaprozin.

iii. To understand the experimental operation process of oxaprozin synthesis.

Ⅱ. Experimental principle

i. Drug introduction

Oxaprozin, also known as 4, 5-diphenyl-2-oxazopropionic acid, bemegride, has the molecular formula $C_{18}H_{15}NO_3$. This drug is mainly used for the treatment of rheumatoid arthritis, rheumatoid arthritis, osteoarthritis, ankylosing spondylitis, neck, shoulder and wrist syndrome, scapulohumeral periarthritis, gout and trauma and postoperative analgesia.

The chemical structure formula is:

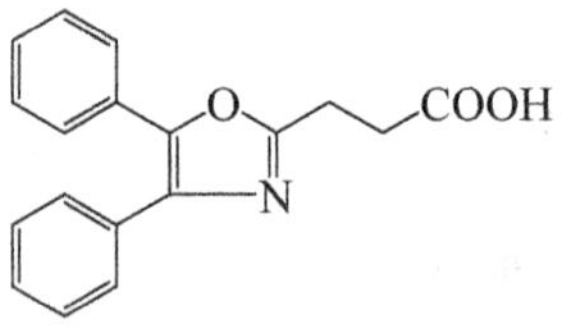

4,5-Diphenyl-2-oxazolepropionic Acid

ii. Synthetic process route

The synthesis of oxaprozin is mainly prepared by using dl-benzoin, succinic acid and acetic anhydride as raw materials. The synthesis process is as follows:

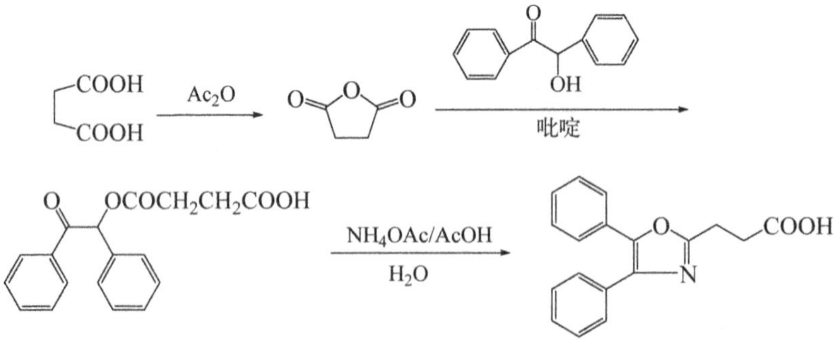

Ⅲ. Main instruments and reagents

i. Main instruments: three necked flask, magnetic stirrer, analytical balance, thermometer, constant pressure drip funnel, condensing tube, drying tube, separation funnel, beaker, rubber tube, conical bottle, suction bottle, Brinell funnel, filter paper, vacuum pump, *etc*.

ii. Main reagents: anhydrous ether, succinic acid, dl-benzoin, ammonium acetate, acetic anhydride, pyridine, glacial acetic acid, methanol, anhydrous ethanol, sodium hydroxide, *etc*.

IV. Experimental steps

i. Synthesis of succinic anhydride

Add 71 g of succinic acid and 120 mL of acetic anhydride into a three necked flask equipped with a magnetic agitator, a thermometer, a reflux condensing tube and a drying tube. Heat to reflux, and maintain for 1 h. After the reaction, pour the reaction solution into a dry beaker and let it stand for 0.5 h. Naturally cool to room temperature, and freeze in the refrigerator. Then filter and dry to obtain the crude product of succinic anhydride. Wash with ether to yield white succinic anhydride.

ii. Synthesis of oxaprozin

Add 40 g of succinic anhydride, 62 g of dl-benzoin and 35 g of pyridine into a three necked flask equipped with a magnetic agitator, a thermometer, a reflux condensing tube and a drying tube. Stir vigorously. Heat to 90-95 ℃, and maintain for 1 h. Add 45 g of ammonium acetate and 150 g of glacial acetic acid into the reaction system. Continue to stir and maintain for 2.5 h. Then add 80 mL of water and continue stir and maintain for 1 h. After the reaction, cool to room temperature. Some crystals will precipitate, then filter. Wash the cake with water and dry to obtain crude product. Recrystallise with methanol to obtain pure white oxaprozin product.

V. Attention or thinking questions

i. What factors affect the yield of the reaction?

ii. Try to write the principle of oxaprozin synthesis reaction.

实验十五 苯佐卡因的合成

一、实验目的

1. 掌握苯佐卡因的合成反应机理。
2. 熟悉苯佐卡因药物合成的基本过程。
3. 掌握氧化、酯化和还原反应的能力及基本操作。
4. 了解苯佐卡因的理化性质及临床用途。

二、实验原理

（一）药物简介

苯佐卡因又名对氨基苯甲酸乙酯，化学式为 $C_9H_{11}NO_2$，是一种重要的有机化合物。为白色结晶性粉末，味微苦而麻。易溶于醇、醚、氯仿，能溶于杏仁油、橄榄油、稀酸，难溶于水。

临床上用于创面、溃疡面、烧伤、皮肤擦裂及痔的镇痛、止痒。苯佐卡因是一种脂溶性表面麻醉剂，与其他几种局麻药，如利多卡因、丁卡因等相比，其作用强度较小，因而在作用于黏膜时不会因麻醉作用而使人感到不适。苯佐卡因是一种脂溶性较强的药物，故而易与黏膜或皮肤的脂层结合，但不易透过而进入体内产生毒性。

化学结构式为：

$$\underset{\text{4-氨基苯甲酸乙酯}}{\begin{array}{c}\text{COOC}_2\text{H}_5 \\ \diagdown\!\!\!\bigcirc\!\!\!\diagup \\ \text{NH}_2\end{array}}$$

（二）合成工艺路线

苯佐卡因的合成，主要通过硝基甲苯被重铬酸钾强氧化剂氧化，随后用铁还原等反应制得，其合成工艺路线具体如下：

$$\underset{\text{NO}_2}{\underset{|}{\bigcirc}}\!\!-\!\text{CH}_3 + Na_2Cr_2O_7 + H_2SO_4 \longrightarrow \underset{\text{NO}_2}{\underset{|}{\bigcirc}}\!\!-\!\text{COOH} + Na_2SO_4 + Cr_2(SO_4)_3$$

$$\underset{NO_2}{\underset{|}{C_6H_4}}-COOH \xrightarrow{C_2H_5OH} \underset{NO_2}{\underset{|}{C_6H_4}}-COOC_2H_5 + H_2O$$

$$\underset{NO_2}{\underset{|}{C_6H_4}}-COOC_2H_5 + Fe + H_2O \longrightarrow \underset{NH_2}{\underset{|}{C_6H_4}}-COOC_2H_5 + Fe_3O_4$$

三、主要仪器和试剂

1. 主要仪器：三颈烧瓶、单口瓶、回流装置、烧杯、磁子、加热套、玻璃棒、橡胶管、分析天平、球形冷凝管、干燥管、抽滤瓶、布氏漏斗、恒压滴液漏斗、滤纸、研钵、真空泵等。

2. 主要试剂：重铬酸钾、硝基甲苯、浓硫酸、活性炭、5%碳酸钠、铁粉、冰醋酸、氯化铵、氯仿、5%氢氧化钠、乙醇、浓盐酸等。

四、实验步骤

（一）对硝基苯甲酸的合成

在配有磁力搅拌器、回流冷凝管及恒压滴液漏斗的三颈烧瓶中，加入重铬酸钾 23.6 g、水 50 mL，剧烈搅拌，待重铬酸钾溶解后，加入硝基甲苯 8 g，滴加浓硫酸 32 mL，滴加完毕后，升温，保持反应液微沸腾，反应 90 min。冷却后，将反应液倒入冷水 80 mL 中，抽滤，滤饼用水洗涤 3 次，每次 15 mL。将滤饼转移到烧瓶中，加入 5%硫酸 35 mL，在沸水浴上加热 10 min，并不断搅拌，冷却后抽滤，滤饼溶于热的 5%氢氧化钠溶液 70 mL 中，在 50℃ 左右抽滤，滤液加入活性炭 0.5 g 脱色，约 10 min，趁热抽滤，冷却，在充分搅拌下，将滤液缓慢倒入 15%硫酸 50 mL 中，抽滤，充分洗涤，干燥得到对硝基苯甲酸。

（二）对硝基苯甲酸乙酯的合成

在配有磁力搅拌器、回流冷凝管、干燥管及恒压滴液漏斗的三颈烧瓶中，加入对硝基苯甲酸 6 g、无水乙醇 24 mL 以及浓硫酸 2 mL，振摇混合均匀，加热回流 80 min 后，稍微冷却后，将反应液全部倒入水 100 mL 中，抽滤，滤饼转移至研钵中，研细，加入 5%碳酸钠溶液 10 mL，研磨 5 min，抽滤，用充分洗涤，干燥，得到对硝基苯甲酸乙酯。

（三）对氨基苯甲酸乙酯的制备（还原）

在配有磁力搅拌器、回流冷凝管、干燥管及恒压滴液漏斗的三颈烧瓶中，加入水 35 mL、乙酸 2.5 mL 以及处理好的铁粉 8.6 g，剧烈搅拌，将反应体系升温至 95～98℃，反应 5 min 后，稍冷，加入对硝基苯甲酸乙酯 6 g 和 95%乙醇 35 mL，在搅拌状态下，回流反应 90 min 后，分 3 次加入温热的碳酸钠饱和溶液（由碳酸钠 3 g 和水 30 mL 配成），搅拌 5 min，趁热抽滤，滤液经冷却后，有晶体析出，抽滤，固体用稀的乙醇洗涤，干燥，得对氨基苯甲酸乙酯的粗品。

（四）精制

将对氨基苯甲酸乙酯粗品置于装有冷凝器的圆底烧瓶中，加入 10～15 倍量的 50%乙醇，在水浴上加热溶解。稍冷，加活性炭进行脱色，升温回流 30 min，趁热抽滤。将滤液趁热转移至烧杯中，自然冷却，待结晶完全析出后，抽滤，用少量 50%乙醇洗涤两次，压干，干燥，得对氨基苯甲酸乙酯纯品。

五、注意事项与思考题

（一）注意事项

1. 抽滤过程中的布氏漏斗需要预热，否则会在布氏漏斗上析出晶体。
2. 氧化反应时，在用 5%氢氧化钠处理滤饼时，温度应保持在 50℃左右，若温度过低，对硝基苯甲酸钠会析出而被滤去。
3. 酯化反应对无水的要求特别严格，反应中所涉及的仪器等均需做无水处理。须在无水条件下进行，如有水进入反应系统中，收率将会降低。
4. 对硝基苯甲酸乙酯及少量未反应的对硝基苯甲酸均溶于乙醇，但均不溶于水。
5. 还原反应中，因铁粉比重大，沉于瓶底，必须将其搅拌起来，才能使反应顺利进行，故充分激烈搅拌是铁还原反应的重要因素。
6. 反应中所用的铁粉需预处理方法为：称取铁粉 10 g 置于烧杯中，加入 2%盐酸 25 mL，在石棉网上加热至微热，抽滤，水洗至 pH 5～6，烘干，备用。

（二）思考题

氧化反应结束后，将对硝基苯甲酸从混合物中分离出来的原理是什么？

Experiment 15 Synthesis of benzocaine

I. Purpose of the experiment

i. To master the synthesis mechanism of benzocaine.

ii. To familiar with the basic process of benzocaine synthesis.

iii. To master the ability and basic operation of oxidation, esterification and reduction reactions.

iv. To understand the physicochemical properties and clinical uses of benzocaine.

II. Experimental principle

i. Drug introduction

Benzocaine, also known as ethyl *p*-aminobenzoate, with the molecular formula $C_9H_{11}NO_2$, is an important organic compound. It is a white crystalline powder. It is slightly bitter and numbing taste. It is easily soluble in alcohol, ether, chloroform, soluble in almond oil, olive oil, dilute acid, difficult to dissolve in water.

It is used for wound surface, ulcer surface, burn, skin rub crack and hemorrhoids analgesia, itching. Benzocaine is a fat soluble topical anesthetic. Compared with several other local anesthetics, such as lidocaine, *etc.*, its effect is less intense, so it does not cause discomfort when acting on mucous membranes due to the anesthetic effect. Benzocaine is a drug with strong fat solubility, so it is easy to combine with mucous membrane or the lipid layer of the skin, but does not easily penetrate into the body to produce toxicity.

The chemical structure formula is:

p-COOC₂H₅ benzene NH₂

4-ethyl aminobenzoic acid

ii. Synthetic process route

Benzocaine is synthesized mainly through nitrotoluene oxidized by strong oxidant potassium dichromate and then reduced by iron. The synthesis process is as follows:

$$\text{p-O}_2\text{N-C}_6\text{H}_4\text{-CH}_3 + Na_2Cr_2O_7 + H_2SO_4 \longrightarrow \text{p-O}_2\text{N-C}_6\text{H}_4\text{-COOH} + Na_2SO_4 + Cr_2(SO_4)_3$$

$$\underset{NO_2}{C_6H_4}\text{-COOH} \xrightarrow{C_2H_5OH} \underset{NO_2}{C_6H_4}\text{-COOC}_2H_5 + H_2O$$

$$\underset{NO_2}{C_6H_4}\text{-COOC}_2H_5 + Fe + H_2O \longrightarrow \underset{NH_2}{C_6H_4}\text{-COOC}_2H_5 + Fe_3O_4$$

III. Main instruments and reagents

i. Main instruments: three necked flask, single-mouth flask, reflux device, beaker, magneton, heating sleeve, glass rod, rubber tube, drying tube, analytical balance, spherical condensing tube, drying tube, suction bottle, Brinell funnel, constant pressure dripping funnel, filter paper, mortar, vacuum pump, *etc*.

ii. Main reagents: potassium dichromate, nitrotoluene, concentrated sulfuric acid, activated carbon, sodium carbonate, iron powder, ice acetic acid, ammonium chloride, chloroform, sodium hydroxide, ethanol, concentrated hydrochloric acid, *etc*.

IV. Experimental steps

i. Synthesis of *p*-nitrobenzoic acid

Add 23.6 g of potassium dichromate and 50 mL of water into a three necked flask equipped with a magnetic agitator, a reflux condensing tube and a constant pressure dripping funnel. Stir vigorously until the potassium dichromate dissolves, Add 8 g of nitrotoluene and 32 mL of concentrated sulfuric acid. After the addition, heat the reaction liquid to maintain a gentle boil for 90 min. After cooling, pour the reaction liquid into 80 mL of cold water. Filter and wash the filter cake with water for 3 times, 15 mL each time. Transfer the filter cake to the flask. Add 35 mL of 5% sulfuric acid. Heat for 10min in the boiling water bath while stirring constantly. Then filter after cooling. Dissolve the filter cake in 70 mL of hot 5% sodium hydroxide solution. Filter at about 50℃. Added 0.5 g of activated carbon to decolorization the filtration for about 10min. Filter while hot. Then cool it. While thoroughly stirring, slowly pour the filtrate into 50 mL of 15% sulfuric acid and filter. Then wash thoroughly. And dry to obtain p-nitrobenzoic acid.

ii. Synthesis of ethyl p-nitrobenzoate

Add 6 g of *p*-nitrobenzoic acid, 24 mL of anhydrous ethanol and 2 mL of concentrated sulfuric acid into a three necked flask equipped with a magnetic agitator, a reflux condensing tube, a drying tube and a constant pressure drip funnel. Shake to mix evenly. Heat to reflux for 80 min. After slightly cooling, pour all the reaction liquid into 100 mL of water and filter. Transfer the filter cake to a mortar and grind finely. Add 10 mL of 5% sodium carbonate solution. Grind for 5 min and filter. Wash thoroughly. Dry to obtain p-nitrobenzoate.

iii. Preparation of ethyl p-aminobenzoate (reduction)

Add 35 mL of water, 2.5 mL of acetic acid and 8.6 g of treated iron powder into a three-necked flask equipped with a magnetic stirrer, a reflux condensing tube, a drying tube and a constant pressure drip funnel. Stir vigorously. Heat the reaction system to 95-98℃, and maintain the reaction for 5 min. Cold slightly. Add 6 g of *p*-nitroethyl benzoate and 35 mL of 95% ethanol. Reflux the reaction for 90 min while stirring. Add warm sodium carbonate saturated solution (made of 3 g of sodium carbonate and 30mL of water) in 3 times. Stir for 5 min. Filter while hot, and after cooling. Crystals will precipitate and filter. Wash the solid with dilute ethanol and dry to obtain the crude product of ethyl p-aminobenzoate.

iv. Refining

Place the crude ethyl *p*-aminobenzoate in a round-bottomed flask equipped with a condenser. Add 10~15 times the amount of 50% ethanol to dissolve it in a water bath. After slightly cooling, add activated carbon for decolorization. Heat to reflux for 30 min, and filter while hot. Transfer the filtrate to the beaker and cool naturally. After the crystallization, drain the filtrate. Wash the solid twice with a small amount of 50% ethanol. Press and dry to obtain p-aminobenzoate pure product.

Ⅴ. Attention or thinking questions

i. Attention

(i) The Buchner funnel used during filtration needs to be preheated, otherwise crystals may precipitate on the funnel.

(ii) During the oxidation reaction, the temperature should be maintained at about 50℃ when treating the cake with 5% sodium hydroxide. If the temperature is too low, sodium p-nitrobenzoate will precipitate and be filtered.

(iii) Esterification reaction is very strict to anhydrous requirements, and the instruments involved in the reaction need to do anhydrous treatment. It must be carried out without water. If water enters the reaction system, the yield will be reduced.

(iv) Ethyl p-nitrobenzoic acid and a small amount of unreacted p-nitrobenzoic acid are soluble in ethanol, but not soluble in water.

(v) In the reduction reaction, due to the significant ratio of iron powder, which is sunk at the bottom of the bottle, it must be stirred up in order to make the reaction proceed smoothly, so sufficiently intense stirring is an important factor in the ferric acid reduction reaction.

(vi) The pretreatment method of the iron powder used in the reaction is as follows: weigh 10 g of iron powder and place it in a beaker, add 25 mL 2% hydrochloric acid, heat it on asbestos net until it is slightly hot, filter it, wash it to pH 5-6, dry it and set aside.

ii. Thinking questions

What is the principle of separating p-nitrobenzoic acid from the mixture after the oxidation reaction is complete?

实验十六 对乙酰氨基酚（扑热息痛）的合成及鉴定

一、实验目的

1. 掌握对乙酰氨基酚的重结晶精制操作方法。
2. 熟悉芳香族伯胺类药物的重氮化-偶合反应。
3. 熟悉酚类化合物与三氯化铁特殊颜色反应。
4. 了解对乙酰氨基酚的理化性质及临床用途。

二、实验原理

（一）药物简介

对乙酰氨基酚，又名扑热息痛，商品名称有百服宁、必理通、泰诺、醋氨酚等。能溶于乙醇、丙酮和热水，难溶于水，不溶于石油醚及苯。无气味，味苦。饱和水溶液 pH 5.5～6.5。

该药物国际非专有药名为 Paracetamol，是最常用的非抗炎解热镇痛药，解热作用与阿司匹林相似，镇痛作用较弱，无抗炎、抗风湿作用，是乙酰苯胺类药物中最好的品种，特别适合于不能应用羧酸类药物的患者，用于感冒、牙痛等症。对乙酰氨基酚也是重要的有机合成中间体，过氧化氢的稳定剂，照相化学药品。

对乙酰氨基酚为乙酰苯胺类解热镇痛药。通过抑制环氧化酶，选择性抑制下丘脑体温调节中枢前列腺素的合成，导致外周血管扩张、出汗而达到解热的作用，其解热作用强度与阿司匹林相似，通过抑制前列腺素等的合成和释放，提高痛阈而起到镇痛作用，属于外周性镇痛药，作用较阿司匹林弱，仅对轻、中度疼痛有效，本药物无明显抗炎作用。

化学结构式为：

对乙酰氨基酚

（二）合成工艺路线

对乙酰氨基酚的合成，主要通过对硝基苯酚经过还原反应、酰化反应等制备，其合成工艺路线具体如下：

三、主要仪器和试剂

1. 主要仪器：分析天平、三颈烧瓶、抽滤瓶、回流冷凝管、搅拌子、磁力搅拌器、200℃温度计、布氏漏斗、滤纸、水浴锅、烧杯、橡胶管、锥形瓶、真空泵等。
2. 主要试剂：对硝基苯酚、0.5%亚硫酸氢钠、乙酸酐、活性炭等。

四、实验步骤

（一）扑热息痛的合成

在装有磁力搅拌器、温度计、回流冷凝管以及干燥管的三颈烧瓶中，加入对氨基苯酚 10.6 g、水 30 mL 以及乙酸酐 12 mL，剧烈搅拌，升温至 80℃，保温反应 30 min，冷却至室温，有晶体析出，抽滤，滤饼用冷水充分洗涤，抽干，干燥，得到白色粗产物，即为对乙酰氨基酚。

（二）扑热息痛的精制

在锥形瓶中加入对乙酰氨基酚粗产物，加水（按每克粗产物加水 5 mL 的比例），加热使完全溶解，稍冷后，加入适量活性炭，煮沸 10 min，在抽滤瓶中先加入亚硫酸氢钠 0.5 g，趁热过滤，滤液放冷析晶，抽滤，滤饼用 0.5%亚硫酸氢钠溶液 5 mL，分 2 次洗涤，抽干，干燥，得白色对乙酰氨基酚纯品。

（三）对乙酰氨基酚的定性鉴别

1. 取对乙酰氨基酚少量→逐滴加水振摇使溶解→滴加三氯化铁试液→应出现蓝紫色。
2. 取对乙酰氨基酚约 0.1 g→加稀盐酸 5 mL→放冷→分取 0.5 mL→滴加亚硝酸钠试液 5 滴→摇匀→加水 3mL 稀释→加碱性 β-萘酚试液 2 mL→振摇→应不出现红色。
3. 取对乙酰氨基酚约 0.1 g→加稀盐酸 5 mL→置水浴中加热 40 min→放冷→分取 0.5 mL→滴加亚硝酸钠试液 5 滴→摇匀→加水 3 mL 稀释→加碱性 β-萘酚试液 2 mL→振摇→应出现红色。

五、注意事项与思考题

（一）注意事项

1. 对氨基苯酚的纯度，对目标物产量、质量等，有着重要的影响。
2. 酰化反应中，有水存在，乙酸酐可选择性的酰化氨基而不与酚羟基作用。若以乙酸代替乙酸酐，则难以控制氧化副反应，反应时间长，产品质量差。
3. 加亚硫酸氢钠，可防止对乙酰氨基酚被空气氧化，但亚硫酸氢钠浓度不宜过高，否则会影响产品质量。

（二）思考题

1. 酰化反应为何选用乙酸酐，而不用乙酸作酰化剂？
2. 加亚硫酸氢钠的目的是什么？
3. 对乙酰氨基酚中的特殊杂质是何物？是怎样产生的？

Experiment 16 Synthesis and identification of acetaminophen(paracetamol)

Ⅰ. Purpose of the experiment

i. To master the operation method of recrystallization refining of acetaminophen.
ii. To familiar with the diazotization and coupling reaction of aromatic primary amines.
iii. To familiar with the special color reaction of phenolic compounds and ferric chloride.
iv. To understand the physicochemical properties and clinical uses of acetaminophen.

Ⅱ. Experimental principle

i. Drug introduction

Acetaminophen, also known as Paracetamol, is marketed under brand names such as bufferin, panadol, Tylenol, acetaminophen and so on. It is soluble in ethanol, acetone and hot water, It is very insoluble in water. It is insoluble in petroleum ether and benzene. It is odorless and bitter taste, with a saturated aqueous solution pH of 5.5-6.5.

The international non-proprietary drug name is Paracetamol. It is the most commonly used non-anti-inflammatory antipyretic and analgesic drug. Its antipyretic effect is similar to aspirin, with weak analgesic effect and no anti-inflammatory and antirheumatic effect. It is the best variety of acetanilide drugs, especially suitable for patients who cannot use carboxylic acid drugs, such as colds, toothaches and other diseases. Acetaminophen is also an important intermediate in organic synthesis, a stabilizer of hydrogen peroxide, and a photographic chemical.

Paracetamol is an acetanilide antipyretic analgesic. By inhibiting cyclooxygenase, it selectively inhibits the synthesis of prostaglandins in the thermoregulatory center of the hypothalamus, leading to peripheral vascular dilation and sweating, thus achieving an antipyretic effect. Its antipyretic effect is similar to that of aspirin. By inhibiting the synthesis and release of prostaglandins, it raises the pain threshold and plays an analgesic effect only for mild and moderate pain effective. This drug has no obvious anti-inflammatory effect.

The chemical structure formula is:

NHCOCH$_3$

⎯⎯⎯

OH

4-acetamido phenol

ii. Synthetic process route

The synthesis of paracetamol is mainly prepared through reduction reaction and acylation reaction of *p*-nitrophenol. The synthesis process is as follows:

$$\underset{\underset{OH}{}}{\overset{NO_2}{\bigcirc}} \xrightarrow{Fe, HCl} \underset{\underset{OH}{}}{\overset{NH_2}{\bigcirc}} \xrightarrow{(CH_3CO)_2O} \underset{\underset{OH}{}}{\overset{NHCOCH_3}{\bigcirc}}$$

III. Main instruments and reagents

i. Main instruments: analytical balance, three-neck flask, filter flask, condensing tube, stirrer, magnetic stirrer, Brinell funnel, filter paper, water bath, beaker, rubber tube, conical flask, 200 ℃ thermometer vacuum pump, *etc*.

ii. Main reagents: *p*-nitrophenol, 0.5%sodium bisulfite, acetic anhydride, actirated carbon, *etc*.

IV. Experimental steps

i. the synthesis of paracetamol

Add 10.6 g of *p*-amino-phenol, 30 mL of water and 12 mL of acetic anhydride into a three necked flask equipped with a magnetic agitator, a thermometer, a reflux condensing tube and a drying tube. Stir vigorously. Heat to 80 ℃, and maintain for 30 min. Cool to room temperature, and allow crystals to precipitate. Filter and wash the filter cake thoroughly with cold water. Dry and obtain the white crude product.

ii. paracetamol refining

Add the crude product of acetaminophen in the conical bottle. Add water (according to the ratio of 5 mL per gram of crude product). Heat to completely dissolve. Slightly cold. Add the appropriate amount of activated carbon. Boil for 10 min. Add 0.5 g of sodium bisulfite in the filter bottle. Filter while hot. Cool the filtrate for crystallization and filter. Wash the filter cake with 5 mL of 0.5% sodium bisulfite solution in 2 times. Dry to obtain the white pure acetaminophen.

iii. Qualitative identification of acetaminophen

(i) Take a small amount of acetaminophen → add water dropwise and shake to dissolve → add a few drops of ferric chloride test solution → a blue-purple colour should appear.

(ii) Take about 0.1 g of paracetamol → add 5 mL of dilute hydrochloric acid → cool → take 0.5 mL → add 5 drops of sodium nitrite solution → shake well → dilute with 3 mL of water → add 2 mL of alkaline β-naphthol solution → shake → no red colour should appear.

(iii) Take about 0.1 g of paracetamol → add 5 mL of diluted hydrochloric acid → heat in a water bath for 40 min → cool → divide 0.5 mL → add 5 drops of sodium nitrite test solution → shake well → add 3 mL of water to dilute → add 2 mL of basic β-naphthol test solution → shake → Red color should appear.

V. Attention or thinking questions

i. Attention

(i) The purity of *p*-aminophenol has an important effect on the yield and quality of the target.

(ii) In the acylation reaction, the presence of water, acetic anhydride can selectively acylate the amino group without interacting with the phenol hydroxyl group. If acetic acid is used instead of acetic anhydride, it is difficult to control the side reactions of oxidation, the reaction time is long and the product quality is poor.

(iii) The addition of sodium bisulfite can prevent acetaminophen by air oxidation, but the concentration of sodium bisulfite should not be too high. Otherwise it will affect the quality of the product.

ii. Thinking questions

(i) Why is acetic anhydride chosen for acylation instead of acetic acid as acylation agent?

(ii) What is the purpose of adding sodium bisulfite?

(iii) What is the special impurities in acetaminophen? How are they product?

实验十七　亚胺-154 的合成

一、实验目的

1. 掌握亚胺-154 药物合成反应中缩合、环合反应基本的操作过程。
2. 熟悉亚胺-154 药物的理化性质及临床用途。
3. 了解调节溶液酸碱度的方法。

二、实验原理

（一）药物简介

亚胺-154 又名抗癌 161、宁癌-154，熔点为 290~292℃。其为白色针状结晶。难溶于水及乙醇，碱中不稳定，亚胺-154 为抗肿瘤药物，主治恶性淋巴瘤，对胃癌、肺癌等有一定的缓解作用，对肝癌、网状细胞肉瘤也有缓解作用，也可用于银屑病的治疗。

化学结构式为：

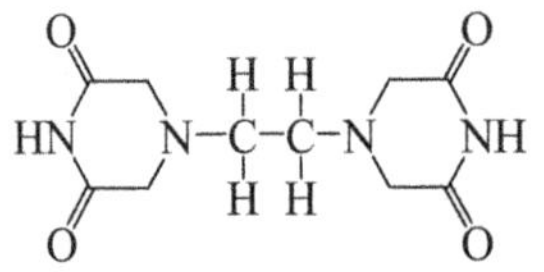

1,2-双（3,5-二氧哌-1-嗪）乙烷

（二）合成工艺路线

亚胺-154 药物的合成，主要通过氯乙酸、乙二胺盐酸盐、甲酰胺等原料进行制备，其合成工艺路线如下：

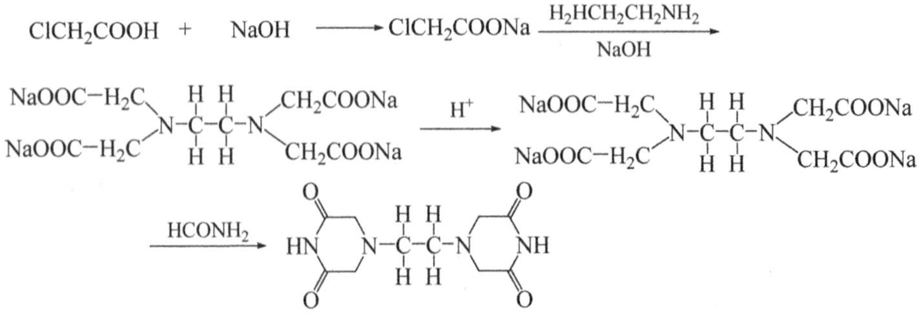

$$ClCH_2COOH + NaOH \longrightarrow ClCH_2COONa \xrightarrow[NaOH]{H_2HCH_2CH_2NH_2}$$

合成路线续图

三、主要仪器和试剂

1. 主要仪器：温度计、恒压滴液漏斗、分析天平、三颈烧瓶、搅拌子、冷凝管、磁力搅拌器、烧杯、玻璃棒、布氏漏斗、抽滤瓶、滤纸、加热套、橡胶管、油浴锅、沸石、真空泵等。

2. 主要试剂：氯乙酸、氢氧化钠、乙二胺盐酸盐、氢氧化钠、活性炭、浓盐酸、甲酰胺、乙醇等。

四、实验步骤

（一）乙二胺四乙酸的合成

在装有温度计、磁力搅拌器、冷凝管以及恒压滴液漏斗的三颈烧瓶中，加入氯乙酸 22.5 g，水 45 mL 溶解，搅拌至完全溶解。另将氢氧化钠 22 g 溶于水 60 mL 中，再加入乙二胺盐酸盐 6.6 g，搅拌混匀后，置于恒压滴液漏斗中，在搅拌下滴加到氯乙酸溶液中（1~2 min）。滴加完毕后，升温至 105℃，pH 约为 9 时，搅拌保温反应 2 h。于前 30 min 内，测定反应液的 pH。当 pH 低于 9 时，补加少许 30%氢氧化钠，使 pH 维持在 9 左右。反应 2 h 后，加入活性炭进行脱色，抽滤。滤液用盐酸调节 pH 至 1，低温静置，有固体析出，抽滤，固体用水洗涤，干燥，得乙二胺四乙酸。

（二）乙二胺四乙酰亚胺的合成

在装有温度计、磁力搅拌器及恒压滴液漏斗的三颈烧瓶中，加入乙二胺四乙酸 14.6 g、甲酰胺 26 g，加热至 140℃左右，保温反应 90 min 后，随后升温至 160℃，再次保温反应 4 h。反应过程中逸出的气体的 pH 缓慢上升，从 pH 3 逐渐上升至 pH 8~9 时，即为反应终点，此时，趁热将反应液倒入冷水中，有固体晶体析出，抽滤。结晶分别用水、乙醇洗涤，烘干，得白色的乙二胺四乙酰亚胺晶体。

五、注意事项与思考题

（一）注意事项

试写出本反应的反应机理。

（二）思考题

1. 本反应中，在乙二胺和氯乙酸钠缩合反应中，为什么控制 pH 9 左右？
2. 在乙二胺四乙酸与甲酰胺环合反应中，最初逸出的气体为什么 pH 约为 3，而当结束时 pH 变为 8~9？

Experiment 17 Synthesis of Ethylimine

Ⅰ. Purpose of the experiment

i. To master the basic operation process of condensation and cyclation reaction in ethylimine drug synthesis reaction.

ii. To familiar with the physicochemical properties and clinical uses of ethylimine.

iii. To understand how to adjust the pH of the solution.

Ⅱ. Experimental principle

i. Drug introduction

Ethylimine, also known as anti-cancer 161 and Ning-cancer 154, has a melting point of 290-292℃. It is a white acicular crystal. It is insoluble in water and ethanol, and unstable in alkali. Ethylimine is an antitumor drug for malignant lymphoma. It has a certain remission effect on gastric cancer and lung cancer, as well as liver cancer and reticular cell sarcoma. It is also used in the treatment of psoriasis.

The chemical structure formula is:

$$\underset{O}{\overset{O}{\underset{\|}{HN}}}\hspace{-2pt}\underset{O}{\overset{}{\underset{}{}}}\hspace{-2pt}N-\underset{H}{\overset{H}{C}}-\underset{H}{\overset{H}{C}}-N\underset{O}{\overset{}{\underset{}{}}}\hspace{-2pt}\underset{O}{\overset{O}{\underset{\|}{NH}}}$$

Ethylenediamine Tetraacetylimide

ii. Synthetic process route

The synthesis of ethylimine drug is mainly prepared by chloroacetic acid, ethylene diamine hydrochloride, formamide and other raw materials. The synthesis process is as follows:

$$ClCH_2COOH + NaOH \longrightarrow ClCH_2COONa \xrightarrow[NaOH]{H_2HCH_2CH_2NH_2}$$

$$\begin{array}{c} NaOOC-H_2C \\ NaOOC-H_2C \end{array}\hspace{-4pt}N-\underset{H}{\overset{H}{C}}-\underset{H}{\overset{H}{C}}-N\hspace{-4pt}\begin{array}{c} CH_2COONa \\ CH_2COONa \end{array} \xrightarrow{H^+} \begin{array}{c} NaOOC-H_2C \\ NaOOC-H_2C \end{array}\hspace{-4pt}N-\underset{H}{\overset{H}{C}}-\underset{H}{\overset{H}{C}}-N\hspace{-4pt}\begin{array}{c} CH_2COONa \\ CH_2COONa \end{array}$$

$$\xrightarrow{HCONH_2} \underset{O}{\overset{O}{HN}}\hspace{-2pt}N-\underset{H}{\overset{H}{C}}-\underset{H}{\overset{H}{C}}-N\underset{O}{\overset{O}{NH}}$$

Ⅲ. Main instruments and reagents

i. Main instruments: thermometer, constant pressure drip funnel, analysis balance, three necked flask, stirrer, condensing tube, magnetic stirrer, beaker, glass rod, Brinell funnel, filter bottle, filter paper, heating sleeve, rubber tube, oil bath, zeolite, vacuum pump, ethanol, *etc*.

ii. Main reagents: chloroacetic acid, sodium hydroxide, ethylenediamine hydrochloride, sodium hydroxide, activated carbon, concentrated hydrochloric acid, formamide, *etc*.

Ⅳ. Experimental steps

i. Synthesis of ethylene diamine tetraacetic acid

Add 22.5 g of chloroacetic acid and 45 mL of water into a three necked flask fitted with a thermometer, a magnetic agitator, a condensing tube and a constant pressure drip funnel. Stir until completely dissolve. Dissolve 22 g of sodium hydroxide of in 60 mL of water, and then add 6.6 g of ethylenediamine hydrochloride. After mixing well, place the mixture in a constant pressure drip funnel, and then add this solution dropwise to the chloroacetic acid solution (about 1-2 min) under stirring. After the addition, raise the temperature to 105 ℃ with pH about 9. Stir and maintain for 2 h. During the first 30 min, measure the pH of the reaction solution. When the pH drops below 9, add a little 30% sodium hydroxide to maintain the pH around 9. After 2 h of reaction, add activated carbon for decolorization and filter. Acidify the filtrate to pH 1 with hydrochloric acid, allow to stand at low temperature and collect the, precipitated solid by filtration. Wash the solid with water and dry to obtain ethylenediamine tetraacetic acid.

ii. Synthesis of ethylenediamine tetraacetylimide

Add 14.6 g of ethylene diamine tetraacetic acid and 26 g of formamide into a three necked flask equipped with a thermometer, a magnetic agitator and a constant pressure drip hopper. Heat to about 140 ℃ for 90 min. Then heat to 160 ℃ for 4 h. During the reaction, the pH of the gas escaping rises slowly from 3 to 8~9, which is the reaction end point. At this time, pour the reaction liquid into cold water, while hot, resulting in the precipitation of solid crystals, which are then filtered. Wash the crystals with water and ethanol. Dry to obtain white ethylenediamine tetraacetylimide crystals.

Ⅴ. Attention or thinking questions

i. Attention

Try to write the reaction mechanism of this reaction.

ii. Thinking questions

(i) In this reaction, why is the pH controlled around 9 during the condensation reaction of ethylene diamine and chloroacetic acid?

(ii) In the reaction of ethylene diamine tetraacetic acid and formamide, why is the pH of the gas initially escaped about 3, while it changes to 8-9 the end?

实验十八　丙戊酸钠的合成

一、实验目的

1. 掌握丙戊酸钠药物的合成，熟悉烃化、水解、脱羧等反应。
2. 掌握丙戊酸钠成盐条件及精制方法。
3. 熟悉减压蒸馏等基本操作。
4. 了解丙戊酸钠药物的理化性质及临床用途。

二、实验原理

（一）药物简介

丙戊酸钠，化学式为 $C_8H_{15}O_2Na$，白色粉末，是一种不含氮的广谱抗癫痫药。其为原发性大发作和失神小发作的首选药，对部分性发作（简单部分性和复杂部分性及部分性发作继发大发作）疗效不佳。对婴儿良性肌阵挛癫痫、婴儿痉挛有一定疗效，对肌阵挛性失神发作，需加用乙琥胺或其他抗癫痫药才有效。对难治性癫痫可以试用。本药物除用于抗癫痫外，还可用于治疗热性惊厥、运动障碍、舞蹈症、卟啉症、精神分裂症、带状疱疹引发的疼痛、肾上腺功能紊乱，以及预防酒精戒断综合征。

化学结构式为：

$$\begin{array}{c} H_3CH_2CH_2C \\ \diagdown \\ C-COONa \\ \diagup H \\ H_3CH_2CH_2C \end{array}$$

2-丙基戊酸钠

（二）合成工艺路线

$$CH_3CH_2CH_2OH \xrightarrow[2h]{H_2SO_4, NaBr} CH_3CH_2CH_2Br + Na_2SO_4 + H_2O$$

$$H_2C\begin{array}{c}COOC_2H_5\\COOC_2H_5\end{array} \xrightarrow[C_2H_5ONa, 3h]{2CH_3CH_2CH_2Br} \begin{array}{c}H_3CH_2CH_2CCH_2CH_2CH_3\\ \diagdown C \diagup \\ H_3CH_2CH_2CCH_2CH_2CH_3\end{array} \xrightarrow[4h]{KOH, HCl}$$

$$\begin{array}{c}H_3CH_2CH_2CCOOH\\ \diagdown C \diagup \\ H_3CH_2CH_2CCOOH\end{array} \xrightarrow{180℃} \begin{array}{c}H_3CH_2CH_2C\\ \diagdown \\ C-COOH \\ \diagup \\ H_3CH_2CH_2C\end{array} \xrightarrow{NaOH} \begin{array}{c}H_3CH_2CH_2C\\ \diagdown \\ C-COONa \\ \diagup H\\ H_3CH_2CH_2C\end{array}$$

三、主要仪器和试剂

1. 主要仪器：温度计、冷凝管、油浴锅、三颈烧瓶、搅拌子、磁力搅拌器、玻璃棒、布氏漏斗、分析天平、抽滤瓶、滤纸、干燥管、加热套、烧杯、橡胶管、锥形瓶、真空泵等。
2. 主要试剂：11.5%溴化钠、正丙醇、浓硫酸、5%碳酸钠、无水碳酸钠、丙二酸二乙酯、钠、乙醇、浓盐酸、氢氧化钠、乙酸乙酯等。

四、实验步骤

（一）1-溴正丙烷的合成

在装有磁力搅拌器、温度计、回流冷凝管的三颈烧瓶中，加入 11.5%溴化钠溶液 68 mL，冷却至 0℃，快速搅拌，加入正丙醇 45 mL，再缓慢滴加浓硫酸 54 mL，约 1 h 内加完，升温回流，保温反应 2 h。进行蒸馏，收集 60～110℃的馏分，得到 1-溴正丙烷粗产物，用 1-溴正丙烷体积的 1/8 量浓硫酸、1/2 量水以及 1/2 量 5%碳酸钠溶液，将该粗产物依次充分洗涤，随后，无水碳酸钠干燥，即得到 1-溴正丙烷产物。

注：

1. 本反应中，11.5%溴化钠溶液的配制过程中，即使有部分溴化钠不溶解，也不需要加热使之全部溶解，因为反应体系冷却后，会再次析出，当反应回流加热时，会完全溶解。

2. 在 1-溴正丙烷的制备中，氢卤酸置换醇羟基是一个可逆反应，通过蒸馏或缓慢滴加浓硫酸，可使反应体系中有足够浓度的氢溴酸，从而使反应平衡向右移动，提高收率。

（二）二丙基丙二酸二乙酯的合成

在配有磁力搅拌器、温度计、回流冷凝管的三颈烧瓶中，加入丙二酸二乙酯 20 mL、质量分数为 17%的乙醇钠溶液 120 mL，快速搅拌，缓慢升温至 70℃，缓慢滴加上步制得的 1-溴正丙烷，滴加结束后，保温回流反应 3 h。随后，将回流装置转换成蒸馏装置，进行蒸馏，回收乙醇，当烧瓶中有固体析出时，停止蒸馏，室温冷却，静置 1 h，由溴化钠固体生成，抽滤。母液继续蒸馏至无液滴馏出（油浴的温度约 110℃），停止蒸馏，冷却，得到棕红色油状液体，即为二丙基丙二酸二乙酯，将其全部倒入烧杯中，待用。

注：

1. 本反应中，乙醇钠的制备方法：在 50 mL 烧杯中加入无水甲苯 25 mL，称量烧杯，用镊子将金属钠取出，擦去附着的油，将其迅速放入二甲苯中，用减重法称取金属钠 20.4 g，用小刀切碎，置于二甲苯中备用。

在装有搅拌、回流冷凝器、干燥管的 250 mL 三颈烧瓶中，加入乙醇 120 mL，分批用镊子取出金属钠，擦去二甲苯，迅速加至乙醇中，反应放出热量，使乙醇沸腾，当金属钠全部变成乙醇钠且溶于乙醇中时，冷却备用。

2. 切割金属钠需小心，切割过程中，所用的纸张、刀具等需经酒精处理后，再经水处理，废物方可扔掉。

3. 蒸馏回收乙醇时，应将反应体系中乙醇完全除尽，否则乙醇混入产品中，使产品实际含量下降，将会影响下步反应的投料配比。

4. 回收乙醇中有固体析出时，冷却放置时间延长，固体析出将增多，滤除后将有利于回收完全。

（三）二丙基丙二酸的合成

在配有磁力搅拌器、温度计、回流冷凝管的三颈烧瓶中，加入上步制得的二丙基二酸二乙酯、乙醇 44 mL、0.75 mol/L 氢氧化钠溶液 85 mL，快速搅拌，升温至回流，保温反应 4 h。蒸馏回收乙醇至无液滴馏出。将其自然冷却，用浓盐酸酸化，反应瓶中残留物至 pH 1，低温静置，有淡黄色晶体析出，抽滤，得到粗产物，即为二丙基丙二酸。

注：
1. 注意回收乙醇的油浴温度，在 100℃左右为宜。温度过高，产物颜色将变深。
2. 在使用浓盐酸酸化过程中，应小心控制浓盐酸的用量，酸如果过量，将形成无机盐的过饱和溶液，析晶后会分为两层，上层为产物，下层为无机盐，需进行分离。

（四）α-正丙基戊酸的合成

在配有磁力搅拌器、温度计、冷凝管及 CO_2 吸收装置的三颈烧瓶中，加入上步制备的二丙基丙二酸粗产物，加入 3 粒沸石，将反应体系，升温至 180℃，待反应物全部熔化后，无 CO_2 逸出时，停止加热，将反应装置转换成蒸馏装置，进行减压蒸馏，收集 112～114℃的馏分，得到淡黄色 α-正丙基戊酸液体。

注：
1. 反应中放出的 CO_2 气体为弱酸性，故应用碱液吸收。使用时，注意漏斗与碱液保持一定的间隙，避免引起倒吸。
2. 检查是否有 CO_2 气体逸出，可用湿润 pH 试纸在排气管末端检验是否变为橙红色。
3. 减压蒸馏系统应保持密封，磨口接头处均涂上真空油脂。蒸馏前，首先检查抽气泵的效率，如果装置漏气，应仔细检查各塞子和橡胶管的连接处是否紧密，如接收瓶和安全吸收系统有漏气现象，可用熔融的固体石蜡进行密封。

（五）丙戊酸钠的合成

1. 成盐反应

将上步所得 α-正丙基戊酸置于三颈烧瓶中，加入搅拌子，快速搅拌，通过恒压滴液漏斗缓慢滴加氢氧化钠溶液调节 pH 至 8～9，注意不能出现固体沉淀，将反应体系升温，浓缩至干，得到粗产物，即为 α-正丙基戊酸钠。

2. 精制

向 α-正丙基戊酸钠粗产物中，加入 1.5 倍量的乙酸乙酯，升温至回流，使固体全部溶解，静置，冷却，有固体晶体析出，抽滤，干燥，得到丙戊酸钠纯品。

注：
1. 用氢氧化钠调节 pH 至 8～9 的过程中，若碱过量，将会出现沉淀，可以用浓盐酸回调其酸碱度。
2. 重结晶所加的乙酸乙酯的量是粗产物重量的 1.5 倍，为重量与体积之比。

五、注意事项与思考题

1. 本反应中制备乙醇钠溶液时，需注意哪些问题？
2. 实验中，在 1-溴正丙烷的制备中，为什么要控制滴加浓硫酸的速度？
3. 实验中蒸馏所得 1-溴正丙烷分别用浓硫酸、水及碳酸钠溶液各洗一次，目的是什么？洗涤时，液体分上、下层，所需产品在哪一层？
4. 实验处理中，不小心酸化过度产生无机盐，与产品混在一起，用何种方法进行分离纯化？

Experiment 18 Synthesis of sodium valproate

I. Purpose of the experiment

i. To master the synthesis of sodium valproate, familiar with alkylation, hydrolysis, decarboxylation and other reactions.

ii. To master the salting conditions and refining methods of sodium valproate.

iii. To familiar with vacuum distillation and other basic operations.

iv. To understand the physicochemical properties and clinical uses of sodium valproate.

II. Experimental principle

i. Drug introduction

Sodium valproate, with the molecular formula $C_8H_{15}O_2Na$, is a white powder. It is a nitrogen-free broad-spectrum antiepileptic. It is the first choice for primary grand mal and absence minor seizures. It is less effective for partial seizures (simple partial and complex partial and partial secondary grand mal). It has certain efficacy for benign infantile myoclonic epilepsy and infantile spasm. But for myoclonic seizures, it must be combined ethosuximide or other antiepileptic drugs to be effective. It can be tried for refractory epilepsy. In addition to antiepileptic use, this drug can also be used to treat febrile convulsions, movement disorders, chorea, porphyria, schizophrenia, pain caused by shingles, adrenal dysfunction, and the prevention of alcohol withdrawal syndrome.

The chemical structure formula is:

$$\begin{array}{c} H_3CH_2CH_2C \\ \diagdown \\ H_3CH_2CH_2C \end{array} \hspace{-6pt} \begin{array}{c} H \\ C-COONa \end{array}$$

Sodium Valproate

ii. Synthetic process route

$$CH_3CH_2CH_2OH \xrightarrow[2h]{H_2SO_4,\ NaBr} CH_3CH_2CH_2Br + Na_2SO_4 + H_2O$$

$$H_2C\begin{array}{c}COOC_2H_5\\COOC_2H_5\end{array} \xrightarrow[C_2H_5ONa,\ 3h]{2CH_3CH_2CH_2Br} \begin{array}{c}H_3CH_2CH_2C\\ \diagdown \\ H_3CH_2CH_2C\end{array}\!\!C\!\!\begin{array}{c}CH_2CH_2CH_3\\ \diagup \\ CH_2CH_2CH_3\end{array} \xrightarrow[4h]{KOH,\ HCl}$$

$$\begin{array}{c}H_3CH_2CH_2C\\ \diagdown \\ H_3CH_2CH_2C\end{array}\!\!C\!\!\begin{array}{c}COOH\\ \diagup \\ COOH\end{array} \xrightarrow{180\,^\circ\!C} \begin{array}{c}H_3CH_2CH_2C\\ \diagdown \end{array}\!\!C\text{-COOH} \xrightarrow{NaOH} \begin{array}{c}H_3CH_2CH_2C\\ \diagdown \\ H_3CH_2CH_2C\end{array}\!\!\begin{array}{c}H\\ C\text{-COONa}\end{array}$$

III. Main instruments and reagents

i. Main instruments: thermometer, condensing tube, oil bath, three-necked flask, stirrer, magnetic stirrer, glass rod, Brinell funnel, analytical balance, filter bottle, filter paper, drying tube, heating sleeve, beaker, rubber tube, conical flask, vacuum pump, *etc.*

ii. Main reagents: 11.5% sodium bromide, *n*-butanol, concentrated sulfuric acid, 5% sodium carbonate, anhydrous sodium carbonate, diethyl malonate, sodium, ethanol, concentrated hydrochloric acid, sodium hydroxide, ethyl acetate, *etc.*

IV. Experimental steps

i. Synthesis of 1-bromo-n-propane

Add 68 mL of 11.5% sodium bromide solution into a three necked flask equipped with a magnetic stirrer, a thermometer and a reflux condensing tube. Cool to 0℃. Stir rapidly. Add 45 mL of *n*-propanol. Then slowly add 54 mL of concentrated sulfuric acid in about 1 h. Heat to reflux, and maintain the reaction for 2 h. Distill and collect the crude product of 1-bromo-*n*-propane at 60-110 ℃. Wash the crude product with 1/8 volume of 1-bromo-*n*-propane concentrated sulfuric acid, 1/2 volume of water and 1/2 volume of 5% sodium carbonate solution. Dry with anhydrous sodium carbonate to obtain a 1-bromo-n-propane product.

Note:

(i) In the preparation of 11.5% sodium bromide solution, it is not necessary to heat to dissolve all sodium bromide, even if some remains undissolved, because it will precipitate again after the reaction system cools, and will completely dissolve when the reaction is heated to reflux.

(ii) In the preparation of 1-bromopropane, the hydrohalic acid displacement of the alcohol hydroxyl group is a reversible reaction. By distilling or slowly adding concentrated sulfuric acid, a sufficient concentration of hydrobromic acid can be maintained in the reaction system, thus shifting the reaction equilibrium to the right and increasing the yield.

ii. Synthesis of diethyl dipropyl malonate

Add 20 mL of diethyl malonate and 120 mL of sodium ethanol solution (mass fraction 17%) into a three necked flask equipped with a magnetic agitator, a thermometer and a reflux condensing tube. Stir rapidly and slowly heat to 70℃. Slowly add the previously prepared 1-bromo-n-propane. After the addition, maintain the reaction at reflux for 3 hours. Then, convert the reflux apparatus to a distillation apparatus, distill to recover ethanol, and stop distillation when solids begin to precipitate in the flask. Cool to room temperature, let stand for 1 hour to allow sodium bromide solids to form, and then filter. Continue distilling the mother liquor until no liquid droplets are distilled (the oil bath temperature is about 110°C), stop distillation, and cool to obtain a brown-red oily liquid, which is diisopropyl malonate. Pour it all into a beaker for later use.

Note:

(i) In this reaction, the preparation method of sodium ethanol: add 25 mL of anhydrous toluene into a 50 mL beaker. Weigh the beaker. Remove the metal sodium with tweezers. Wipe off the attached oil. Put it into xylene quickly. Weigh 20.4 g metal sodium by weight reduction method.

Chop it with a knife. Put it in xylene for use.

Add 120 mL of ethanol into a 250 mL three necked flask equipped with a magnetic stirring, a reflux condenser and a drying tube. Remove sodium metal with tweezers in batches. Wipe off xylene. Quickly add to ethanol. Reaction releases heat. Make the ethanol boiling. When all the sodium metal is into sodium ethanol and dissolv in ethanol, cool for use.

(ii) Cutting metal sodium should be carried out carefully. During the cutting process, the paper and tools should be treated with alcohol and then treated with water before the waste can be thrown away.

(iii) When distilling and recovering ethanol, ethanol in the reaction system should be completely removed, otherwise ethanol will be mixed into the product, resulting in a decrease in the actual content of the product, which will affect the feeding ratio of the next reaction.

(iv) When solids are precipitated in the recovered ethanol, if the cooling time is prolonged, solids will be precipitated more, which will be conducive to complete recovery after filtration.

iii. Synthesis of dipropyl malonic acid

Add the previously prepared diethyl dipropyl dicarboxylate, 44 mL of ethanol and 85 mL of sodium hydroxide solution (0.75 mol/L) into a three necked flask equipped with a magnetic stirrer, a thermometer and a reflux condensing tube. Stir rapidly. Heat to reflux, and maintain the reaction for 4 h. Distill to recover ethanol until no liquid droplets are distilled. Allow it to cool naturally, acidify with concentrated hydrochloric acid until the pH of the residue in the reaction flask is 1, and let it stand at low temperature to allow light yellow crystals to precipitate. Filter to obtain the crude product, which is diisopropyl malonate.

Note:

(i) Pay attention to the oil bath temperature of ethanol recovery at about 100℃. If the temperature is too high, the product will become darker.

(ii) In the process of using concentrated hydrochloric acid acidification, the amount of concentrated hydrochloric acid should be carefully controlled. If the acid is excessive, the supersaturated solution of inorganic salts will be formed, which will be divided into two layers after crystallization.

iv. Synthesis of α-n-valproic acid

Add the crude product of dipropyl malonic acid prepared in the previous step into the three necked round-bottom flask equipped with a magnetic agitator, a thermometer, a condensing tube. Add three zeolite. Heat the reaction system to 180℃. After all the reactants have melted and no CO_2 is escaping escaped, stop heating. Convert the reaction device to a distillation device for vacuum distillation. Collect the fraction at 112-114 ℃ to obtain light yellow α-n-propylvaleric acid liquid.

Note:

(i) The CO_2 gas released in the reaction is weakly acidic, so it should be absorbed by lye. When using, pay attention to the funnel and lye to maintain a certain gap, to avoid causing suction.

(ii) To check for escaping CO_2 gas, use moist pH paper at the end of the exhaust pipe to see if it turns orange-red.

(iii) The reduced pressure distillation system should be kept sealed, and vacuum grease should be applied to the ground glass joints. Before distillation, first check the efficiency of the vacuum pump. If there is a leak in the apparatus, carefully check whether the connections of each stopper and rubber tube are tight. If there is a leak in the receiving bottle and safety absorption system, it can be sealed with molten solid paraffin.

v. Synthesis of sodium valproate

(i) Salting reaction

Place the obtained *α-n*-propylvaleric in a three necked flask. Add a stirring agent. Stir rapidly. Slowly add sodium hydroxide solution through a constant pressure drip funnel until the pH reaches 8-9, ensuring no solid precipitatate. Heat forms the reaction system and concentrate. Dry to obtain the crude product, which is *α-n*-propylvaleric sodium.

(ii) Refine

Add 1.5 times the amount of ethyl acetate into the *α-n*-valproate sodium crude product. Heat to reflux to dissolve all solids. After standing and cooling, some solid crystals precipitate. After filtrating and drying obtain pure, sodium valproate.

Note:

(i) During the process of adjusting the pH to 8-9 with sodium hydroxide, if the base is in excess, a precipitate will form, which can be adjusted back to the desired acidity with concentrated hydrochloric acid.

(ii) The amount of ethyl acetate added by recrystallization is 1.5 times the weight of the crude product, which is the ratio of weight to volume.

Ⅴ. Attention or thinking questions

i. What issues should be noted when preparing sodium ethanol solution in this reaction?

ii. In the experiment, why is it necessary to control the rate of adding concentration sulfuric acid during the preparation of 1-bromo-*n*-propane?

iii. In the experiment, why is the distilled 1-bromopropane washed once each with concentrated sulfuric acid, water, and sodium carbonate solution? Which layer contains the desired product during washing when the liquid separates into upper and lower layers?

iv. In the experimental process, if excessive acidification leads to the formation of inorganic salts mixed with the product, what method can be used for separation and purification?

实验十九　苯妥英钠的合成

一、实验目的

1. 掌握安息香缩合反应的反应机理。
2. 熟悉三价铁作为氧化剂的使用条件。
3. 了解安息香缩合反应的应用。
4. 了解苯妥英钠的理化性质及临床用途。

二、实验原理

（一）药物简介

苯妥英钠，又名大伦丁钠，分子式为 $C_{15}H_{13}N_2NaO_2$，白色粉末，无臭、味苦。熔点 290~299℃。置于空气中能吸收水分和二氧化碳，析出苯妥英。易溶于乙醇、水，不溶于氯仿和乙醚。

苯妥英钠为抗癫痫药，用于治疗癫痫大发作，也可用于三叉神经痛以及某些类型的心律不齐。苯妥英钠对大脑皮层运动区有高度选择性抑制作用，一般认为，是通过稳定脑细胞膜的功能，增加脑内抑制性神经递质 5-羟色胺以及 γ-氨基丁酸的作用，来防止异常放电的传播，从而具有抗癫痫作用。

苯妥英钠的抗神经痛的作用机制，可能与本药物作用与中枢神经系统、降低突触传递或降低引起神经元放电的短暂刺激有关，还可对心房与心室的异位节律有抑制作用，也可加速房室的传导，降低心肌自律性，具有抗心律失常作用。

化学结构式为：

5,5-二苯基-2,4-咪唑烷二酮钠

（二）合成工艺路线

苯妥英钠的合成，以维生素 B_1、苯甲醛等作为原料，其合成路线具体如下：

三、主要仪器和试剂

1. 主要仪器：三颈烧瓶、抽滤瓶、分析天平、冷凝管、干燥管、搅拌子、磁力搅拌器、布氏漏斗、滤纸、油浴锅、加热套、烧杯、橡胶管、锥形瓶、真空泵等。

2. 主要试剂：维生素 B_1、NaOH、苯甲醛、冰醋酸、三氯化铁、95%盐酸、乙醇、尿素等。

四、实验步骤

（一）安息香的合成

在装有磁力搅拌器、温度计、回流冷凝管的三颈烧瓶中，加入维生素 B_1 17.5 g、水 35 mL，搅拌溶解，随后加入 95%乙醇 150 mL，用冰浴冷却，缓慢加入 3 mol/L 氢氧化钠 40 mL，反应体系颜色呈深黄色或淡棕色，接着加入苯甲醛 100 mL。缓慢升温至 60~70℃，回流反应 90 min，反应结束后，自然冷却，有白色晶体析出，抽滤，用冷水充分洗涤，干燥，得粗产物，可以用 95%乙醇重结晶，可得安息香纯品。

（二）联苯甲酰的合成

在装有磁力搅拌器、温度计、回流冷凝管以及干燥管的三颈烧瓶中，加入 $FeCl_3·6H_2O$ 4.5 g、乙酸 50 mL 以及水 25 mL，升温至水沸腾，保温搅拌 10 min。加入上步反应制备的安息香 10.6 g 后，继续加热回流反应 1 h。反应结束后，冷却，加入水 200 mL，再将反应体系升温至沸腾，冷却至室温，有黄色晶体析出，抽滤，得粗产物，接着用 95%乙醇进行重结晶，趁热抽滤，冷却后，析出黄色晶体，抽滤，得到联苯甲酰纯品。

（三）苯妥英钠的合成

在装有磁力搅拌器、温度计、回流冷凝管以及干燥管的三颈烧瓶中，加入 5.5 g 上步制备的联苯甲酰以及 95%乙醇 20 mL，快速搅拌，升温至溶解。同时，在烧杯中配置氢氧化钠溶液（4 g NaOH、12 mL 水），随后，将其加入三颈烧瓶中，再加尿素 2.0 g。升温至回流，保温反应 50 min，反应结束后，冷却，将反应液倒入烧杯中，接着加入 250 mL 的水，用 6 mol/L 盐酸调节酸碱度，至 pH 4，有白色固体析出。抽滤，滤饼用水充分洗涤，抽干，干燥，得粗产物。可以用 95%乙醇重结晶，得到苯妥英钠纯品。

五、注意事项与思考题

（一）注意事项

熟悉重结晶的实验操作过程。

（二）思考题

写出安息香缩合反应的反应机理？

Experiment 19   Synthesis of phenytoin sodium

I. Purpose of the experiment

i. To master the reaction mechanism of benzoin condensation reaction.
ii. To familiarize yourself with the use conditions of iron trivalent as oxidant.
iii. To understand the application of benzoin condensation reaction.
iv. To understand the physicochemical properties and clinical uses of phenytoin sodium.

II. Experimental principle

i. Drug introduction

Phenytoin sodium, also known as great Lundin sodium, has the molecular formula $C_{15}H_{13}N_2NaO_2$. It is a white powder. It is odorless and bitter taste. Its melting point is 290-299℃. When exposed to air, it can absorb water and carbon dioxide and precipitate phenytoin. It is soluble in ethanol, water, insoluble in chloroform and ether.

Phenytoin sodium is an antiepileptic suitable for the treatment of grand mal seizures. It is also used for trigeminal neuralgia and some types of arrhythmia. Phenytoin sodium has a highly selective inhibitory effect on the motor area of the cerebral cortex. It is generally believed that by stabilizing the function of the membrane of the brain cells, increasing the inhibitory neurotransmitter 5-hydroxytryptamine and gamma-aminobutyric acid in the brain, to prevent the transmission of abnormal discharge, so as to have an anti-epileptic effect.

The mechanism of anti neuralgia action of phenytoin sodium may be related to the effect of phenytoin on the central nervous system, the reduction of synaptic transmission or the reduction of transient stimulation that causes neuronal discharge. Phenytoin can also inhibit the ectopic rhythm of the atrium and ventricle, accelerate the atrioventricular conduction, reduce myocardial automata, and have an antiarrhythmic effect.

The chemical structure formula is:

5, 5-diphenyl-2, 4-imidazolidine diketone sodium

ii. Synthetic process route

The synthesis of phenytoin sodium is mainly prepared by using vitamin B_1 and benzaldehyde as raw materials. The synthesis route is as follows:

III. Main instruments and reagents

i. Main instruments: three necked flask, filter flask, analysis balance, condensing tube, drying tube, stirrer, magnetic stirrer, Brinell funnel, filter paper, oil bath, heating sleeve, beaker, rubber tube, conical flask, vacuum pump, *etc.*

ii. Main reagents: vitamin B_1, NaOH, benzaldehyde, glacial acetic acid, ferric chloride, hydrochloric acid, 95% ethanol, urea, *etc.*

IV. Experimental steps

i. Synthesis of Benzoin

Add 17.5 g of vitamin B_1 and 35 mL of water into a three necked flask equipped with a magnetic stirrer, a thermometer and a reflux condensing tube. Stir to dissolve. Add 150 mL of 95% ethanol. Cool with an ice bath. Slowly add 40 mL of 3 mol/L sodium hydroxide. The color of the reaction system is dark yellow or light brown. Then add 100mL of benzaldehyde. After the reaction, cool naturally. Some white crystals will precipitate. Filter and wash thoroughly with cold water. Dry to obtain the crude products, which can be recrystallized with 95% ethanol to obtain pure Benzoin.

ii. Bibenzoyl synthesis

Add 4.5 g of $FeCl_3 \cdot 6H_2O$, 50 mL of acetic acid and 25 mL of water into a three-neck flask equipped with a magnetic stirrer, a thermometer, a reflux condensing tube and a drying tube. Heat until the water boils and maintain stirring 10 min. After adding 10.6 g of benzoin prepared in the previous step, continue heating reflux reaction for 1 h. After the reaction, cool down. Add 200 mL of water, then heat the reaction mixture to boiling. Cool to room temperature. Yellow crystals will precipitate. Filter to obtain the crude product, then recrystallise using 95% ethanol. Filter while hot. After cooling, yellow crystals will precipitate. Filter to obtain pure biphenylcarboxylic acid.

iii. The synthesis of phenytoin sodium

Add 5.5 g of benzoyl prepared in the previous step and 20 mL of 95% ethanol into a three necked flask fitted with a magnetic stirrer, a thermometer, a reflux condensing tube and a drying tube to the flask. Stir rapidly and heat until dissolved. Meanwhile, prepare a sodium hydroxide solution (4 g NaOH in 12 mL of water) in a beaker. Then add it to a three necked flask. Add 2.0 g of urea. Heat to reflux and maintain the reaction for 50 minutes. After the reaction, cool down. Pour the reaction mixture into a beaker, then add 250 mL of water. Adjust the pH to 4 using 6 mol/L hydrochloric acid. A white solid will precipitate. Filter and wash the filter cake thoroughly with water. Dry to obtain the crude product. It can be recrystallised from 95% ethanol to yield pure phenytoin sodium.

V. Attention or thinking questions

i. Attention

Familiar with the experimental operation process of recrystallization.

ii. Thinking questions

Write down the reaction mechanism of Benzoin condensation reaction.

实验二十 硝苯地平的合成

一、实验目的

1. 掌握硝苯地平环合反应的种类、特点及操作条件。
2. 熟悉硝化剂的种类和不同应用范围。
3. 了解硝化反应的种类、特点及操作条件。
4. 了解硝苯地平的理化性质及临床用途。

二、实验原理

（一）药物简介

硝苯地平别名尼非地平、利心平、硝苯啶，其化学式为 $C_{17}H_{18}N_2O_6$，黄色结晶固体，无臭，无味，遇光不稳定。在丙酮或氯仿中易溶，在乙醇中略溶，在水中几乎不溶。

硝苯地平是一种二氢吡啶类钙拮抗剂，用于预防和治疗冠心病心绞痛，特别是变异型心绞痛和冠状动脉痉挛所致心绞痛。对呼吸功能没有不良影响，故适用于有呼吸道阻塞性疾病的心绞痛患者，其疗效优于 β 受体拮抗剂。同时，还适用于各种类型的高血压，对顽固性、重度高血压也有较好疗效。由于能降低后负荷，对顽固性充血性心力衰竭亦有良好疗效，宜于长期服用。

化学结构式为：

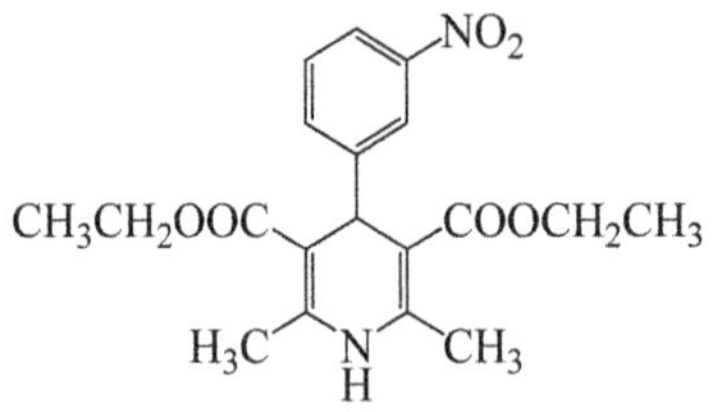

1,4-二氢-2,6-二甲基-4-(2-硝基苯基)-3,5-吡啶二羧酸二甲酯

（二）合成工艺路线

硝苯地平的合成，以苯甲醛、硝酸钾以及乙酰乙酸乙酯等原料，发生硝化反应、环合反应等制得，其合成工艺路线具体如下：

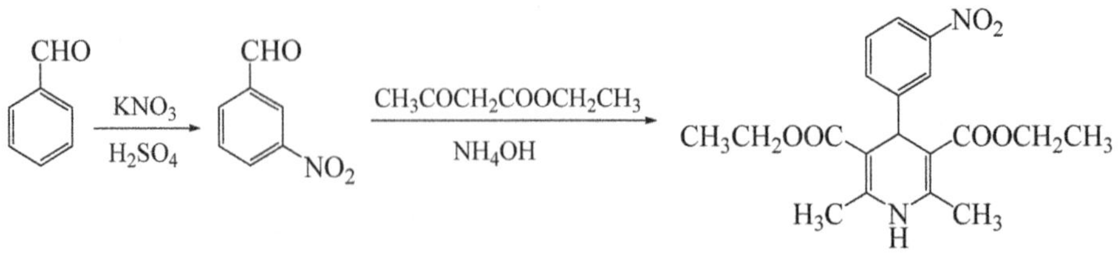

三、主要仪器和试剂

1. 主要仪器：温度计、恒压滴液漏斗、分析天平、三颈烧瓶、搅拌子、磁力搅拌器、烧

杯、玻璃棒、布氏漏斗、抽滤瓶、滤纸、加热套、橡胶管、真空泵、油浴锅、沸石等。

2. 主要试剂：硝酸钾、浓硫酸、苯甲醛、5%碳酸钠、乙酰乙酸乙酯、饱和甲醇氨、95%乙醇等。

四、实验步骤

（一）硝化反应

在配有磁力搅拌器、回流冷凝管、干燥管及恒压滴液漏斗的三颈烧瓶中，加入硝酸钾 11 g 及浓硫酸 40 mL。用冰盐浴将反应体系冷至 0 ℃以下，在快速搅拌下，缓慢滴加苯甲醛 10 g，在 1 h 内滴加完毕，滴加过程中，控制反应温度在 0～2 ℃之间。滴加完毕后，控制反应温度在 0～5 ℃之间，持续保温反应 90 min。将反应物缓慢倾入约 200 mL 冰水中，边倒边搅拌，析出黄色固体，抽滤。滤饼移至研钵中，研细，随后，加入 5% 碳酸钠溶液 20 mL，研磨 5 min，抽滤，用冰水洗涤多次，压干，得到间硝基苯甲醛。

（二）环合反应

在配有磁力搅拌器、回流冷凝管、干燥管及恒压滴液漏斗的三颈烧瓶中，分别加入间硝基苯甲醛 5 g、乙酰乙酸乙酯 9 mL、饱和甲醇氨溶液 30 mL 以及沸石几粒，升温回流，保温反应 5 h 后，然后改为蒸馏装置，蒸出甲醇，直至有结晶析出为止，抽滤，滤饼用 95%乙醇充分洗涤，压干，得黄色结晶性粉末粗产物。可以用 95%的乙醇进行重结晶，得到硝苯地平纯品。

五、注意事项与思考题

（一）注意事项

1. 本反应中所用的甲醇氨饱和溶液应新鲜配制。
2. 试写出本反应的反应机理。

（二）思考题

浓硫酸在本反应中的作用是什么？

Experiment 20 Synthesis of Nifedipine

Ⅰ. Purpose of the experiment

i. To master the types, characteristics and operating conditions of nifedipine cyclization reaction.

ii. To familiar with the types and different application ranges of nitrification agents.

iii. To understand the types, characteristics and operating conditions of nitrification reaction.

iv. To understand the physicochemical properties and clinical applications of nifedipine.

Ⅱ. Experimental principle

i. Drug introduction

Nifedipine, as known as nifedipine, nifedipir, with the molecular chemical formula $C_{17}H_{18}N_2O_6$, is yellow crystalline solid. It is odorless and tasteless. It is unstable in light. It is readily soluble in acetone or chloroform, slightly soluble in ethanol, and almost insoluble in water.

Nifedipine is a dihydropyridine calcium antagonist used for the prevention and treatment of angina pectoris of coronary heart disease, especially variant angina pectoris and angina caused by coronary spasm. It has no adverse effect on respiratory function, so it is suitable for patients with angina pectoris suffering from respiratory obstructive disease, and its efficacy is better than that of β-receptor antagonists. At the same time, it is also suitable for various types of hypertension. It has a good effect on refractory and severe hypertension. Because it can reduce afterload, it also has a good effect on refractory congestive heart failure and is suitable for long-term use.

The chemical structure formula is:

1, 4-dihydro-2, 6-dimethyl-4 -(2-nitrophenyl)-3, 5-pyridine dicarboxylate dimethyl ester

ii. Synthetic process route

The synthesis of nifedipine is mainly prepared by nitrification and cyclization reaction of benzaldehyde, potassium nitrate, ethyl acetoacetate and other raw materials. The synthesis process is as follows:

III. Main instruments and reagents

i. Main instruments: thermometer, constant pressure drip funnel, analysis balance, three necked flask, round-bottom flask, stirrer, magnetic stirrer, beaker, glass rod, Brinell funnel, filter bottle, filter paper, heating sleeve, rubber tube, vacuum pump, oil bath, zeolite, *etc.*

ii. Main reagents: potassium nitrate, concentrated sulfuric acid, benzaldehyde, 5% sodium carbonate, ethyl acetoacetate, methanol amine, 95% ethanol, *etc.*

IV. Experimental steps

i. Nitration reaction

Add 11 g of potassium nitrate and 40 mL of concentrated sulfuric acid in to a three necked flask equipped with a magnetic agitator, a reflux condensing tube, a drying tube and a constant pressure drip funnel. Cool the reaction system to below 0℃ with an ice salt bath. Under rapidly stirring, slowly add 10 g of benzaldehyde within 1 h. During the addition, control the reaction temperature between 0 and 2℃. After the addition, control the reaction temperature between 0 and 5℃. Maintain the reaction for 90 minutes. Pour the reactant slowly into about 200 mL of ice water. Stir while pouring. Extract the yellow solid and filter. Move the cake to a mortar and ground fine. Add 20 mL of 5% sodium carbonate solution. Grind for 5 min and filter. Wash with ice water for several times. Press dry to obtain *m*-nitro benzaldehyde.

ii. cyclization reaction

Add 5 g of *m*-nitro benzaldehyde, 9 mL of ethyl acetoacetate, 30 mL of methanol ammonia saturated solution and a few grains of zeolite into a three-neck flask equipped with a magnetic agitator, a reflux condensing tube, a drying tube and a constant pressure drip funnel. Heat to reflux and maintain the reaction for 5 h. Then switch to a distillation device to distil off the methanol until crystallization occurs. Fully wash the filter cake with 95% ethanol. Press dry to obtain yellow crystalline powder as crude product. Recrystallization with 95% ethanol can yield pure nifedipine.

V. Attention or thinking questions

i. Attention

(i) The methanol-ammonia saturated solution used in this reaction should be prepared fresh.

(ii) Try to write the reaction mechanism of this reaction.

ii. Thinking questions

What is the role of concentrated sulfuric acid in this reaction?

实验二十一　维生素 K_3 的合成

一、实验目的

1. 掌握维生素 K_3 氧化及加成反应的特点。
2. 了解维生素 K_3 的制备方法。
3. 了解在药物合成中常用的氧化剂。

二、实验原理

(一) 药物简介

维生素 K_3，又名甲萘醌，是一种有机化合物，化学式为 $C_{11}H_8O_2$，在临床上属于促凝血药，可以用于治疗维生素 K 缺乏所引起的出血性疾病，如新生儿出血、肠道吸收不良所致维生素 K 缺乏及低凝血酶原血症等。

本药为维生素营养补充剂，维生素 K 为肝脏合成原酶的必需物质，并参与凝血因子Ⅶ、Ⅸ和Ⅹ的合成，维持动物的血液凝固生理过程。缺乏维生素 K 可致上述凝血因子合成障碍，影响凝血过程而引起出血。维生素 K 也为动物机体内的维生素 K 依赖羧化作用体系所必需，是骨骼素合成过程中不可缺少的因子。

此外，在高能化合物代谢和氧化磷酸化过程中，以及与其他脂溶性维生素代谢的方面均起重要作用，并具有利尿、增强肝脏解毒功能，参与膜的结构，降低血压的功能。

化学结构式为：

2-甲基-1,4-萘醌

(二) 合成工艺路线

维生素 K_3 的合成，主要通过 2-甲基萘，用重铬酸钾强氧化剂氧化等反应制备，其合成工艺路线如下：

三、主要仪器和试剂

1. 主要仪器：分析天平、温度计、恒压滴液漏斗、三颈烧瓶、回流冷凝管、干燥管、搅拌子、磁力搅拌器、布氏漏斗、滤纸、水浴锅、真空泵、烧杯、橡胶管、锥形瓶等。
2. 主要试剂：2-甲基萘、重铬酸钾、亚硫酸氢钠、浓硫酸、乙醇、丙酮等。

四、实验步骤

在装有磁力搅拌器、温度计、回流冷凝管以及恒压滴液漏斗的三颈烧瓶中，加入 2-甲基萘 7 g、丙酮 15 g，剧烈搅拌，使其完全溶解。同时，在另一只烧杯中，加入重铬酸钾 35 g、水 52 mL 以及浓硫酸 22.8 mL，搅拌混匀，于 38～40℃时，缓慢滴加至三颈烧瓶中，滴加结束后，在 40℃时，保温反应 30 min，接着升高温度至 60℃，保温反应 45 min。反应结束后，将反应物倒入大量水中，有固体析出，抽滤，滤饼用水洗涤，压紧，抽干，得甲萘醌粗产物。

在装有磁力搅拌器、温度计、回流冷凝管以及干燥管的三颈烧瓶中，加入上步反应得到的甲萘醌粗产物、亚硫酸氢钠 4.4 g 以及水 7 mL，快速搅拌，升高温度，于 38～40℃加入乙醇 11 mL，搅拌 30 min，反应结束后，将反应液倒入烧杯中，冷却至室温，放置冰箱中，降温至 5℃左右，有晶体析出，抽滤，滤饼用少量冷的乙醇进行洗涤，抽干，得到维生素 K_3 粗产物。

上步制得的维生素 K_3 粗产物，放入锥形瓶中，随后加入 4 倍量 95%乙醇及亚硫酸氢钠 0.5 g，在 70℃以下溶解，加入约为维生素 K_3 粗产物 1%量的活性炭。在水浴温度为 68～70℃时，保温脱色反应 30 min，随后，趁热抽滤，滤液冷至 5℃以下，有固体晶体析出，抽滤，滤饼用少量冷的乙醇，进行充分洗涤，抽干，得到维生素 K_3 的纯品。

五、注意事项与思考题

（一）注意事项

1. 配制混合氧化剂时，需将浓硫酸缓慢加入重铬酸钾的水溶液中，同时要不断搅拌。
2. 氧化剂加入反应液后，要保持温度 38～40℃。
3. 在反应结束的母液倒入大量水中（一般为母液 10 倍体积）时，缓慢加入，同时要不断搅拌。

（二）思考题

1. 氧化反应中升高温度时，对产物产生什么影响？
2. 本反应中，加成反应中加入乙醇的目的是什么？
3. 在药物合成中，常用的氧化剂包括哪些？

Experiment 21   Synthesis of vitamin K₃

I. Purpose of the experiment

i. To master the characteristics of vitamin K_3 oxidation and addition reaction.

ii. To understand the preparation method of vitamin K_3.

iii. To understand the oxidants commonly used in drug synthesis.

II. Experimental principle

i. Drug introduction

Vitamin K_3, also known as menadione, is an organic compound, with the molecular formula $C_{11}H_8O_2$. It is a coagulant in clinical practice. It can be used to treat hemorrhagic diseases caused by vitamin K deficiency, such as neonatal hemorrhage, vitamin K deficiency caused by intestinal malabsorption and hypoprothrombinemia.

Vitamin K, as a nutritional supplement, is essential for the synthesis of proenzymes in the liver. It is involved in the synthesis of coagulation factors VII, IX and X to maintain the physiological process of blood clotting in animals. Lack of vitamin K can cause the above coagulation factor synthesis disorders, and affect the coagulation process and cause bleeding. Vitamin K is also required for the vitamin K-dependent carboxylation system in animals. And it is an indispensable factor in the synthesis of skeletal hormone.

In addition, it plays an important role in the metabolism of high-energy compounds and oxidative phosphorylation, as well as in the metabolism of other fat-soluble vitamins. It also has the function of diuresis, enhancing liver detoxification, participating in membrane structure, and lowering blood pressure.

The chemical structure formula is:

2-methyl-1, 4-naphthoquinone

ii. Synthetic process route

The synthesis of vitamin K_3 is mainly prepared through the reaction of 2-methyl naphthalene and oxidation with potassium dichromate strong oxidant. The synthesis process is as follows:

III. Main instruments and reagents

i. Main instruments: analytical balance, thermometer, constant pressure drip funnel, three necked flask, drying tube, reflux condensing tube, stirrer, magnetic stirrer, Brinell funnel, filter paper, water bath, vacuum pump, beaker, rubber tube, conical flask, *etc.*

ii. Main reagents: 2-methyl naphthalene, potassium dichromate, sodium bisulfite, concentrated sulfuric acid, ethanol, acetone, *etc.*

IV. Experimental steps

Add 7 g of 2-methyl naphthalene and 15 g of acetone into a three necked flask fitted with a magnetic agitator, a thermometer, a reflux condensing tube and a constant pressure drip funnel. Stir vigorously to dissolve completely. At the same time, add 35 g of potassium dichromate, 52 mL of water and 22.8 mL of concentrated sulfuric acid into another beaker. Mix well. And slowly add the mixture to the three-neck flask at 38-40℃. After the addition, maintain the temperature at 40℃ for 30 min. Then is raise to 60℃ and maintain the reaction for 45 min. After the reaction, pour the reactant into a large amount of water, causing solid to precipitate. Filter. Wash the filter cake with water. Press dry to obtain crude menadione.

Add the crude menadione product obtained from the previous step, 4.5 g of sodium bisulfite and 7 mL of water into a three necked flask equipped with a magnetic stirrer, a thermometer, a reflux condensing tube and a drying tube. Stir quickly. After raising the temperature, add 11 mL of ethanol at 38-40℃. Stir for 30 min. Pour the reaction liquid into the beakers after the reaction. Cool to room temperature. Plac in the refrigerator to cool to about 5℃ causing crystals to precipitate. Filter. Wash the filter cake with a small amount of cold ethanol. Dry to obtain vitamin K_3 crude products.

Put the crude product of vitamin K_3 prepared in the above step into a conical bottle. Add 4 times the amount 95% ethanol and 0.5 g of sodium bisulfite. Dissolve below 70℃. Add about 1% of the amount of the crude vitamin K_3 product of activated carbon. When the temperature of the water bath is about 68-70℃, maintain decolorization reaction for 30 min. Filter while hot. Cool the filtrate to below 5℃. There are solid crystals precipitation. Then filter. Fully wash the filter cake with a small amount of cold ethanol. Dry to obtain vitamin K_3 pure products.

V. Attention or thinking questions

i. Attention

(i) When mixing the oxidizer, the concentrated sulfuric acid should be slowly added to the aqueous solution of potassium dichromate, while constantly stirring.

(ii) After the oxidant is added to the reaction liquid, the temperature should be kept at 38-40℃.

(iii) When the mother liquor after the reaction is poured into a large amount of water (generally 10 times the volume of mother liquor), slowly add it and stir it constantly.

ii. Thinking questions

(i) What is the effect on the product when the temperature is raised in the oxidation reaction?

(ii) In this reaction, what is the purpose of adding ethanol into the addition reaction?

(iii) What oxidants are commonly used in drug synthesis?

实验二十二　巴比妥酸的合成

一、实验目的

1. 掌握实验中由尿素和丙二酸二乙酯缩合合成六元杂环化合物的实验方法。
2. 熟悉本反应中涉及的回流、结晶等基本操作方法。
3. 了解巴比妥酸的理化性质及临床用途。

二、实验原理

（一）药物简介

巴比妥酸又称为丙二酰脲，化学式为 $C_4H_4N_2O_3$，呈白色结晶性粉末，易溶于热水和稀酸，溶于乙醚，微溶于冷水。水溶液呈强酸性。可以与金属反应生成盐类。丙二酰脲亚甲基上两个氢原子被烃基取代后的若干衍生物，称为巴比妥类药物，是一类重要的镇静催眠药物。

巴比妥酸作为合成巴比妥、苯巴比妥和维生素 B_{12} 等药品的中间体，也可用作聚合催化剂和制取染料的原料。巴比妥类药物是巴比妥酸的衍生物，具有镇静和催眠作用。

化学结构式为：

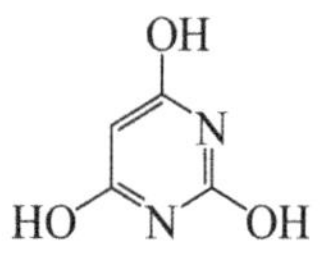

2,4,6-嘧啶三酮

（二）合成工艺路线

巴比妥酸的合成，以丙二酸二乙酯及尿素为原料进行制备，具体合成工艺路线如下：

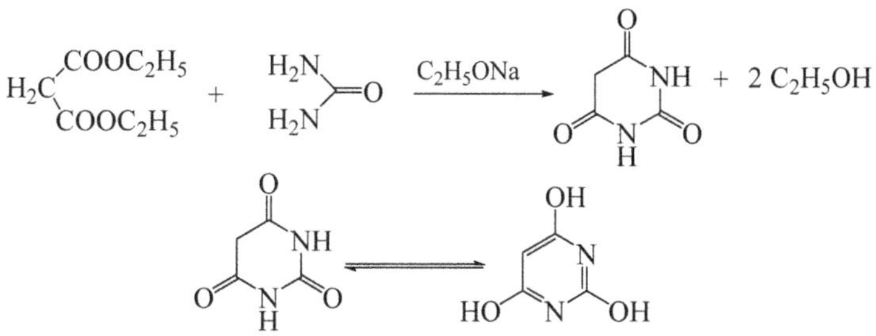

三、主要仪器和试剂

1. 主要仪器：分析天平、锥形瓶、加热套、磁力搅拌器、三颈烧瓶、回流冷凝管、干燥管、分液漏斗、温度计、水浴锅、烧杯、玻璃棒、真空泵等。
2. 主要试剂：钠、丙二酸二乙酯、尿素、无水乙醇、浓盐酸等。

四、实验步骤

在装有温度计、恒压滴液漏斗、磁力搅拌器及干燥管的三颈烧瓶中,加入无水乙醇 20 mL、金属钠 1 g,待金属钠完全溶解后,加入丙二酸二乙酯 6.5 mL,剧烈搅拌后,缓慢滴加干燥的尿素 2.4 g 以及无水乙醇 1.2 mL 配成的溶液,快速搅拌,回流反应 1.5 h,有固体产生。冷却后,加入热水 30 mL 使固体完全溶解,再加入浓盐酸 2 mL 调节 pH 至 3,得到澄清溶液。抽滤除去少量的杂质。其滤液用冰水进行冷却,析出晶体,抽滤,滤饼用少量冰水洗涤 3 次,得到白色晶体,干燥,得到巴比妥酸。

五、注意事项与思考题

1. 反应对无水的要求特别严格,反应中所涉及的仪器等均需做无水处理。
2. 在处理金属钠的过程中,避免钠块过长时间暴露在空气中,在空气中迅速吸水转化为氢氧化钠而使丙二酸二乙酯皂化。
3. 尿素在投料前,需预处理,在 110 ℃ 烘箱中干燥 45 min 后才能够使用。

Experiment 22 Synthesis of barbiturates

I. Purpose of the experiment

i. To master the experimental method of synthesis of six-membered heterocyclic compounds by condensation of urea and diethyl malonate.

ii. To familiar with basic operation methods such as reflux and crystallization involved in this reaction.

iii. To understand the physicochemical properties and clinical uses of barbiturates.

II. Experimental principle

i. Drug introduction

Barbiturate, also known as malonyl urea, has the molecular chemical formula of $C_4H_4N_2O_3$. It is a white crystalline powder. It is easily soluble in hot water and diluted acid, soluble in ether, slightly soluble in cold water. The aqueous solution is strongly acidic. It can react with metals to form salts. Several derivatives of malonylurea, where the two hydrogen atoms on the methylene group are replaced by hydrocarbon groups, are known as barbiturates, which are an important class of sedative-hypnotic drugs.

Barbiturates are mainly used as intermediates in the synthesis of drugs such as barbiturates, phenobarbital and vitamin B_{12}. They are also used as polymerization catalysts and raw materials for making dyes. Barbiturates are derivatives of barbiturates that have sedative and hypnotic effects.

The chemical structure formula is:

2,4, 6-pyrimidine triketone

ii. Synthetic process route

The synthesis of barbituric acid is mainly prepared by using diethyl malonate and urea as raw materials. The specific synthesis process is as follows:

III. Main instruments and reagents

i. Main instruments: analysis balance, conical flask, heating sleeve, magnetic stirrer, three necked flask, reflux condensing tube, drying tube, separator funnel, thermometer, water bath, beaker, glass rod, vacuum pump, *etc*.

ii. Main reagents: sodium, diethyl malonate, urea, anhydrous ethanol, concentrated hydrochloric acid, *etc*.

IV. Experimental steps

Add 20 mL of anhydrous ethanol and 1 g of sodium metal into a three necked flask equipped with a thermometer, a constant pressure drip funder, a magnetic agitator and a drying tube. After the sodium metal is completely dissolved, add 6.5 mL of diethyl malonate. Stir vigorously. Slowly add 2.4 g of dried urea and 1.2 mL of the solution of anhydrous ethanol. Stir rapidly. Reflux the reaction for 1.5 h, during which solid will form. After cooling, add 30 mL of hot water to completely dissolve the solid. Add 2 mL of concentrated hydrochloric acid to adjust the pH to 3 to obtain a clarified solution. Filter to remove small amounts of impurities. Cool the filtrate with ice water to precipitate crystals. Then filter. Wash the filter cake 3 times with a small amount of ice water to obtain white crystal. After drying obtain barbiturate.

V. Attention or thinking questions

i. The requirement for anhydrous conditions is particularly strict; all instruments involved in the reaction must be treated to be anhydrous.

ii. During the handling of sodium metal, avoid prolonged exposure of sodium blocks to air, as they quickly absorb moisture from the air and convert to sodium hydroxide, which can saponify diethyl malonate.

iii. Urea must be pre-treated before use; it should be dried in an oven at 110°C for 45 minutes.

实验二十三　乳酸米力农的合成

一、实验目的

1. 掌握乳酸米力农药物的合成反应机理。
2. 熟悉乳酸米力农药物的合成实验操作过程。
3. 了解乳酸米力农药物的理化性质及临床用途。

二、实验原理

（一）药物简介

米力农为灰白色固体，分子式为 $C_{12}H_9N_3O$。临床上用于治疗慢性充血性心力衰竭、顽固性心力衰竭。

本药品是磷酸二酯酶抑制剂，其为氨力农的同类药物，作用机理与氨力农相同。口服和静注均有效，兼有正性肌力作用和血管扩张作用。其作用较氨力农强 10～30 倍。耐受性较好。本药品正性肌力作用主要是通过抑制磷酸二酯酶，使心肌细胞内环磷酸腺苷浓度增高，细胞内钙增加，心肌收缩力加强，心排血量增加。而与肾上腺素受体或心肌细胞 Na^+、K^+、ATP 酶无关，其血管扩张作用可能是直接作用于小动脉所致，从而可降低心脏前、后负荷，降低左心室充盈压，改善左室功能，增加心脏指数，但对平均动脉压和心率无明显影响。

米力农的心血管效应与剂量有关，小剂量时，主要表现为正性肌力作用，当剂量加大，逐渐达到稳态的最大正性肌力效应时，其扩张血管作用也可随剂量的增加而逐渐加强。本药品对伴有传导阻滞的患者较安全，但本药品口服时不良反应较重，不宜长期应用。

化学结构式为：

2-甲基-6-氧-1,6-二氢-3,4'-双吡啶-5-甲腈

（二）合成工艺路线

米力农的合成，以 4-甲基吡啶和乙酰氯等为原料，制得中间体 1-(4-吡啶基)-2-丙酮后，与原甲酸三乙酯缩合，得到 1-乙氧基-2-（4-吡啶基）乙烯基甲基酮，随后，在碱催化作用下与氰基乙酰胺环合制得产物，其合成工艺路线具体如下：

三、主要仪器和试剂

1. 主要仪器：三颈烧瓶、分析天平、磁力搅拌器、温度计、恒压滴液漏斗、冷凝管、干燥管、分液漏斗、烧杯、橡胶管、锥形瓶、抽滤瓶、布氏漏斗、滤纸、真空泵等。

2. 主要试剂：4-甲基吡啶、二氯甲烷、乙酰氯、饱和碳酸钠溶液、饱和亚硫酸氢钠溶液、氢氧化钠、无水硫酸镁、原甲酸三乙酯、乙酸酐、乙酸、甲醇钠、氰乙酰胺、无水乙醇、乳酸、活性炭等。

四、实验步骤

（一）1-（4-吡啶基）-2-丙酮的合成

在配有磁力搅拌器、温度计、冷凝管、干燥管以及恒压滴液漏斗的 250 mL 三颈烧瓶中，加入 4-甲基吡啶 13 mL、二氯甲烷 40 mL，在冰盐浴条件下冷却，缓慢滴加乙酰氯 19 mL，在滴加过程中，注意反应条件的控制，保证反应体系的温度小于 10℃，滴加结束后，升温至 30℃，保温反应 24 h。反应结束后，冰浴冷却，滴加饱和碳酸钠溶液调节 pH 至 7～8，静置分层，水层用二氯甲烷充分萃取 3 次（每次 50 mL），合并有机层，减压浓缩。浓缩后加入饱和亚硫酸氢钠溶液 30 mL，室温搅拌 3 h。用二氯甲烷充分萃取 3 次（每次 50 mL），合并有机层，用无水硫酸镁充分干燥，过夜。回收溶剂后，蒸去未反应完的 4-甲基吡啶。

水层用 6.25 mol/L 氢氧化钠溶液调节 pH 至 13，室温下，搅拌反应 2 h。反应结束后，加入水 40 mL，用二氯甲烷萃取 2 次（每次 25 mL），用无水硫酸镁充分干燥，过夜。减压蒸馏，得浅黄色液体。

（二）米力农的制备

在配有磁力搅拌器、温度计、冷凝管、干燥管以及恒压滴液漏斗的 100 mL 三颈烧瓶中，加入 1-（4-吡啶基）-2-丙酮 10 g，快速搅拌，加入原甲酸三乙酯 18 mL、乙酸酐 19 mL 以及乙酸 18.5 mL，升温至 40℃，搅拌保温反应 5 h。反应结束后，加入无水乙醇，减压蒸去低沸点溶剂，得深红色油状物，待用。

在无水甲醇 150 mL 中加入甲醇钠 70 g、氰乙酰胺 8 g 以及上一步待用深红色的油状物，升温至回流，保温反应 2 h。反应结束后，冷却，有固体生成，抽滤，滤饼用甲醇充分洗涤，随后用适量水溶解，加入适量活性炭，升温、搅拌、脱色，抽滤，滤液用乙酸调节 pH 至 6.5～7.0，有固体析出，抽滤，固体用乙醇重结晶，得浅黄色米力农纯品。

（三）乳酸米力农的合成

将上述所制得的米力农中，加入 95%乙醇 70 mL 以及等量的乳酸，升温搅拌至全部溶解，加入少许活性炭，升温至回流，保温反应 30 min，反应结束后，趁热过滤，缓慢降温，待溶液析晶，放置冰箱过夜，抽滤，用 95%乙醇 5 mL 洗涤滤饼，干燥，得产物，即为乳酸米力农。

五、注意事项与思考题

（一）注意事项
试写出本反应的反应机理。

（二）思考题
本反应中，常用的环合试剂有哪些？

Experiment 23 Synthesis of Milrinone lactate

I. Purpose of the experiment

i. To master the synthetic reaction mechanism of Milrinone lactate.

ii. To familiar with the synthesis experiment and operation process of Milrinone lactate.

iii. To understand the physicochemical properties and clinical applications of Milrinone lactate.

II. Experimental principle

i. Drug introduction

Milrinone is a grayish solid, with the molecular formula $C_{12}H_9N_3O$. It has clinical using for chronic congestive heart failure, refractory heart failure.

This drug is a phosphodiesterase inhibitor, which is similar to amrinone. It has the same mechanism of action as amrinone. Both oral and intravenous injections are effective, with positive inotropic and vascular dilatation effects. Its effect is 10-30 times stronger than amrinone. The tolerance is good. The positive inotropic effect of this drug is mainly through inhibiting phosphodiesterase, which can increase the concentration of cyclic adenosine phosphate in cardiomyocytes, increase intracellular calcium, strengthen myocardial contractility and increase cardiac output. However, it is not related to adrenergic receptors or myocardial Na^+, K^+, ATPase. Its vasodilation effect may be caused by direct action on arterioles, thus reducing cardiac preload and afterload, reducing left ventricular filling pressure, improving left ventricular function, and increasing cardiac index. But it has no significant effect on mean arterial pressure and heart rate.

The cardiovascular effect of Milrinone is dose-dependent. And it mainly shows positive inotropic effects at small dose. When the dose is increased and the maximum positive inotropic effect is reached gradually, the vasodilation effect of Milrinone can also be gradually enhanced with the increase of dose. This drug is safe for patients with conduction block. But the adverse reactions of this drug are serious when taken orally. So it is not suitable for long-term use.

The chemical structure formula is:

2-methyl-6-oxy-1, 6-dihydro-3, 4'-dipyridine-5-methyl nitrile

ii. Synthetic process route

Milrinone synthesis, mainly through 4-methylpyridine and acetyl chloride reagents, preparation of intermediate 1-(4-pyridinyl) -2-acetone, and triethyl orthoformate condensation, to obtain 1-

ethoxy-2-(4-pyridinyl) vinyl methyl ketone, then, under the catalysis of alkali and cyanoacetamide ring synthesis product, the synthesis process is as follows:

III. Main instruments and reagents

i. Main instruments: three necked flask, analysis balance, magnetic stirrer, thermometer, constant pressure drip funnel, condensing tube, drying tube, separation funnel, beaker, rubber tube, conical bottle, suction bottle, Brinell funnel, filter paper, vacuum pump, *etc.*

ii. Main reagents: 4-methylpyridine, dichloromethane, acetyl chloride, saturated sodium carbonate solution, saturated sodium bisulfite solution, sodium hydroxide, anhydrous magnesium sulfate, triethyl orthoformate, acetic anhydride, acetic acid, sodium methanol, cyanamide, anhydrous ethanol, lactic acid, activated carbon, *etc.*

IV. Experimental steps

i. Synthesis of 1-(4-pyridyl)-2-acetone

Add 13 mL of 4-methylpyridine and 40 mL of dichloromethane into a 250 mL three necked flask equipped with a magnetic agitator, a thermometer, a condensing tube, a drying tube and a constant pressure drip funnel. Cool in an ice-salt bath. Slowly add 19 mL of acetyl chloride. During the addition, pay attention to the control of reaction conditions. Ensure the temperature of the reaction system remains below 10 ℃. After the addition, raise the temperature to 30 ℃ and maintain the reaction for 24 h. After the reaction, cool in the ice bath. Add the saturated sodium carbonate solution in drops to adjust the pH to 7-8. Let it stand to separate layers, and extract the aqueous layer with dichloromethane three times (50 mL each time). Combine the organic layers, and concentrate under reduced pressure. After concentration, add 30 mL of saturated sodium bisulfite solution and stir at room temperature for 3 hours. Extract with dichloromethane three times (50 mL each time). Combine the organic layers. Dry thoroughly with anhydrous magnesium sulfate overnight. After recovering the solvent, evaporate the unreacted 4-methylpyridine.

Adjust the pH of the water layer with sodium hydroxide solution (6.25 mol/L) to 13. Stir at room temperature for 2 h. After the reaction, add 40 mL of water and extracte with methylene chloride two times (25 mL each time). Then, dry thoroughly with anhydrous magnesium sulfate overnight. Distil under reduced pressure to abtain a yellowish liquid.

ii. Preparation of Milrinone

Add 10 g of 1-(4-pyridyl)-2-acetone into a 100 mL three necked round-bottom flask equipped with a magnetic agitator, a thermometer, a condensing tube, a drying tube and constant pressure drip hopper. Stir quickly. Add 18 mL of triethyl orthoformate, 19 mL of acetic anhydride and 18.5 mL of acetic acid. Heat to 40 ℃ and maintain stirring for 5 h. After the reaction, add anhydrous ethanol and solvent the low boiling solvent under pressure to obtain a dark red oily substance.

Add 70 g of sodium methanol, 8 g of cyanoacetamide and the dark red oily material used in the previous step into 150 mL of anhydrous methanol. Heat to reflux and maintain the reaction for 2 h. After the reaction, cool down, and a solid will form Filter and Fully wash the filter cake with methanol. Then dissolve in an appropriate amount of water. Add the appropriate amount of activated carbon. Heat and stir. Decolorizate and filter. Adjust the pH to 6.5-7.0 with acetic acid. There are solids precipitation. Filter again, and recrystallize the solid with ethanol to obtain pure milrinone as a light yellow product.

iii. Synthesis of Milrinone lactate

Add 70 mL of 95% ethanol and the same amount of lactic acid to the milinone prepared above. Heat and stir until fully dissolved. Add a little activated carbon. Heat to reflux and maintain the reaction for 30 min. After the reaction, filter while hot. Slowly cool down. Wait for the solution to crystallize. Place it in the refrigerator overnight. Filter and wash the filter cake with 5 mL 95% ethanol. Dry to obtain the product, which is Milrinone lactate.

Ⅴ. Attention or thinking questions

i. Attention

Try to write the reaction mechanism of this reaction.

ii. Thinking questions

What are the commonly used cyclization reagents in this reaction?

实验二十四　美沙拉秦的合成

一、实验目的

1. 掌握美沙拉嗪中硝化反应的机理。
2. 熟悉硝基还原反应的操作步骤。
3. 熟悉美沙拉秦药物的相关理化性质及临床用途。

二、实验原理

(一) 药物简介

美沙拉嗪又名马沙拉嗪，化学式为 $C_7H_7NO_3$，用于治疗溃疡性结肠炎。对肠壁的炎症有显著的抑制作用。美沙拉嗪可以抑制引起炎症的前列腺素的合成和炎性介质白三烯的形成，从而对肠黏膜的炎症起显著抑制作用。美沙拉秦以剂量依赖方式，抑制前列腺素的合成，减少前列腺素 E2 在结肠黏膜的释放。

此外，美沙拉秦还可以抑制中性粒细胞的脂肪氧化酶活性。高剂量时美沙拉秦能够抑制中性粒细胞的某些功能，如迁移、脱粒、吞噬及氧自由基的合成。同时，美沙拉秦还可以抑制在炎症过程中起重要作用的血小板活动因子的合成。对有炎症的肠壁的结缔组织效果更佳。用于溃疡性结肠炎、溃疡性直肠炎和克罗恩病。

化学结构式为：

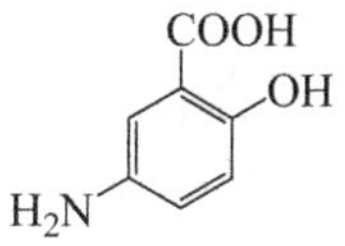

5-氨基-2-羟基苯甲酸

(二) 合成工艺路线

美沙拉秦药物的合成方法种类较多，本实验采用的是水杨酸硝化还原法，进行制备，具体合成工艺路线如下：

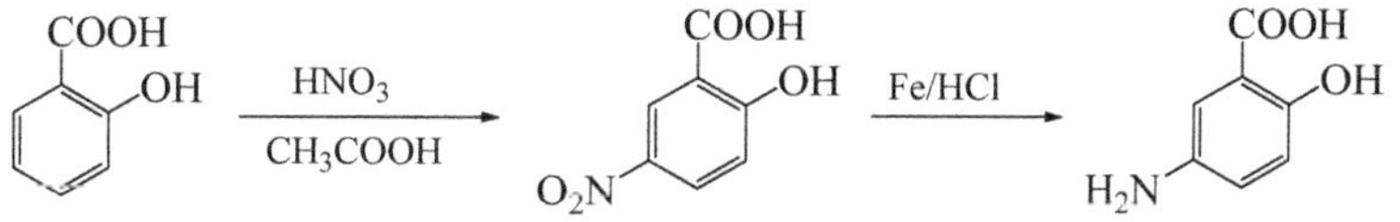

三、主要仪器和试剂

1. 主要仪器：温度计、分析天平、磁力搅拌器、恒压滴液漏斗、三颈烧瓶、烧杯、量筒、加热套、玻璃棒、抽滤瓶、布氏漏斗、滤纸、真空泵等。
2. 主要试剂：水杨酸、硝酸、冰醋酸、铁粉、浓盐酸、保险粉、40%硫酸、亚硫酸氢钠、活性炭、50%氢氧化钠等。

四、实验步骤

（一）5-硝基水杨酸的合成

在配有磁力搅拌器、温度计、回流冷凝管和恒压滴液漏斗的三颈烧瓶中，分别加入水杨酸 13.8 g 及水 35 mL，剧烈搅拌下，升温至 50℃，使其完全溶解。随后，缓慢滴加 68%硝酸 18 mL 及冰醋酸 1.8 mL 的混合液。继续升温至 70～80℃，保温反应 2 h。反应结束后，将反应液倒入水 120 mL 中，放置冰箱 4℃放置，过夜，有晶体析出，抽滤，采用热水进行重结晶，过滤，得到浅黄色固体粗产物，即为 5-硝基水杨酸。

注：
1. 反应中，如果水杨酸没有完全溶解，需要再加入少量水，直至溶解。
2. 本反应中使用混合酸进行硝化反应，反应过程较为剧烈，所以滴加的速度不能太快。

（二）美沙拉秦的合成

在配有磁力搅拌器、温度计、回流冷凝管的三颈烧瓶中，加入浓盐酸 3 mL 和水 35 mL，快速搅拌，升温至 60℃，加入铁粉 2.3 g，加热回流反应 5 min。随后，加入 5-硝基水杨酸约 1.5 g，继续剧烈搅拌 5 min 后，将剩余的 5-硝基水杨酸约 4.5 g 及铁粉 6.7 g 分 3 次加入，每次加入时间间隔 5 min。升温至 100℃后，继续保温反应 1.5 h，反应结束后，趁热用 50%氢氧化钠溶液调节 pH 至 11～12，减压抽滤，滤饼用水洗涤 3 次，每次用水 10 mL。向滤液中加入保险粉 0.7 g 后，用 40%硫酸调节 pH 至 3～4。冷却析出晶体，抽滤，干燥，即得到美沙拉嗪的粗产物。

美沙拉嗪粗产物的精制：将上步得到的粗产物 3 g，溶解在热水 35 mL 中，加入亚硫酸氢钠 0.4 g、活性炭 1 g，加热回流 5 min，趁热过滤，合并滤液和洗液。迅速冷却至 5℃，保温 1 h 后，过滤，冰水洗涤两次，干燥，得到白色结晶。

五、注意事项与思考题

1. 硝化反应对反应的条件要求较高，如何有效控制硝化反应条件？
2. 简述常见的硝化反应的硝化剂类型？
3. 在反应过程中，保险粉的主要作用是什么？

Experiment 24　Synthesis of Mesalazine

I. Purpose of the experiment

i. To master the mechanism of nitration reaction in mesalazine.

ii. To familiar with the operation steps of nitro-reduction reaction.

iii. To familiar with the physicochemical properties and clinical uses of mesalazine.

II. Experimental principle

i. Drug introduction

Mesalazine, also known as Masalazine, with the molecular formula $C_7H_7NO_3$, is used for ulcerative colitis. It has a significant inhibitory effect on intestinal wall inflammation. Mesalazine can inhibit the synthesis of inflammatory prostaglandins and the formation of inflammatory mediums leukotriene, which can significantly inhibit the inflammation of intestinal mucosa. Mesalazine can inhibit prostaglandin synthesis and reduce the release of prostaglandin E_2 in human colon mucosa in a dose-dependent manner.

Mesalazine can also inhibit the activity of fat oxidase in neutrophils. At high doses, mesalazine inhibits certain functions of human neutrophils, such as migration, threshing, phagocytosis, and oxygen free radical synthesis. Mesalazine also inhibits the synthesis of platelet activating factors that play an important role in inflammation. It works better on the connective tissue of the inflamed intestinal wall. It is used in ulcerative colitis, ulcerative proctitis and Crohn's disease.

The chemical structure formula is:

5-Amino Salicylic Acid

ii. Synthetic process route

Mesalazine has a variety of synthesis methods, This experiment adopts the salicylic acid nitration reduction method for preparation. The specific synthesis process is as follows:

III. Main instruments and reagents

i. Main instruments: thermometer, analytical balance, magnetic stirrer, constant pressure drip funnel, three necked flask, beaker, measuring cylinder, heating sleeve, glass rod, suction bottle, Brinell funnel, filter paper, vacuum pump, *etc.*

ii. Main reagents: salicylic acid, nitric acid, ice acetic acid, iron powder, concentrated hydrochloric acid, insurance powder, 40% sulphuric acid, sodium bisulfite, activated carbon, 50% sodium hydroxide, *etc.*

IV. Experimental steps

i. Synthesis of 5-nitrosalicylic acid

Add 13.8 g of salicylic acid and 35 mL of water into a three necked flask equipped with a magnetic stirrer, a thermometer, a reflux condensing tube and a constant pressure drip funnel. Under stirring vigorously, heat to 50℃ to ensure completely dissolution. Slowly add a mixture of 18 mL of 68% nitric acid and 1.8 mL of glacial acetic acid. Continue heating to 70-80℃ and maintain the reaction for 2 h. After the reaction, pour the reaction liquid into 120 mL of water. Place in the refrigerator at 4℃ overnight. There is crystal precipitation. Then filtr. After recrystallizing with hot water, filter to obtain a light yellow solid crude product, which 5-nitrosalicylic acid.

Note:

(i) In the reaction, if the salicylic acid is not completely dissolved, add a small amount of water until it dissolves.

(ii) In this reaction, mixed acid is used for nitrification. The reaction process is relatively intense, so the rate of drip cannot be too fast.

ii. Synthesis of Mesalazine

Add 3 mL of concentrated hydrochloric acid and 35 mL of water into a three necked flask equipped with a magnetic stirrer, a thermometer and a reflux condensing tube. Stir rapidly and heat to 60℃. Add 2.3 g of iron powder. Heat to reflux for 5 min. Then, add about 1.5 g of 5-nitrosalicylic acid. After stirring vigorously for 5 min, add the remaining about 4.5 g of 5-nitrosalicylic acid and 6.7 g of iron powder in 3 times, with a 5 min interval between each addition. After the reaction, adjust the pH with 50% sodium hydroxide solution while hot to 11-12. Extracte the filter under pressure. Wash the filter cake with water 3 times, 10 mL each time. Add 0.7 g of insurance powder into the filtrate. Adjust the pH with 40% sulfuric acid to 3-4. The crystal is precipitated by cooling, filtering and drying. Obtain the crude product of mesalazine.

Refining of mesalazine crude products: dissolve 3 g of the crude products obtained in the previous step in 35 mL of hot water. Add 0.4 g of sodium bisulfite and 1 g of activated carbon. Heat to reflux for 5 min. Filter while hot. Combine filtrate and lotion. Quickly cool to 5℃, and maintain for 1 h. Then filter. Wash with ice water twice, and dry to obtain white crystals.

V. Attention or thinking questions

i. How to effectively control the conditions of nitrification, which requires higher conditions for the reaction?

ii. Describe the nitrification agent types of common nitrification reactions.

iii. What is the main function of insurance powder in the reaction process?

实验二十五　曲尼司特的合成

一、实验目的

1. 掌握曲尼司特药物的合成反应机理。
2. 熟悉曲尼司特药物的合成实验操作过程。
3. 了解曲尼司特药物的理化性质及临床用途。

二、实验原理

（一）药物简介

曲尼司特，别名为利喘贝、肉桂氨茴酸等。分子式为 $C_{18}H_{17}NO_5$，是一种淡黄色或淡黄绿色结晶或结晶性粉末，无臭，无味。在二甲基甲酰胺中易溶，在甲醇中微溶，在水中不溶。

曲尼司特药物在 20 世纪 70 年代首先由日本江田等人研制成功，其国外商品名为 Tranilast 或 Rizaben，国内于 1983 年由南京药学院制药厂首先研制开发成功。

曲尼司特作为抗变态反应药，为过敏反应介质阻释剂，有稳定肥大细胞和嗜碱性粒细胞膜作用。适用于支气管哮喘、过敏性鼻炎、特应性皮炎等的治疗。

本药为新型抗变态反应药物，能稳定肥大细胞和嗜碱性粒细胞的细胞膜，阻止脱颗粒，从而抑制组胺和 5-羟色胺等过敏反应介质的释放，对支气管哮喘、过敏性鼻炎等有较好的治疗作用。与色甘酸钠的对照研究表明，曲尼司特口服有效，对被动皮肤过敏反应的抑制作用，在口服后 30～60 min 达最大效应，240 min 后消失，而色甘酸钠几无抑制作用。静脉注射后两药均于 5 min 后达最大效应，色甘酸钠作用较强，但 60 min 后消失，而曲尼司特 120 min 后仍有显著作用。与色甘酸钠的不同点是，色甘酸钠仅抑制反应素抗体介导的过敏反应，曲尼司特尚能抑制局部过敏坏死反应。此外，尚有降低血中 IgE 水平、抑制抗原抗体反应、减少外周血中嗜酸粒细胞的绝对计数、调节胶原合成代谢等作用。

化学结构式为：

2-{[3-（3,4-二甲氧苯基)-1-氧代-2-丙烯基]氨基}苯甲酸

（二）合成工艺路线

曲尼司特的合成，主要通过丙二酸亚异丙酯、邻氨基苯甲酸、藜芦醛等原料制备，其合成工艺路线具体如下：

三、主要仪器和试剂

1. 主要仪器：三颈烧瓶、分析天平、磁力搅拌器、温度计、恒压滴液漏斗、冷凝管、干燥管、分液漏斗、烧杯、橡胶管、锥形瓶、抽滤瓶、布氏漏斗、滤纸、真空泵等。
2. 主要试剂：丙二酸亚异丙酯、邻氨基苯甲酸、藜芦醛、哌啶、乙腈、乙醇、10%氢氧化钠等。

四、实验步骤

在配有磁力搅拌器、温度计、冷凝管、干燥管以及恒压滴液漏斗的 100 mL 三颈烧瓶中，加入丙二酸亚异丙酯 3.6 g 以及邻氨基苯甲酸 3.4 g 的研磨混合物，随后加入乙腈 15 mL，快速搅拌，升温至回流，保温反应 1 h。减压蒸馏除去乙腈，加入藜芦醛 4.2 g，加入哌啶，至回流时完全溶解，加热回流反应 1 h。反应结束后，将反应体系冷却至室温，分批缓慢加入 10%氢氧化钠 30 mL，适当补加哌啶，然后进行常压蒸馏，利用哌啶共沸脱水，随后，减压继续浓缩。将反应体系中的残余物倒入水中，有黄色固体析出，抽滤，干燥，得粗产物。粗产物可以用乙醇进行重结晶，得到浅黄色曲尼司特纯品。

五、注意事项与思考题

（一）注意事项

试写出本反应的反应机理。

（二）思考题

1. 本反应中加入氢氧化钠的作用是什么？
2. 本反应中用到吡啶与水形成共沸脱水，还有哪些试剂可以与水能形成共沸脱水？

Experiment 25 Synthesis of Tranilast

I. Purpose of the experiment

i. To master the synthetic reaction mechanism of tranilast.

ii. To familiar with the experimental operation process of synthesis of tranilast.

iii. To understand the physicochemical properties and clinical uses of tranilast.

II. Experimental principle

i. Drug introduction

Tranilast, also known as rizaben, cinnamomic anisic acid, *etc.* has the molecular formula $C_{18}H_{17}NO_5$. It is a yellowish or yellowish green crystal or crystalline powder. It is odorless and tasteless. It is soluble in dimethylformamide, slightly soluble in methanol, insoluble in water.

Tranilast was first developed by Eda et al in Japan in the 1970s. Its foreign product was named Tranilast or Rizaben. In China, it was first developed by Nanjing Pharmaceutical University Pharmaceutical Factory in 1983.

Tranilast acts as an antiallergic drug, an inhibitor of anaphylaxis medium. It has the effect of stabilizing mast cells and basophil membrane action. It is suitable for the treatment of bronchial asthma, allergic rhinitis, atopic dermatitis, *etc.*

As a new antiallergic drugs, it can stabilize the cell membranes of mast cells and basophils. Then prevent degranulation, thus inhibiting the release of histamine and 5-hydroxytryptamine and other allergic reaction mediators. This drug has a good therapeutic effect on bronchial asthma and allergic rhinitis. A comparative study with cromolyn sodium showes that the oral administration of tranilast is effective. The inhibitory effect on passive skin allergic reactions reach the maximum effect 30 to 60 min after oral administration, and disappear after 240 min, while cromolyn sodium has little inhibitory effect. After intravenous injection, both drugs reach the maximum effect 5 min later. Cromolyn sodium has a stronger effect, but disappear 60 min later, while the effect of tranilast is still significant 120 min later. Unlike cromolyn sodium, cromolyn sodium only inhibits allergic reactions mediated by reactin antibodies, while tranilast inhibits local allergic necrosis reactions. In addition, it can reduce Ig E level in blood, inhibit antigen-antibody response, reduce the absolute count of eosinophilic granulocytes in peripheral blood, regulate collagen anabolism and so on.

The chemical structure formula is:

2-{ [3-(3,4-Dimethoxyphenyl)-1-oxo-2-propenyl] amino } benzoic Acid

ii. Synthetic process route

The synthesis of tranilast is mainly prepared by isopropyl malonate, *o*-aminobenzoic acid, veratrol and other raw materials. The synthesis process is as follows:

III. Main instruments and reagents

i. Main instruments: three necked flask, analysis balance, magnetic stirrer, thermometer, constant pressure drip funnel, condensing tube, separation funnel, beaker, rubber tube, drying tube, conical bottle, suction bottle, Brinell funnel, filter paper, vacuum pump, *etc.*

ii. Main reagents: isopropyl malonate, *o*-aminobenzoic acid, veratrol, piperidine, acetonitrile, ethanol, 10% sodium hydroxide, *etc.*

IV. Experimental steps

Add a grinding mixture of 3.6 g isopropyl malonate and 3.4 g *o*-aminobenzoic acid into a 100 mL three necked flask equipped with a magnetic agitator, a thermometer, a condensing tube, a drying tube and a constant pressure drip hopper. Add 15 mL of acetonitrile. Stir rapidly. Heat to reflux and maintain the reaction for 1 h. Remove acetonitrile by vacuum distillation. Add 4.2 g of veratrol and piperidine. Completely dissolve at reflux. Heat to reflux for 1 h. After the reaction, cool the reaction system to room temperature. Slowly add 30 mL of 10% sodium hydroxide in batches. Appropriately supplement piperidine. Then carry out atmospheric distillation. Piperidine is used as azeotropic dehydration. Continue concentration after decompression. Pour the residue of the reaction system into water. There is yellow solid precipitation. Filter and dry to obtain the crude product. The crude product can be recrystallized with ethanol to obtain the pure yellow tranilast product.

V. Attention or thinking questions

i. Attention

Try to write the reaction mechanism of this reaction.

ii. Thinking questions

(i) What is the function of adding sodium hydroxide in this reaction?

(ii) In this reaction, pyridine is used to form azeotropic dehydration with water. What other reagents can form azeotropic dehydration with water?

附 录

附录01 常用的溶剂的纯化处理措施、方法一览表

名称	纯化处理方法
石油醚 (petroleum)	石油醚为轻质石油产品,是低分子量的烷烃类混合物,按其沸程收集不同馏分。沸程为30~150℃,一般把温度相差30℃左右的馏分收集在一起,如通常有30~60℃、0~90℃、90~120℃以及120~150℃等沸程规格的石油醚。 石油醚中含有少量不饱和烃杂质,其沸点与烷烃相近,用蒸馏的方法是不能分离的,通常可用浓硫酸和高锰酸钾溶液洗去。首先,将石油醚用相当其体积10%的浓硫酸,洗涤2~3次,再用10%硫酸加入高锰酸钾配成的饱和溶液洗涤,直至水层中的紫色不再消失为止。然后用水洗,经无水氯化钙干燥后蒸馏。如要绝对干燥的石油醚,则加入钠丝,与无水乙醚一样的方法进行处理
环己烷 (cyclohexane)	分子式C_6H_{12},沸点80.7℃,熔点6.5℃。环己烷为无色液体,不溶于水,当温度高于57℃时,能与无水乙醇、甲醇、苯、醚、丙酮等混溶。环己烷中含有的杂质主要是苯。作为一般溶剂使用,并不需要特殊处理。若要除去苯,可用冷的浓硫酸与浓硝酸的混合液洗涤数次,使苯硝化后溶于酸层,进而除去,然后用水洗,干燥分馏,加入钠丝保存
正己烷 (n-hexane)	分子式$CH_3(CH_2)_4CH_3$,沸点68.7℃。正己烷为无色挥发性液体,能与醇、醚和三氯甲烷混合,不溶于水。在60~70℃沸程的石油醚中,主要为正己烷,因此在许多方面可以用该沸程的石油醚代替正己烷作溶剂。其纯化方法为先用浓硫酸洗涤数次,继以含0.5 mol/L高锰酸钾的10%硫酸溶液洗涤,再以含0.5 mol/L高锰酸钾的10% NaOH溶液洗涤,最后用水洗,干燥,蒸馏
苯 (benzene)	分子式C_6H_6,沸点80.1℃。普通苯试剂中常含有少量水和噻吩,噻吩的沸点为84℃,与苯接近,不能用蒸馏的方法除去。噻吩的检验方法:取1 mL苯,加入2 mL溶有2 mg吲哚醌的浓硫酸,振荡片刻,若酸层呈蓝绿色,即表示有噻吩存在。噻吩和水的除去:将苯装入分液漏斗中,加入相当于苯体积1/7的浓硫酸,振摇使噻吩磺化,弃去酸液,再加入新的浓硫酸,重复操作几次,直到酸层呈现无色。将上述无噻吩的苯依次用水、10%碳酸钠溶液和水洗至中性,再用氯化钙干燥,进行蒸馏,收集80℃的馏分,最后用金属钠脱去微量的水得无水苯
甲苯 (toluene)	分子式$CH_3C_6H_5$,沸点110.6℃。普通甲苯中可能含有少量甲基噻吩,欲除去甲基噻吩,可以用浓硫酸(甲苯:浓硫酸=10:1)振摇30 min(温度不能超过30℃),除去酸层,然后依次用水、10%碳酸钠溶液、水洗至中性,然后用无水氯化钙干燥,进行蒸馏,收集110℃的馏分
二甲苯 (xylene)	分子式$(CH_3)_2C_6H_4$,相对分子质量106.17,无色透明液体,商品为邻、对、间二甲苯三种异构体的混合物。能与乙醇、乙醚、三氯甲烷等有机溶剂相混合,不溶于水,沸点137~144℃。该品易燃,应远离火种。吸入或接触皮肤有害,对皮肤有刺激性
吡啶 (pyridine)	分子式C_5H_5N,沸点115.5℃。吡啶吸水力强,能与水、醇和醚任意混合。与水形成恒沸溶液,沸点为94℃。目前市售的分析纯吡啶含量为99%,可供一般实验用。如要制得无水吡啶,可将吡啶与粒状氢氧化钾(钠)一同加热回流,然后隔绝潮气蒸出。干燥的吡啶吸水性很强,应保存于含有氧化钡、分子筛或氯化钙的容器中
碘甲烷 (iodomethane)	分子式CH_3I,为无色液体,见光游离出碘后,变褐色。纯化可用硫代硫酸钠或亚硫酸钠的稀溶液,反复洗至无色,然后用水洗,用无水氯化钙干燥,蒸馏,沸点42~42.5℃。碘甲烷应盛放于棕色瓶中,避光保存
二氯甲烷 (dichloromethane)	分子式CH_2Cl_2,沸点40℃。使用二氯甲烷比氯仿安全,因此常用它代替氯仿,作为比水重的萃取剂。普通的二氯甲烷,一般都能直接作萃取剂用。如需纯化,可用5%碳酸钠溶液洗涤,再用水洗涤,然后用无水氯化钙干燥,蒸馏收集40~41℃的馏分,保存在棕色瓶中
氯仿 (chloroform)	分子式$CHCl_3$,沸点61.7℃。氯仿在日光下易氧化成氯气、氯化氢和光气(剧毒),故氯仿应贮于棕色瓶中。市场上供应的氯仿,多用1%乙醇作稳定剂,以消除产生的光气。氯仿中乙醇的检验,可用碘仿反应;游离氯化氢的检验可用硝酸银的醇溶液。除去乙醇可将氯仿用其1/2体积的水振摇数次,分离下层的氯仿,用氯化钙干燥24 h,然后蒸馏。另一种纯化方法,是将氯仿与少量浓硫酸一起振动2~3次。每200 mL氯仿用10 mL浓硫酸,分去酸层,以后的氯仿用水洗涤,干燥,然后蒸馏,除去乙醇后的无水氯仿,应保存在棕色瓶中并避光存放,以免光化作用产生光气
四氯化碳 (carbon tetrachloride)	分子式CCl_4,沸点76.8℃。目前四氯化碳,主要由二硫化碳经氯化制得,因此普通四氯化碳中含有二硫化碳(含量约4%)。纯化方法为将1000 mL四氯化碳与120 mL 50%氢氧化钾水溶液混合,再加100 mL乙醇,剧烈摇动半小时(温度50~60℃)。必要时,可用半量氢氧化钾溶液和乙醇重复处理一次。然后分出四氯化碳,先用水洗,再用少量浓硫酸洗至无色,最后再以水洗,用无水氯化钙干燥,蒸馏即得。四氯化碳不能用金属钠干燥,否则会发生爆炸

续表

名称	纯化处理方法
1,2-二氯乙烷 (1, 2-dichloroethane)	1,2-二氯乙烷为无色油状液体，具有芳香气味，与水可形成共沸物（含量为81.5%，沸点72℃）。与乙醇、乙醚和三氯甲烷相混溶，是重结晶和提取的较为常用的溶剂。可用五氧化二磷（20 g/L）加热回流2 h，常压蒸馏纯化
无水乙醇 (absolute ethanol)	分子式 C_2H_5OH，沸点78.5 （1）纯度98%～99%乙醇的纯化 第一种方法：利用苯、水和乙醇形成低共沸混合物的性质，将苯加入乙醇中，进行分馏，在64.9℃时，蒸出苯、水、乙醇的三元恒沸混合物，多余的苯，在68.3℃与乙醇形成二元恒沸混合物被蒸出，最后蒸出乙醇。工业上多采用此方法 第二种方法：用生石灰脱水。于100 mL 95%乙醇中加入，新鲜的块状生石灰20 g，回流3～5 h，然后进行蒸馏 （2）纯度99%以上乙醇的纯化 第一种方法：在500 mL 99%乙醇中，加入7 g金属钠，待反应完毕，再加入27.5 g邻苯二甲酸二乙酯或25 g草酸二乙酯，回流2～3 h，然后进行蒸馏。金属钠虽能与乙醇中的水作用，产生氢气和NaOH，但所生成的NaOH又与乙醇发生如下的平衡反应： $$NaOH + C_2H_5OH \rightleftharpoons H_2O + C_2H_5ONa$$ 因此单独使用金属钠，不能完全除去乙醇中的水，须加入过量的高沸点酯，如邻苯二甲酸二乙酯与生成的NaOH作用，抑制上述平衡水的形成，可得到99.95%的乙醇 第二种方法：在60 mL 99%乙醇中，加入5 g镁和0.5 g碘，待镁溶解生成醇镁后，再加入900 mL 99%乙醇，回流5 h后，蒸馏，可得到99.9%乙醇 检验乙醇是否含有水分，常用的方法有：①在一支干净试管中，加入制得的无水乙醇，随即加入少量的无水硫酸铜粉末，如果变为蓝色，则表明乙醇中含有水分；②在另一支干净试管中，加入制得的无水乙醇，随即加入几粒干燥的高锰酸钾，如果呈现紫红色，则表明乙醇中含有水分。由于乙醇具有非常强的吸湿性，所以在操作时，动作要迅速，尽量减少转移次数，以防止空气中的水分进入。同时所用仪器必须在实验前干燥
甲醇 (methanol)	分子式 CH_3OH，沸点64.96℃。普通未精制的甲醇，含有0.02%丙酮和0.1%水。而工业甲醇中这些杂质的含量达0.5%～1%，为了制得纯度达99.9%以上的甲醇，可将甲醇用分馏柱分馏，收集64℃的馏分，再用镁去水（与制备无水乙醇相同）。甲醇有毒，处理时应在通风柜中进行，防止吸入其蒸气
正丁醇 (butanol)	分子式 $CH_3CH_2CH_2CH_2OH$，分子量74.12。溶于乙醇、乙醚、苯，微溶于水，与水可形成共沸物，共沸点92℃（含水量37%）。该品易燃，空气中爆炸极限1.4%～11.2%，工作场所空气中允许浓度150 mg/m^3
甘油 (glycerol)	分子式 $HOCH_2CH(OH)CH_2OH$，称丙三醇（1, 2, 3-trihydroxypropane），分子量92.09，无色无臭黏稠液体，略有甜味。有强吸湿性，能吸收硫化氢、氢氰酸、二氧化硫。能与水、乙醇相混溶，1份该化合物能溶于11份乙酸乙酯、500份乙醚，不溶于苯、二硫化碳、三氯甲烷、四氯化碳、石油醚等。易脱水，失水生成双甘油和聚甘油。氧化生成甘油和甘油酸等。在0℃下，凝固形成闪光的斜方结晶。该化合物与铬酸酐、氯酸钾、高锰酸钾等强氧化剂接触，能引起燃烧或爆炸
聚乙二醇 (polyethylene glycol)	分子式 $HOCH_2(CH_2OCH_2)nCH_2OH$，为平均分子量200～6000以上的乙二醇高聚物的总称。溶于水、乙醇和许多有机溶剂。聚乙二醇400、聚乙二醇600
无水乙醚 (diethyl ether)	分子式 $(C_2H_5)_2O$，沸点34.51℃。普通乙醚常含有2%乙醇和0.5%水，久藏的乙醚常含有少量过氧化物 （1）过氧化物的检验和除去　制备无水乙醚时，首先必须检验有无过氧化物，否则，容易发生爆炸。检验方法：在干净的试管中加入2～3滴浓硫酸、1 mL 2%碘化钾溶液（若碘化钾溶液已被空气氧化，可用稀亚硫酸钠溶液滴到黄色消失）和1～2滴淀粉溶液，混合均匀后加入乙醚，振摇，如果出现蓝色，即表示有过氧化物存在。除去方法：除去过氧化物，可用新配制的硫酸亚铁稀溶液（配制方法是60 g $FeSO_4 \cdot 7H_2O$、100 mL水和6 mL浓硫酸）。把乙醚置于分液漏斗中，加入新配制的硫酸亚铁溶液，充分振荡混合后，弃去水层，此操作可以重复数次，至无过氧化物为止 （2）醇和水的检验和除去方法 检验方法：乙醚中放入少许高锰酸钾粉末与一粒NaOH。放置后，NaOH表面附有棕色树脂，即证明有醇存在。水的存在用无水硫酸铜检验 除去方法：先用无水氯化钙除去大部分水，再经金属钠干燥。其方法是，将100 mL乙醚放在干燥锥形瓶中，加入20～25 g无水氯化钙，盖好，放置一天以上，并间断摇动，然后蒸馏，收集33～37℃的馏分。加入1 g钠丝放于盛乙醚的瓶中，放置至无气泡发生即可使用；放置后，若钠丝表面已变黄变粗时，须再蒸一次，然后再压入钠丝

续表

名称	纯化处理方法
四氢呋喃 (tetrahydrofunan, THF)	分子式 C_4H_8O，沸点 67℃。四氢呋喃与水能混溶，并常含有少量水分及过氧化物。处理四氢呋喃时，应先用小量进行试验，确定其中只有少量水和过氧化物，作用不过于剧烈时，方可进行纯化。四氢呋喃中的过氧化物，用酸化的碘化钾溶液来检验。如过氧化物较多，应另行处理为宜 如要制得无水四氢呋喃，可先用无机干燥剂干燥后，再用少量金属钠，在隔绝潮气下回流，除去其中的水和过氧化物，以二苯甲酮作指示剂，变为蓝色后蒸馏，收集 66℃的馏分。精制后的液体，加入钠丝并在氮气氛围下保存。如需较久放置，应加 0.025% 4-甲基-2,6-二叔丁基苯酚作抗氧剂
乙二醇二甲醚 (dimethoxy ethane)	分子式 $CH_3OCH_2CH_2OCH_3$，又名二甲氧基乙烷，沸点 82～83℃。无色液体，有乙醚气味，能溶于水和碳氢化合物，对某些不溶于水的有机化合物是很好的惰性溶剂，并可促使芳香族碳氢化合物与钠反应。其化学性质稳定，溶于水、乙醇、乙醚和氯仿。纯化时，先用钠丝干燥。在氮气下加氢化锂铝蒸馏；或者用无水氯化钙干燥数天，过滤，加金属钠蒸馏，可加入氢化锂铝保存，用前蒸馏
二氧六环 (dioxane)	分子式 $C_4H_8O_2$，又称 1,4-二氧六环、二噁烷，与水互溶，无色，易燃，能与水形成共沸物（含量为 81.6%，沸点 87.8℃）。普通品中含有少量二乙醇缩醛与水。久贮的二氧六环中，可能含有过氧化物，要注意除去，然后再处理。纯化方法：可加入 10%浓盐酸回流 3 h，同时慢慢通入氮气，以除去生成的乙醛。冷却后，加入粒状氢氧化钾，直至其不再溶解；分去水层，再用粒状氢氧化钾干燥一天；过滤，在其中加入金属钠回流数小时，蒸馏。可压入钠丝保存
丙酮 (acetone)	分子式 CH_3COCH_3，沸点 56.21℃。普通丙酮含有少量的甲醇、乙醛以及水等杂质，不可能利用简单蒸馏把这些杂质分离。其纯化方法有如下两种：①在 250 mL 丙酮中加入 2.5 g 高锰酸钾进行回流，若高锰酸钾紫色很快消失，再加入少量高锰酸钾继续回流，至紫不褪为止。然后将丙酮蒸出，用无水碳酸钾或无水硫酸钙干燥，过滤后蒸馏，收集 55～56.5℃的馏分。用此法纯化丙酮时，须注意丙酮中含还原性物质不能太多，否则会过多消耗高锰酸钾和丙酮，使处理时间增长。②将 100 mL 丙酮装入分液漏斗中，先加入 4 mL10%硝酸银溶液，再加入 3.6 mL 1mol/L NaOH 溶液，振摇 10 min，分出丙酮层，再加入无水硫酸钾或无水硫酸钙进行干燥，最后蒸馏收集 55～56.5℃馏分。此法比方法①要快，但硝酸银较贵，只宜小量纯化用
苯甲醛 (benzaldehyde)	分子式 C_6H_5CHO，为带有苦杏仁味的无色液体，能与乙醇、乙醚、氯仿相混溶，微溶于水，由于在空气中易氧化成苯甲酸，使用前需经蒸馏，沸点 64～65℃/1.60 kPa（12 mmHg）或 179℃/101.0 kPa（760 mmHg）。苯甲醛低毒，但对皮肤有刺激，触及皮肤可用水洗
乙酸乙酯 (ethyl acetate)	分子式 $CH_3COOCH_2CH_3$，沸点 77.06℃。乙酸乙酯一般含量为 95%～98%，含有少量水、乙醇和乙酸。纯化方法如下：①于 1000 mL 98%乙酸乙酯中加入 100 mL 乙酸酐，10 滴浓硫酸，加热回流 4 h，除去乙醇和水等杂质，然后进行蒸馏。馏出液用 20～30g 无水碳酸钾振荡，再蒸馏，产物沸点为 77℃，纯度可达 99.7%以上。②先用等体积 5%碳酸钠溶液洗涤，然后用饱和氯化钙溶液洗，最后用无水碳酸钾干燥蒸馏。如需进一步干燥，可再与五氧化二磷回流 0.5 h，过滤，防潮蒸馏
乙腈 (acetonitrile)	分子式 CH_3CN，沸点 81.5℃。乙腈是惰性溶剂，可用于化学反应及重结晶。乙腈与水、醇、醚可任意混溶，与水生成共沸物（含乙腈 84.2%，沸点 76.7℃）。市售乙腈常含有水、不饱和腈、醛和胺等杂质，三级以上的乙腈含量应高于 95%。纯化方法：可将乙腈用无水碳酸钾干燥，过滤，再与五氧化二磷加热回流（20 g/L），直至无色，用分馏柱分馏。乙腈可贮存于放有分子筛（2A）的棕色瓶中。乙腈有毒，常含有游离氢氰酸
N，N-二甲基甲酰胺 (dimethyl formide, DMF)	分子式 $HCON(CH_3)_2$，沸点 153℃。无色液体，与多数有机溶剂与水任意混合。化学和热稳定性好，对有机和无机化合物的溶解性能较好。纯化时常用硫酸钙、硫酸镁、氧化钡、硅胶或分子筛干燥，然后减压蒸馏，收集 46℃/4800Pa（36 mmHg）的馏分。二甲基甲酰胺见光可慢慢分解为二甲胺和甲醛，因此纯化后的 N，N-二甲基甲酰胺要避光贮存。其中游离胺，可用 2,4-二硝基氟苯产生颜色来检查
二甲基亚砜 (dimethyl sulfoxide, DMSO)	分子式 $(CH_3)_2SO$，沸点 189℃，熔点 18.5℃。二甲基亚砜为无色、无嗅、微苦、易吸湿的液体，是一种优异的非质子极性溶剂，常压下加热至沸腾会发生部分水解。市售二甲基亚砜含水量约为 1%，一般先减压蒸馏，再用 4A 型分子筛干燥，或用氢氧化钙（10 g/L）干燥，搅拌 4～8 h，再减压蒸馏，蒸馏时温度不宜高于 90℃，否则会发生歧化反应，生成二甲砜及二甲硫醚。沸点 189℃/101.0 kPa（760 mmHg）或 71～72℃/2.80 kPa（21 mmHg）
苯胺 (aniline)	市售苯胺，经过氢氧化钾（钠）干燥，要除去含硫的杂质，可在少量氯化锌存在下，氮气保护下减压蒸馏，沸点 77～78℃/2.0 kPa（15 mmHg）或 184.4℃/101.0 kPa（760 mmHg）。在空气中或光照下苯胺颜色变深，应密封贮存于避光处。苯胺稍溶于水，能与乙醇、氯仿和大多数有机溶剂混溶。可与酸成盐，苯胺盐酸盐熔点 198℃。吸入苯胺蒸气或经皮肤吸收会引起中毒症状

续表

名称	纯化处理方法
冰醋酸 (acetic acid or glacial acetic acid)	将市售冰醋酸，在 4℃ 下慢慢结晶，并在冷却下迅速过滤，压干。少量的水可用五氧化二磷回流干燥几小时除去（10 g/L），熔点 16~17℃，沸点 117~118℃/101.0 kPa（760 mmHg）。冰醋酸对皮肤有腐蚀作用，触及皮肤或溅到眼睛时，要用大量水冲洗
乙酸酐 (acetic anhydride)	分子式 $(CH_3CO)_2O$，沸点 139~141℃/101.0 kPa（760 mmHg）。对皮肤有严重腐蚀作用，使用时需使用防护眼镜及手套
亚硫酰氯 (thionyl chloride)	分子式 $SOCl_2$，又称氯化亚砜，为无色或微黄色液体，有刺激性，遇水强烈分解。纯化时使用硫黄处理，操作较为方便，效果较好。将硫黄（20 g/L）在搅拌下，加入亚硫酰氯中，加热，回流 4.5 h，用分馏柱分馏，得无色纯品，沸点 78~79℃。氯化亚砜对皮肤与眼睛有很大刺激性，操作中要小心防护
尿素 (carbamide)	分子式 H_2NCONH_2，简称脲，分子量 60.06，无色柱状结晶或白色结晶性粉末。1 g 该品溶于 1 mL 水、10 mL 95%乙醇、1 mL 95%沸乙醇、20 mL 无水乙醇、6 mL 甲醇、2 mL 甘油，几乎不溶于乙醚、三氯甲烷。熔点 132.7℃。该品有刺激性，使用时避免吸入粉尘，避免与眼睛和皮肤接触。光照受热易分解，避光密闭保存
盐酸苯肼 (phenyl hydrazine hydrochloride)	分子式 $C_6H_5NHNH_2·HCl$，分子量 144.60，无色而有光泽的片状结晶，见光变黄。易溶于水，溶于乙醇，不溶于乙醚。熔点 243~246 ℃（微变棕色）。本品有毒，吸入、口服或皮肤接触时有害，对机体有不可逆性操作的可能。接触皮肤后，应立即用大量指定的液体冲洗。密封于避光干燥处保存
对甲氧基偶氮苯 (p-methoxyazobenzene)	分子式 $CH_3OC_6H_4N_2C_6H_5$，分子量 212.25。橙红色结晶，易溶于有机溶剂，不溶于水，熔点 54~56℃。使用时注意避免吸入其粉尘，避免与眼睛和皮肤接触，密封保存
苏丹黄 (sudan yellow)	分子式 $C_{16}H_{14}N_4O$，也称苏丹Ⅰ，分子量 248.28，暗红色粉末，熔点 129~134℃。溶于乙醚、苯、二硫化碳等有机溶剂中呈橙黄色，溶于浓硫酸呈深红色，不溶于水、碱溶液。该品具有刺激性，可能致癌，应密封保存
苏丹红 (sudan red)	分子式 $C_{24}H_{20}N_4O$，即苏丹红，分子量 380.45，红色粉末。溶于乙醇、丙酮等有机溶剂。使用时应避免吸入其粉尘，避免与眼睛和皮肤接触。
对氨基偶氮苯 (p-aminoazobene, AAB)	分子式 $C_{12}H_{11}N_3$，分子量 197.24，黄褐色针状结晶。溶于乙醚、乙醇、苯、三氯甲烷，微溶于水。本品可能致癌，应避光保存
对羟基偶氮苯 (p-hydroxy azobenzene)	分子式 $C_{12}H_{10}N_2O$，分子量 198.23，橙黄色结晶或粉末。溶于乙醇。熔点 152~155℃。使用时避免吸入其粉尘，避免与眼睛和皮肤接触
金属钠 (sodium, Na)	原子量 22.99，银白色金属。熔点 97.82℃，沸点 881.4℃，相对密度 0.968。遇水和醇发生反应生成氢气。与水反应剧烈，会发生燃烧、爆炸，一旦起火，一定不能用水灭火，要用指定灭火器。本品有腐蚀性，能引起烧伤。用时保持容器干燥，平时一般储存在煤油中
镁 (magnesium, Mg)	原子量 24.305，带金属光泽，银白色金属。熔点 651℃，沸点 1100℃，相对密度 1.738, 空气中，易被氧化生成暗膜。高度易燃，与水接触时产生易燃气体，故不能用水灭火。燃烧时产生炫目白光，冒白烟。溶于酸，不溶于水
锌粉 (zinc powder, Zn)	原子量 65.38，浅灰色的细小粉末，具强还原性。熔点 419.5℃，沸点 908℃，相对密度 7.140。该品与水接触时，释放出高度易燃气体，在空气中能自动燃烧，万一着火应使用指定灭火设备灭火，而绝不能用水。密封于干燥处保存
无水氯化钙 (anhydrous calcium chloride)	分子式 $CaCl_2$，分子量 110.98，白色固体。极易吸潮，易溶于水、乙醇、丙酮、乙酸。熔点 772℃，沸点>1600℃，相对密度 2.152。该品对眼睛有刺激性，使用时避免吸入粉尘，避免与皮肤接触，密封于干燥处保存
氯化锌 (zinc chloride)	分子式 $ZnCl_2$，分子量 136.30，白色粉末或颗粒，无味，极易潮解。该品溶于 0.25 mL 2%盐酸、1.3 mL 乙醇、2 mL 甘油，极易溶于水（25℃，432 g/100 g；100℃，614 g/100 g），易溶于丙酮。其水溶液呈酸性，pH 约为 4，熔点 290℃，沸点 732℃。该品具有腐蚀性，能引起烧伤。接触皮肤后，应立即用大量清水与 2%碳酸氢钠溶液冲洗

续表

名称	纯化处理方法
氯化铵 (ammonium chloride)	分子式 NH_4Cl，分子量 53.49，无色结晶或粉末，无味，吸潮结块。溶于水（25℃，28.3% w/w）、甘油、甲醇、乙醇，不溶于丙酮、乙醚、乙酸乙酯。加热至 337.8℃升华并分解，该品口服有害，对眼睛有刺激性，使用时应避免吸入粉尘。密封于干燥处保存
无水三氯化铝 (anhydrous aluminum chloride)	分子式 $AlCl_3$，分子量 133.34，无色透明六角晶体，有强盐酸气味，容易潮解，在湿空气中发烟。熔点 194℃（253.3 kPa），177.8℃升华，262℃分解。空气中能吸收水分，一部分水解而放出氯化氢。溶于水（15℃，69.9 g/100 mL），能生成六水合物 $AlCl_3 \cdot 6H_2O$，相对密度 2.40。也能溶于乙醇、乙醚、氯仿、二硫化碳、四氯化碳等有机溶剂。本品具有腐蚀性，能引起烧伤，溶于水能产生大量热，激烈时能燃烧或爆炸，接触皮肤后用大量清水和 2%碳酸氢钠溶液冲洗。密封于干燥处保存
亚硝酸钠 (sodium nitrite)	分子式 $NaNO_2$，分子量 69.0，白色或微黄色结晶，有潮解性。溶于 1.5 份冷水
高锰酸钾 (potassium permanganate)	分子式 $KMnO_4$，分子量 158.03，深紫色或类似青铜色有金属光泽的结晶，无味。能溶于 14.2 份冷水、3.5 份沸水。遇醇和其他有机溶剂或浓酸即分解而释放出游离氧，属强氧化剂，外用有杀菌作用。约 240℃分解。本品与易燃品接触能引起燃烧。密闭于干燥处保存
二水合重铬酸钠 (sodium dichromate dihydrate)	分子式 $Na_2Cr_2O_7 \cdot 2H_2O$，分子量 298.00，红色或橙红色结晶。易潮解，易溶于水，溶液呈酸性（1%溶液 pH 4.0），不溶于乙醇。100℃时失去结晶水。熔点 320℃，相对密度 2.348。本品有毒，对眼睛和皮肤有刺激性，接触皮肤会引起过敏，产生炎症和溃疡，可能致癌，使用时应尽量避免吸入本品的粉尘，接触皮肤后立即用大量指定的液体（5%硫代硫酸钠溶液）冲洗。在使用前应得到专门指导，避免暴露，密封于干燥处保存

附录02 国际原子量一览表

元素符号	名称	原子量	元素符号	名称	原子量	元素符号	名称	原子量	元素符号	名称	原子量
Ac	锕	[227]	Er	铒	167.26	Mn	锰	54.93805	Ru	钌	101.07
Ag	银	107.8682	Es	锿	[254]	Mo	钼	95.94	S	硫	32.066
Al	铝	26.98154	Eu	铕	151.965	N	氮	14.00674	Sb	锑	121.760
Am	镅	[243]	F	氟	18.9984032	Na	钠	22.989768	Sc	钪	44.95591
Ar	氩	39.948	Fe	铁	55.845	Nb	铌	92.90638	Se	硒	78.96
As	砷	74.92159	Fm	镄	[257]	Nd	钕	144.24	Si	硅	28.0855
At	砹	[210]	Fr	钫	[223]	Ne	氖	20.1797	Sm	钐	150.36
Au	金	196.96654	Ga	镓	69.723	Ni	镍	58.6934	Sn	锡	118.710
B	硼	10.811	Gd	钆	157.25	No	锘	[254]	Sr	锶	87.62
Ba	钡	137.327	Ge	锗	72.61	Np	镎	237.0482	Ta	钽	180.9479
Be	铍	9.012182	H	氢	1.00794	O	氧	15.9994	Tb	铽	158.92534
Bi	铋	208.98037	He	氦	4.002602	Os	锇	190.23	Tc	锝	98.9062
Bk	锫	[247]	Hf	铪	178.49	P	磷	30.973762	Te	碲	127.60
Br	溴	79.904	Hg	汞	200.59	Pa	镤	231.03588	Th	钍	232.0381
C	碳	12.011	Ho	钬	164.93032	Pb	铅	207.2	Ti	钛	47.867
Ca	钙	40.078	I	碘	126.90447	Pd	钯	106.42	Tl	铊	204.3833
Cd	镉	112.411	In	铟	114.818	Pm	钷	[145]	Tm	铥	168.93421
Ce	铈	140.115	Ir	铱	192.217	Po	钋	[210]	U	铀	238.0289
Cf	锎	[251]	K	钾	39.0983	Pr	镨	140.90765	V	钒	50.9415
Cl	氯	35.4527	Kr	氪	83.80	Pt	铂	195.08	W	钨	183.84
Cm	锔	[247]	La	镧	138.90550	Pu	钚	[244]	Xe	氙	131.29
Co	钴	58.93320	Li	锂	6.941	Ra	镭	226.0254	Y	钇	88.90585
Cr	铬	51.9961	Lr	铹	[257]	Rb	铷	85.4678	Yb	镱	173.04
Cs	铯	132.90543	Lu	镥	174.967	Re	铼	186.207	Zn	锌	65.39
Cu	铜	63.546	Md	钔	[256]	Rh	铑	102.9055	Zr	锆	91.224
Dy	镝	162.50	Mg	镁	24.3050	Rn	氡	[222]	—	—	—

附录03 不同温度下水的蒸气压一览表

温度/℃	蒸气压/Pa	温度/℃	蒸气压/Pa	温度/℃	蒸气压/Pa
−10	259.46	11	1312.0	32	4754.5
−9	283.32	12	1402.6	33	5030.5
−8	309.46	13	1497.3	34	5319.8
−7	337.59	14	1598.6	35	5623.8
−6	368.12	15	1705.3	36	5941.1
−5	401.05	16	1817.3	37	6275.8
−4	436.79	17	1937.3	38	6619.8
−3	475.45	18	2063.9	39	6991.8
−2	516.78	19	2197.3	40	7375.8
−1	562.11	20	2338.6	41	7778.4
0	610.51	21	2486.6	42	8199.7
1	657.31	22	2646.6	43	8639.7
2	705.31	23	2809.2	44	9101.0
3	758.64	24	2983.9	45	9583.7
4	813.31	25	3167.9	46	10086
5	871.79	26	3361.2	47	10613
6	934.64	27	3565.2	48	11161
7	1001.3	28	3779.9	49	11736
8	1073.3	29	4005.2	50	12334
9	1148.0	30	4242.5	51	12960
10	1228.0	31	4493.2	—	—

附录 04 常见化合物的物理常数一览表

试剂名称	分子量	性状	折光率 n	密度 ρ	熔点/℃	沸点/℃	溶解度 水	溶解度 乙醇	溶解度 乙醚
乙醇	46.07	无色液体	1.3611	0.7893	−117.3	78.5	∞	∞	∞
萘	128.19	白色晶体	1.4003	1.0253	80.55	218	不	溶	热易
对-二氯苯	147.01	白色晶体	1.5285	1.2475	53.1	174	不	∞	溶
正丁醇	74.12	无色液体	1.3993	0.8098	−89.53	117.25	微	∞	∞
正溴丁烷	137.02	无色液体	1.4399	1.2758	−112.4	101.6	不	溶	溶
叔丁醇	74.12	无色固体	1.379	0.7858	25.8	82.4	可溶	溶	溶
叔丁基氯	92.57	无色液体	1.3856	0.8511	−25.4	51~52	微	溶	溶
咖啡因	194.20	无色晶体	—	1.23	234.5	178 升华	微	微	不
对甲基苯乙酮	134.12	无色液体	1.5328	1.0051	−19~24	224~226	不	易	∞
邻甲基苯乙酮	134.12	无色液体	1.5302	1.026	—	214	不	易	∞
间甲基苯乙酮	134.12	无色液体	1.5290	1.007	−9	218~220	不	溶	溶
甲苯	92.14	无色液体	1.4969	0.8669	−95	110.62	不	可溶	可溶
乙酸酐	102.09	无色液体	1.3901	1.0820	−73.1	139.55	易	溶	∞
对叔丁基苯酚	150.21	白色晶体	1.4787	0.908	100~101	236~238	不	溶	溶
苯酚	94.11	白色晶体	1.5418	1.0576[41]	40.9	181.8	微	溶	溶
三苯甲醇	260.34	无色棱晶	—	1.199	164.2	380	不	易	易
溴苯	157.02	无色液体	1.5597	1.4950	−30.82	156	不	易	易
苯甲酸乙酯	150.18	无色液体	1.5007	1.0468	−34.6	213	不	溶	∞
乙醚	74.12	无色液体	1.3526	0.7137	−116.2	34.51	微	∞	∞
联苯	154.21	白色晶体	1.588	0.8660	71	255.9	不	溶	溶
二茂铁	186.04	橙色晶体	—	—	172.5	249 升华	不	溶	溶
乙酰二茂铁	229.08	橙色晶体	—	1.35	84.5	—	不	溶	溶
石油醚（60~90℃）	—	无色液体	—	0.816	—	30~120	不	溶	溶
二氯甲烷	84.93	无色液体	1.4242	1.3266	−95.1	40	微	∞	∞
正己烷	86.18	无色液体	1.3750	0.6603	−95	68.95	不	易	溶
甲醇	32.04	无色液体	1.3288	0.7914	−97.8	64.96	∞	∞	∞
二苯甲醇	184.24	白色晶体	/	1.1108	68~69	2972~98	热微	易	易
二苯酮 (α)/(β)	182.21	无色晶体	1.6077 / 1.6060	1.0976 / 1.108	48.1 / 26	305.9	不	溶	溶
8-羟基喹啉	145.16	白色结晶	—	1.03	75~76	267	不	溶	不
邻硝基苯酚	139.11	黄色晶体	1.5723	1.485	45.3~457	216	微	热易	易
邻氨基苯酚	109.13	白色晶体	—	1.328	174	153 升华	溶	易	溶

续表

试剂名称	分子量	性状	折光率 n	密度ρ	熔点/℃	沸点/℃	溶解度 水	溶解度 乙醇	溶解度 乙醚
丙三醇	92.11	无色液体	1.4746	1.2613	20	290 分解	∞	∞	微
安息香	212.25	白色晶体	—	1.310	137	344	微	热溶	微
VB$_1$	337.21	白色结晶	—	—	246~250	—	微	微	微
苹甲醛	106.13	无色液体	1.5463	1.0415	−26	178.1	微	∞	∞
二苯乙二酮	210.23	无色晶体	—	1.0844	95~96	346~348	不	易	易
醋酸	60.05	无色液体	1.3716	1.0492	16.604	117.9	∞	∞	∞
二苯乙酸	228.25	无色液体	—	—	150	—	微	易	易
烯丙基溴	120.98	无色液体	1.465	1.451	−50	70	不	∞	∞
1-苯基-3-丁烯基-1-醇	148.21	无色液体	—	—	−0.56	243.31	不	溶	溶
乙酸乙酯	88.12	无色液体	1.3723	0.9003	−83.57	77.06	微	∞	∞
环己烯	82.15	无色液体	1.4465	0.8102	−103.5	82.98	不	∞	∞
环己醇	100.16	无色液体	1.4641	0.9624	25.15	161.1	溶	溶	溶
7.7-二氯双环 [4.1.0] 庚烷	165.05	无色液体	1.5014	1.21	—	197~198	不	—	溶
氯仿	119.38	无色液体	1.4459	1.4832	−63.5	61.7	微	∞	∞
四乙基溴化铵	210.16	白色晶体	—	1.397	287	—	溶	溶	氯仿
苯甲酸	122.13	白色晶体	1.504	1.2659	122.4	249	溶	易	易
环己烷	84.16	无色液体	1.4266	0.7786	6.55	80.74	不	∞	∞
正己酸	116.16	无色液体	1.4163	0.9274	−2~−1.5	205	不	溶	溶
丙二酸二乙酯	160.17	无色液体	1.4139	1.0551	−48.9	199.3	微	∞	∞
正丁基丙二酸二乙酯	216.28	无色液体	1.4220	0.983	—	235~240	不	溶	溶
水杨酸	138.12	白色晶体	1.565	1.443	159	2011 升华	微	溶	溶
乙酰水杨酸	180.16	白色结晶	—	1.35	135~136	—	微	微溶	微溶
浓盐酸(37%)	36.46	无色液体	—	1.19	—	—	∞	溶	溶
浓硫酸(96%)	98.08	无色液体	—	1.84	—	—	∞	溶	分解
磷酸	98.00	无色液体	—	1.834	42.35	213	∞	溶	不
浓硝酸(71%)	63.01	无色液体	—	1.42	—	—	∞	溶	溶
氢氧化钠	40.00	白色	1.3576	2.130	318.4	1390	溶	溶	不
氢氧化钾	56.11	白色结晶	/	2.044	360.4	1324	易溶	溶	不
碳酸钠	105.99	白色粉末	1.535	2.532	851	分解	易溶	微溶	不
碳酸氢钠	84.00	白色结晶	1.500	2.159	270 分解	—	易溶	微溶	不
硫酸钠	142.04	单斜	1.480	2.68	884	—	易溶	不	不
亚硫酸氢钠	104.06	单斜	1.526	1.48	分解	—	易溶	微溶	不
硫酸镁	120.37	无色结晶	1.56	2.66	1124 分解	—	易溶	溶	微溶

续表

试剂名称	分子量	性状	折光率 n	密度ρ	熔点/℃	沸点/℃	溶解度		
							水	乙醇	乙醚
氧化钙	56.08	无色立方	1.838	3.25~3.38	2614	2850	分解	不	不
氯化钙	110.99	无色立方	1.52	2.15	782	>1600	易溶	溶	不
溴化钠	102.90	无色立方	1.6412	3.203	747	1390	易溶	微溶	不
氯化铵	53.49	无色立方	1.642	1.527	340 升华	520	易溶	微溶	不
氯化钠	58.44	无色立方	1.5442	2.165	801	1413	易溶	微溶	不
氯化铁	162.21	褐色结晶	—	2.895	306	315 分解	易溶	溶	溶
无水氯铝	133.34	白色六方	/	2.44	190	263.3	水解	微溶	微溶
氧化铝	101.96	无色立方	1.768	3.965	2045	2980	不	不	不
硅胶 G	—	白色粉末	—	—	—	—	不	不	不
CMC	—	白色絮状	—	—	—	—	溶	不	不
碘	129.90	黑紫斜方	—	4.93	113.5	184.35	溶	—	—
锌粉	65.37	蓝色六方	—	7.14	419.47	907	不	不	不
镁粉	24.31	银色立方	—	1.74	650	1117	不	不	不

附录 05 常见干燥剂使用一览表

干燥剂	适合干燥的物质	不适合干燥的物质	吸水量/(g/g)	活化温度
氧化铝	烃、空气、NH_3、Ar_2、He、N_2、O_2、CO_2、SO_2	—	0.2	175℃
氧化钡	有机碱、醇、醛、胺	酸性物质、CO_2	0.1	—
氧化镁	烃、醛、醇、碱性气体、胺	酸性物质	0.5	800℃
氧化钙	醇、氨、氨气	酸性物质、酯	0.3	1000℃
硫酸钙	大多数有机物	—	0.066	235℃
硫酸铜	酯、醇、（适合苯、甲苯干燥）	—	0.6	200℃
硫酸钠	氯代烷烃、氯代芳烃、醛、酮、酸	—	1.2	150℃
硫酸镁	酸、酮、醛、酯、腈	对酸敏感物质	0.8	200℃
氯化钙	氯代烷烃、氯代芳烃、酯、饱和芳香烃、芳香烃、醚	醇、胺、苯酚、醛、酰胺、氨基酸、某些酯和酮	$0.2(1H_2O)$ $0.3(2H_2O)$	250℃
氯化锌	烃	氨、胺、醇	0.2	110℃
氢氧化钾	胺、有机碱	酸、苯酚、酯、酰胺、酸性气体、醛	—	—
氢氧化钠	胺	酸、苯酚、酯、酰胺	—	—
碳酸钾	醇、腈、酮、酯、胺	酸、苯酚	0.2	300℃
金属钠	饱和脂肪烃和芳香烃、醚	酸、醇、醛、酮、胺、酯、氯代有机物、含水过高的物质	—	—
五氧化二磷	烷烃、芳香烃、醚、氯代烷烃、氯代芳烃、腈、酸酐、酯	醇、酸、胺、酮、氟化氢和氯化氢	0.5	—
浓硫酸	惰性气体、HCl、Cl_2、CO、SO_2	基本不能与其他物质接触	—	—
硅胶（6～16目）	绝大部分有机物	氟化氢	0.2	200～350℃
分子筛（0.3 mm）	分子直径>0.3 mm	分子直径<0.3 nm	0.18	117～260℃
分子筛（0.4 nm）	分子直径>0.4 nm	分子直径<0.4 nm，乙醇、H_2S、CO_2、SO_2、乙烯、乙炔、强酸	0.18	250℃
分子筛（0.5 nm）	分子直径>0.5 nm，如支链化合物和有4个碳原子以上的环	分子直径<0.5 nm，如，丁醇、正丁烷到正二十二烷	0.18	250℃

附录06 常见试剂的除水剂一览表

干燥剂	与水形成的化合物	注释
Na	NaOH、H_2	用于烃和醚的去水很出色,不得用于醇和卤代烃
CaH_2	$Ca(OH)_2$,H_2	最佳去水剂之一,比 $LiAlH_4$ 缓慢,但效率高相对较安全,用于烃、醚、胺、酯和更高级的醇(勿用于 C_1,C_2,C_3 醇),不得用于醛和活泼羰基化合物
$LiAlH_4$	LiOH、$Al(OH)_3$、H_2	只使用于惰性溶剂[烃基、芳基卤(不能用于烷基卤)、醚];能与任何酸性氢和大多数功能团(卤素、羟基、硝基等)反应。使用时要小心:多余者可慢慢加入乙酸乙酯加以破坏
BaO、CaO、P_2O_5	$Ba(OH)_2$、$Ca(OH)_2$、HPO_3、H_3PO_4、$H_4P_2O_7$	慢而有效,主要适用于醇类和醚类,但不易用于对强碱敏感的化合物,非常快而且效率高,高度耐酸,建议先预干燥,仅用于惰性化合物(尤其适用于烃、醚、卤代烃、酸、酐)

附录 07 常见溶剂与水形成的二元共沸物一览表

溶剂	沸点/℃	共沸点/℃	含水量/%	溶剂	沸点/℃	共沸点/℃	含水量/℃
氯仿	61.2	56.1	2.5	甲苯	110.5	85.0	20
四氯化碳	77.0	66.0	4.0	正丙醇	97.2	87.7	28.8
苯	80.4	69.2	8.8	异丁醇	108.4	89.9	88.2
丙烯腈	78.0	70.0	13.0	二甲苯	137-40.5	92.0	37.5
二氯乙烷	83.7	72.0	19.5	正丁醇	117.7	92.2	37.5
乙腈	82.0	76.0	16.0	吡啶	115.5	94.0	42
乙醇	78.3	78.1	4.4	异戊醇	131.0	95.1	49.6
乙酸乙酯	77.1	70.4	8.0	正戊醇	138.3	95.4	44.7
异丙醇	82.4	80.4	12.1	氯乙醇	129.0	97.8	59.0
乙醚	35	34	1.0	二硫化碳	46	44	2.0
甲酸	101	107	26	—	—	—	—

附录08 常见有机溶剂间的共沸物一览表

共沸混合物	组分的沸点/℃	共沸物的组成（质量分数）	共沸物的沸点/℃
乙醇-乙酸乙酯	78.3，78.0	30∶70	72.0
乙醇-苯	78.3，80.6	32∶68	68.2
乙醇-氯仿	78.3，61.2	7∶93	59.4
乙醇-四氯化碳	78.3，77.0	16∶84	64.9
乙酸乙酯-四氯化碳	78.0，77.0	43∶57	75.0
甲醇-四氯化碳	64.7，77.0	21∶79	55.7
甲醇-苯	64.7，80.4	39∶61	48.3
氯仿-丙酮	61.2，56.4	80∶20	64.7
甲苯-乙酸	101.5，118.5	72∶28	105.4
乙醇-苯-水	78.3，80.6，100	19∶74∶7	64.9

附录 09 实验室常用酸碱的浓度一览表

试剂名称	密度/(20℃,g/mL)	浓度/(mol/L)	质量分数/%
浓硫酸	1.84	18.0	0.960
浓盐酸	1.19	12.1	0.372
浓硝酸	1.42	15.9	0.704
磷酸	1.70	14.8	0.855
冰醋酸	1.05	17.45	0.998
浓氨水	0.90	14.53	0.566
浓氢氧化钠	1.54	19.4	0.505

附录 10　常见冰盐浴冷却剂一览表

盐		每 100 g 碎冰用盐/g	冷却剂温度降低到/℃
$NaNO_3$		50	−18.5
$NaCl$		33	−21.2
$NaCl$	混合物	40	−26
NH_4Cl		20	
NH_4Cl	混合物	13	−30.7
$NaNO_3$		37.5	
K_2CO_3		33	−46
$CaCl_2 \cdot 6H_2O$		143	−35

附录11 常见冷却剂一览表

冷却剂	最低冷却温度/℃	冷却剂	最低冷却温度/℃
冰	0	三氯甲烷/N_2	63
乙二醇/CO_2	−15	三氯甲烷/CO_2	63
冰(100)/NH_4Cl(25)	−15	乙醇/CO_2	72
冰(100)/NaCl(33)	−21	乙醇/CO_2	77
四氯化碳/N_2	−23	乙醇/CO_2	78
四氯化碳/CO_2	−23	甲醇/N_2	98
冰(100)/EtOH(100)	−30	n-戊烷/N_2	131
乙腈/N_2	−41	N_2	180
冰(100)/$CaCl_2$(150)	−49	—	—

附录 12 常见干燥剂的分类以及使用方法一览表

分类	干燥剂	适用的物质与条件	不适用的物质与条件	干燥特点	使用方法	备注
金属、金属氢化物	Mg	醇类	—	—	无水 MeOH 的制备：MeOH 和 Mg 一起加热回流，然后蒸馏出 MeOH	不要蒸馏到过干
	Na	烷烃、芳烃、醚类	用于卤代烃时，有爆炸的危险，不适用于醇、酯、酸、醛酮、胺类的干燥	干燥能力高，但在表面易覆盖 NaOH，效果下降，脱水能力小	切成薄片或压成丝状，放入待干燥液体中。对 THF 和 Et₂O 也可加入 Ph₂CO 和 Na 回流再进行蒸馏	利水反应生成 H_2 与大量水接触会燃烧。保存和处理时要注意。蒸馏时不要蒸干，用过的 Na 用乙醇分解破坏
	CaH_2	烃类、卤代烃、t-丁醇、三级胺、醚类、二氧六环、THF、吡啶、DMSO 等	醛、酮、羧酸	脱水容量大，处理方便，适用范围广	加入 CaH_2，在 Ar 或 N_2 气流中蒸馏，或者将粒状的 CaH_2 加到液体中进行干燥	和水反应产生 H_2，保存和处理上要注意
	$LiAlH_4$	醚类、乙醚、THF 等	易和酸、胺、硫醇、乙炔等含活泼氢的化合物及酮、酯、酰氯、酰胺、腈、硝基化合物、环氧化物、二硫化物、烯丙醇反应，生成高沸点化合物	能分解待干燥物中的醇、羰基化合物、过氧化物	加入 $LiAlH_4$，在 Ar 或 N_2 气流中蒸馏	$LiAlH_4$ 在 125℃ 时分解，蒸馏时不要蒸干、过量的 $LiAlH_4$ 用氯化铵水溶液或乙酸乙酯分解。保存时不要与水和 CO_2 接触
中性干燥剂	Na_2SO_4 $MgSO_4$ $CaSO_4$	几乎全部溶剂	Na_2SO_4 在 33℃ 以上，$MgSO_4$ 在 48℃ 以上释放出结晶水，因此不适合在以上温度使用	Na_2SO_4 脱水容量大，脱水速度慢；$MgSO_4$ 脱水容量大，脱水速度比 $NaSO_4$ 快 $CaSO_4$ 脱水容量小，但脱水力强，速度快	加到待干燥液体中	$CaSO_4$ 在 235℃ 加热 2~3 h 后可以再生
	$CuSO_4$	乙醇、苯、乙醚等	能和甲醇反应，不用于甲醇干燥	无水物呈白色与结晶水物呈蓝色	加到待干燥液体中	—
	$CaCl_2$	烃类、卤代烃、醚类、中性气体等	醇、胺、氨基酸、酰胺酮、酯、酸等	吸水速度慢，30℃ 以下生成六水合物，脱水容量大，有潮解性	加入到待干燥液体中。加入干燥器、干燥管中使用	—
	氧化铝	烃、醚类、三氯甲烷、苯、吡啶等	—	同时能除去醚类中的过氧化物，处理方便，吸收力大	做成填充柱，让溶剂通过	170℃ 以上加热 6~8 h 可以再生。加热到 800℃ 以上变成活性氧化铝
	硅胶（蓝色）	几乎全部固体和气体物质	—	处理方便、脱水力极强，无水时蓝色，吸水后粉红色	加入干燥器、干燥管中使用	150℃ 以上加热 2~3 h 可以再生
	分子筛	卤代烃、烃类、THF、二噁烷、丙酮、Py、DMSO、DMF、HMPA 等，适用 pH 范围 5~11	对强酸、碱性物质不稳定	随干燥时间长，脱水力显著高，高温时，吸附力也不降低	加入到待干燥溶剂瓶中，结晶的孔径不同种类、根据溶剂进行选择使用	350℃加热 3 h 再生

续表

分类	干燥剂	适用的物质与条件	不适用的物质与条件	干燥特点	使用方法	备注
碱性干燥剂	KOH NaOH	胺类等碱性物质、中性或碱性气体	酸、醛、酮、醇、酯等	脱水速度、脱水力大,易潮解	加到液体中,干燥皿中、干燥管中	—
	N_2CO_3 K_2CO_3	胺类等碱性物质,醇酮酯、腈等	酸	—	加到液体中,适合预干燥	可加热熔化,活化
	CaO	胺类等碱性物质,醇等	酸	脱水速度小,便宜,可大量使用,能吸收CO_2	加到液体、干燥皿、干燥管中。块状可粉碎使用	细的粉末物中,以$Ca(OH)_2$和$CaCO_3$为主,干燥能力低
酸性干燥剂	H_2SO_4	Br_2,中性气体	醇、酚、酮、乙烯等	吸收速度、容量大,吸水浓度降低后,干燥能力急剧下降	加到干燥皿、气体干燥瓶中	—
	P_2O_5	烃、卤烃、酸酐、腈中性气体	碱性物质,酮醇、胺、酰胺、卤化氢、丙酮	吸水速度、吸水能力最大。在表面上形成偏磷酸膜时,效率变低,生成白色粉末,难处理	加到干燥皿、干燥管中,多用于固体、气体干燥	P_2O_5的后处理,用乙醇分解或自然放置让其吸湿潮解

注:THF—四氢呋喃;DMSO—二甲基亚砜;Py—吡啶;DMF—二甲基甲酰胺;HMPA—六甲基磷酰三胺。

附录 13 常见溶剂的提纯、干燥以及贮藏方法一览表

溶剂	沸点/℃（容许沸距/℃）	初步提纯	进一步干燥与提纯	贮藏
戊烷 己烷 环己烷 其他烷烃	36 (2~3) 69 (2.5) 80.7 (1)	必要时，首先用浓硫酸洗涤几次，以除去烯烃，然后水洗，用 $CaCl_2$ 干燥，蒸馏，收集潮湿的前馏分之后的正沸点蒸馏液	几乎没有进一步处理的必要；一定要处理时可利用恒沸蒸馏脱水	500 mL 以内贮藏于带塞的试剂瓶中；大量和长期贮藏时应采用螺旋盖的棕色瓶，向其中加入分子筛是没有意义的
苯 甲苯 邻二甲苯 间二甲苯 对二甲苯	80.1 (0.5) 110.6 (1) 144.5 139 138.3 (1)	$CaCl_2$ 干燥、分馏、弃去前面 5%~10%潮湿的前馏分	重蒸，分去前面 5%的馏分	500 mL 以内贮藏于带塞的试剂瓶中；大量和长期贮藏时应采用螺旋盖的棕色瓶，向其中加入分子筛是没有意义的
二氯甲烷 三氯甲烷 四氯化碳 1,2-二氯乙烷	40 (1) 61.2 (0.5) 76.8 (0.5) 83.5 (1)	水洗，$CaCl_2$ 干燥，蒸馏，弃去前面 5%的潮湿的前馏分	加入 P_2O_5 重蒸；在小量和特殊的情况下可通过氧化铝（碱性，一级活性）直接蒸入反应瓶	500 mL 以内贮藏于带塞的试剂瓶中；大量和长期贮藏时，应采用螺旋盖的棕色瓶，向其中加入分子筛是没有意义的。长期贮藏的三氯甲烷，应放在密闭的瓶中，装满，并保存于黑暗处
乙醚 二异丙基醚	34.5 (1) 68.5 (1)	检查是否含过氧化物。如证实其存在，用 5%偏亚硫酸氢钠溶液洗涤，然后以饱和 NaCl 溶液洗涤，用 $CaCl_2$ 干燥，蒸馏（不能用浓硫酸）	小量：通过相当于其重量 10%的氧化铝（碱性，一级活性）蒸馏反应瓶	装于有螺旋盖的金属容器中，几乎装满，置于阴凉黑暗处，长期贮藏时并加以密封
四氢呋喃 1,2-二甲氧基乙烷（甘醇）	65.5 (0.5) 84 (1)	加 KOH，静置过夜，氧化物试验。如呈阳性，则加入最多 0.4%重量的 $NaBH_4$ 搅拌过夜。加入 CaH_2 蒸馏，但不能蒸干	在氩气保护下加入金属钾蒸馏；少量的通过氧化铝（碱性，一级活性）直接进入反应瓶	盛于干燥的塑料瓶中，加入碱性的活性氧化铝，并用氩气保护；长期贮藏时，必须加以密封
二噁烷	101.5 (1) (mp 11~12)	加 KOH，静置过夜，过氧化物试验。如呈阳性，则加入最多 0.4%重量的 $NaBH_4$ 搅拌过夜。加入 CaH_2 蒸馏，但不能蒸干	加入金属钠，在氩气保护下蒸馏	盛于干燥的塑料瓶中，加入碱性的活性氧化铝，并用氩气保护；长期贮藏时，必须加以密封，最好冷冻，保存于冰箱中
二硫化碳	46.5 (1)	加入少量 P_2O_5，蒸馏；使用水浴，加热	加少量汞，振荡，再加入 P_2O_5 重蒸	不要贮藏于实验室内，极易着火
甲酸	101 (1) (mp 8.3)	分馏，最好稍作减压。加邻苯二甲酸酐，回流后，重蒸能获得进一步干燥。与水恒沸物的沸点 107℃，含水 22.5%	将经过纯化的试剂完全冷冻，再让其温热，熔化总量 10%~20%，倾出液化部分，使用剩下的试剂。全部操作应在脱水的条件下完成	贮藏于有螺旋盖的瓶中
乙酸	118 (0.5) (mp 16.6)	加入总量 5%的醋酐和 2%的 CrO_3 后分馏	—	—

续表

溶剂	沸点/℃（容许沸距/℃）	初步提纯	进一步干燥与提纯	贮藏
吡啶甲基吡啶	115.5 (0.5)	向粗品中加入 KOH，倾泻、分馏	加入 CaO、BaO 或活性很强的碱性氧化铝，重新分馏	加入 5A 分子筛，密闭保存并注明日期
N，N-二甲基甲酰胺 N，N-二甲基乙酰胺 N-甲基吡咯烷酮	153，42/1333Pa 55/2666Pa (1) 166，58～59/1466.3Pa 63/2399.4Pa (1) 202，78～79/1333Pa 96～97/3199.2 Pa (1)	真空分馏，弃去前面和最后各10%的馏分；避免常压蒸馏	加入 CaO，BaO 或氧化铝（碱性，一级活性），搅拌过夜，再次真空分馏	加入新活化的分子筛，贮藏于小瓶中，并注明日期。大量贮藏超过 500 mL 时，考虑到多次开启将有水汽渗入，应加入大量的分子筛
二甲基亚砜	190，50/340Pa 7/1360Pa 84～85/2932.6Pa (1) (mp 8.5)	—	加入 CaH$_2$ 搅拌过夜，然后从中减压分馏；如足够干燥，可通过部分冷冻而进一步提纯	—
六甲基磷酰三胺	235，68～70/133.3Pa 115/1200Pa 126/3999Pa (1) (mp 7)	—	加入 CaH$_2$，于 100℃下减压搅拌 1 h，然后真空分馏	分装于小的（50mL）塑料瓶中，加入活化的 13X 分子筛 或除去了矿物油的 NaH，并以氩气保护
醋酸乙酯 醋酸甲酯	77.1 (0.5) 57 (1)	用活性硫酸钙和（或）水碳酸钾干燥、倾泻，小心蒸馏	加入最多 5%重量的醋酐后分馏	加入 5A 活性分子筛，密闭保存
其他沸点低于 100℃的酯	—	—	分馏	—
乙腈	81.5 (0.5)	顺次以 MgSO$_4$ 和无水 K$_2$CO$_3$ 干燥，倾泄；加 CaH$_2$ 蒸馏	通过 P$_2$O$_5$ 分馏；小量：通过氧化铝（碱性，一级活性）直接蒸入反应瓶	加入 3A 活性分子筛，保存于小瓶中，并注明日期
丙酮	56.2 (0.5)	蒸馏，控制 2℃的收集沸程，以无水硫酸钙干燥，倾泻、重蒸	如用于氧化反应，需在回流下加入足够数量的 KMnO$_4$ 直到紫色不褪为止。蒸馏，干燥，再分馏。通过 NaI 化合物，得到很纯的试剂	加入新活化的 3A 分子筛
2-丁酮	79.5 (0.5)	恒沸蒸馏除去水（沸点 73.5℃）以无水硫酸钙分别干燥馏出的恒沸物和残余部分，倾泻、重蒸	—	加入新活化的 5A 分子筛
甲醇	64.5 (0.5)	即使对于工业级产品，简单蒸馏也已足够	经过预干燥后加入 CaH$_2$ 重蒸，直接蒸入反应瓶	贮藏于小瓶中，加入 3A 活性分子筛
乙醇	78.3 (0.5)	将 95%乙醇与 CaO 一同回流并蒸馏（CaO 的用量至少应达含水量的 1.5 倍）	—	—
异丙醇	82.5 (0.5)	分馏，蒸去恒沸物（沸点 80.3℃）之后，收集正沸物；对恒沸物的处理与95%乙醇相同	—	—

续表

溶剂	沸点/℃ （容许沸距/℃）	初步提纯	进一步干燥与提纯	贮藏
正丙醇	97.2 (0.5)	分馏，除去含水的恒沸物后，收集正馏分	—	—
较高级的醇		收集正馏分	—	—
叔丁醇	82.5 (0.5) (mp 25.8)	水恒沸物的沸点 79.9℃，处理与异丙醇相同	与前述的醇相同，在蒸馏时，需防止产物凝结于冷凝管中导致堵塞	与前述的醇相同，但冷天最好保存于温暖处，以免固化
乙二醇 较高级的二醇	198，68~70/533.2Pa 108~110/3732.4Pa (2)	真空分馏，弃 5%~10%的前馏分。注意，其蒸发潜热很大	溶入1%重量的金属钠，重新分馏	分装于小塑料瓶中，但冷天保存于温暖处，以免固化
硝基甲烷 硝基乙烷	101.3 (1) 115	$CaCl_2$，干燥，倾泻，分馏	加入 4A 分子筛，重蒸	加入 4A 分子筛贮藏

附录 14 标准缓冲液的配制方法一览表

缓冲溶液	配制方法
草酸三氢钾标准缓冲液 （0.05 mol/L）	称取在 54℃干燥 4～5 h 的草酸三氢钾[$KH_3(C_2O_4)_2 \cdot H_2O$] 12.7 g，溶于蒸馏水，在容量瓶中稀释至 1 L（20～25℃，pH 1.68）
25℃饱和酒石酸氢钾溶液	在磨口玻璃瓶中，装入蒸馏水和过量的酒石酸氢钾($KHC_4H_2O_6$)粉末（约 20 g/L），温度控制在 25±5℃，剧烈摇动 20～30 分钟，溶液澄清后，用倾泻法取其清液备用。（25℃，pH 3.56）
邻苯二甲酸氢钾标准缓冲液 （0.05 mol/L）	称取在 115℃干燥 2～3 h 的邻苯二氢钾 10.21 g，溶于蒸馏水，在容量瓶中稀释至 1 L（20℃，pH 4.00；25℃，pH 4.01）
混合磷酸盐标准缓冲液 （0.025 mol/L 磷酸二氢钾和 0.025 mol/L 磷酸氢二钠的混合液）	分别称取在 115℃干燥 2～3 h 的磷酸氢二钠(Na_2HPO_4) 3.549 g 和磷酸二氢钾(KH_2PO_4) 3.402 g，溶于蒸馏水，在容量瓶中稀释至 1 L（20℃，pH 6.88；25℃，pH 6.66）
硼砂标准缓冲液 （0.01 mol/L）	称取硼砂($Na_2B_4O_7 \cdot 10H_2O$) 3.814 g（注意不能烘），溶于蒸馏水，在容量瓶中稀释至 1 L（20℃，pH 9.22；25℃，pH 9.18）
25℃饱和氢氧化钙溶液	在玻璃磨口瓶或聚乙烯塑料瓶中，装入蒸馏水和过量的氢氧化钙[$Ca(OH)_2$]粉末（5～10 g/L），温度控制在 25℃，剧烈摇动 20～30 min，迅速用抽滤法，滤取其清液备用（20℃，pH 12.63；25℃，pH 12.45）

注：用 pH 基准试剂配制；所用水应符合三级水规格；可用于精度为 0.1 的 pH 酸度计校正；所测样品用无二氧化碳的水配成 5%的溶液。

附录 15 常用的缓冲液的配制一览表

缓冲溶液组成	pKa	缓冲液 pH	缓冲溶液配制方法
氨基乙酸-HCl	2.35 (pKa1)	2.3	取氨基乙酸 150 g，溶于 500 mL 水中，加浓 HCl 80 mL，水稀释至 1 L
H_3PO_4-枸橼酸盐	—	2.5	取 $Na_2HPO_4 \cdot 12H_2O$ 113 g 溶于 200 mL 水后，加枸橼酸 387 g，溶解，过滤后，稀释至 1 L
一氯乙酸-NaOH	2.86	2.8	取 200 g 一氯乙酸溶于 200 mL 水中，加 NaOH 40 g 溶解，稀释至 1 L
邻苯二甲酸氢钾-HCl	2.95 (pKa1)	2.9	取 500 g 邻苯二甲酸氢钾溶于 500 mL 水中，加浓 HCl 80 mL，稀释至 1 L
甲酸-NaOH	3.76	3.7	取 95 g 甲酸和 NaOH 40 g 于 500 mL 水中，溶解，稀释至 1 L
NH_4Ac-HAc	—	4.5	取 NH_4Ac 77 g 溶于 200 mL 水中，加冰 HAc 59 mL，稀释至 1 L
NaAc-HAc	4.74	4.7	取无水 NaAc 83 g 溶水中，加冰 HAc 60 mL，稀释至 1 L
NaAc-HAc	4.74	5.0	取无水 NaAc 160 g 溶于水中，加冰 HAc 60 mL，稀释至 1 L
NH_4Ac-HAc	—	5.0	取 NH_4Ac 250 g 溶于水中，加冰 HAc 25 mL，稀释至 1 L
六次甲基四胺-HCl	5.15	5.4	取六次甲基四胺 40 g 溶于 200 mL，加浓 HCl 10 mL 稀释至 1 L
NH_4Ar-HAc	—	6.0	取 NH_4Ac 600 g 溶于水中，加冰 HAc 20 mL，稀释至 1 L
HAc-NaAc	—	6.0	取醋酸钠 54.6 g，加醋酸液(1 mol/L)20 mL 溶解，加水稀释 500 mL 即得
NaAc-H_3PO_4盐	—	8.0	取无水 NaAc 50 g 和 $Na_2HPO_4 \cdot 12H_2O$ 50 g，溶于水中，稀释至 1 L
NH_3-NH_4Cl	9.26	9.2	取 NH_4Cl 54 g 溶于水中，加浓氨水 63 mL，稀释至 1 L
NH_3-NH_4Cl	9.26	9.5	取 NH_4Cl 54 g 溶于水中，加浓氨水 126 mL，稀释至 1 L
NH_3-NH_4Cl	9.26	10.0	取 NH_4Cl 54 g 溶于水中，加浓氨水 350 mL，稀释至 1 L

附录 16 常用酸碱试剂的含量与密度一览表

试剂	密 度（ρ）	含 量/%	浓 度/(mol/L)
盐酸	1.18～1.19	36～38	11.6～12.4
硝酸	1.39～1.40	65.0～67.0	14.4～14.9
硫酸	1.83～1.84	93.6～95.7	17.5～18.0
磷酸	1.69	85.5	14.7
高氯酸	1.68	70.2	11.7
冰醋酸	1.05	99.0（分析纯、化学纯）	17.4
氢氟酸	1.13	40	22.5
氢溴酸	1.49	47.0	8.6
氨水	0.88～0.90	34.31～27.33	17.8～14.4
乙酸	1.05	40.2	7.03
浓氢氧化钠	1.43	40	14
稀氢氧化钠	1.09	8	2
$Ca(OH)_2$（饱和）	—	0.15	—

附录 17 常用酸碱指示剂以及配制方法一览表

指示剂名称	变色 pH 范围	颜色变化	溶液配制方法
甲基紫（第一变色范围）	0.13～0.5	黄～绿	0.1%或 0.05%水溶液
苦味酸	0.0～1.3	无色～黄	0.1%水溶液
甲基绿	0.1～2.0	黄 绿～浅蓝	0.05%水溶液
孔雀绿（第一变色范围）	0.13～2.0	黄～浅蓝～绿	0.1%水溶液
甲酚红（第一变色范围）	0.2～1.8	红～黄	0.04 g 指示剂溶于 100 mL 50%乙醇中
甲基紫（第二变色范围）	1.0～1.5	绿～蓝	0.1%水溶液
百里酚蓝（麝香草酚蓝）（第一变色范围）	1.2～2.8	红～黄	0.1 g 指示剂溶于 100 mL 20%乙醇中
甲基紫（第三变色范围）	2.0～3.0	蓝～紫	0.1%水溶液
茜素黄 R（第一变色范围）	1.9～3.3	红～黄	0.1%水溶液
二甲基黄	2.9～4.0	红～黄	0.1 或 0.01 g 指示剂溶于 100 mL 90%乙醇中
甲基橙	3.1～4.4	红～橙黄	0.1%水溶液
溴酚蓝	3.0～4.6	黄～蓝	0.1g 指示剂溶于 100 mL 20%乙醇中
刚果红	3.0～5.2	蓝紫～红	0.1%水溶液
茜素红 S（第一变色范围）	3.7～5.2	黄～紫	0.1%水溶液
溴甲酚绿	3.8～5.4	黄～蓝	0.1 g 指示剂溶于 100 mL 20%乙醇中
甲基红	4.4～6.2	红～黄	0.1 或 0.2 g 指示剂溶于 100 mL60%乙醇中
溴酚红	5.0～6.8	黄～红	0.1 或 0.04 g 指示剂溶于 100 mL 20%乙醇中
溴甲酚紫	5.2～6.8	黄～紫红	0.1 g 指示剂溶于 100 mL 20%乙醇中
溴百里酚蓝	6.0～7.6	黄～蓝	0.05 g 指示剂溶于 100 mL 20%乙醇中
中性红	6.8～8.0	红～亮黄	0.1 g 指示剂溶于 100 mL 60%乙醇中
酚红	6.8～8.0	黄～红	0.1 g 指示剂溶于 100 mL 20%乙醇中
甲酚红	7.2～8.8	亮黄～紫红	0.1 g 指示剂溶于 100 mL 50%乙醇中
百里酚蓝(麝香草酚蓝)（第二变色范围）	8.0～9.0	黄～蓝	参看第一变色范围
酚酞	8.2～10.0	无色～紫红	0.1 g 指示剂溶于 100 mL 60%乙醇中
百里酚酞	9.4～10.6	无色～蓝	0.1 g 指示剂溶于 100 mL 90%乙醇中
茜素红 S（第二变色范围）	10.0～12.0	紫～淡黄	参看第一变色范围
茜素黄 R（第二变色范围）	10.1～12.1	黄～淡紫	0.1%水溶液
孔雀绿（第二变色范围）	11.5～13.2	蓝绿～无色	参看第一变色范围
达旦黄	12.0～13.0	黄～红	溶于水、乙醇

附录 18　常用试剂的配制方法一览表

试剂名称	浓度/(mol/L)	配制方法
硫化钠 Na_2S	1	称取 200 g $Na_2S·9H_2O$、40 g 的 NaOH 溶于适量水中，稀释至 1 L 混匀
硫化铵 $(NH_4)_2S$	3	通 H_2S 于 200 mL 浓 $NH_3·H_2O$ 中直至饱和，然后再加 200 mL 浓 $NH_3·H_2O$，最后加水稀释至 1 L
氯化亚锡 $SnCl_2$	0.25	称取 56.4 g $SnCl_2·2H_2O$ 溶于 100 mL 液 HCl 中，加水稀释至 1 L，在溶液中放几颗纯锡粒
氯化铁 $FeCl_3$	0.5	称取 135.2 g $FeCl_3·6H_2O$ 溶于 100 mL 6 mol/L HCl，加水稀释至 1 L
三氯化铬 $CrCl_3$	0.1	称取 26.7 g $CrCl_3·6H_2O$ 溶于 30 mL 6 mol/L HCl，加水稀释至 1 L
硝酸亚汞 $Hg_2(NO_3)_2$	0.1	称取 56 g $Hg_2(NO_3)_2·2H_2O$ 溶于 250 mL 6 mol/L HNO_3 中，加水稀释至 1 L，并置入金属汞少许
硝酸铅 $Pb(NO_3)_2$	0.25	称取 83 g $Pb(NO_3)_2$ 溶于适量水中，加入 15 mL 6 mol/L NO_3，加水稀释至 1 L
硝酸铋 $Bi(NO_3)_2$	0.1	称取 48.5 g $Bi(NO_3)_2·2H_2O$ 溶于 250 mL 1 mol/L HNO_3 中，加水稀释至 1 L
硫酸亚铁 $FeSO_4$	0.25	称取 69.5 g $FeSO_4·7H_2O$ 溶于适量水中，加入 5 mL 18 mol/L H_2SO_4，再加水稀释至 1 L，并置入小铁钉数枚
Cl_2 水	Cl_2 饱和水溶液	将 Cl_2 通入水中至饱和为止（用时临时配制）
Br_2 水	Br_2 饱和水溶液	在带有良好磨口塞的玻璃瓶内，将市售的 Br_2 约注入 1L 水中，经常剧烈震荡，每次震荡之后微开塞子，使积聚的 Br_2 蒸汽放出。在储存瓶底总有过量的溴。将 Br_2 水倒入试剂瓶时，剩余的 Br_2 应留在储存瓶中，而不倒入试剂瓶（倾倒 Br_2 或 Br_2 水时，应在通风橱中进行，将凡士林涂在手上或戴橡皮手套操作，以防 Br_2 蒸汽灼伤）
I_2 水	~0.005	将 I_2 溶解在尽可能少的水中，待完全溶解后（充分搅动）再加水稀释至 1L
淀粉溶液	~0.5%	称取易溶淀粉 1 g 和 $HgCl$ 25 g（作防腐剂）置于烧杯中，加水少许调成薄浆，然后倾入 200 mL 沸水中
亚硝酰铁氰化钠	3	称取 3 g $Na_2[Fe(CN_3)NO]·2H_2O$ 溶于 100 mL 水中
奈斯勒试剂	—	称取 115 g HgI_2 和 80 g KI 溶于足量的水中，稀释至 500 mL，然后加入 500 mL 6 mol/L NaOH 溶液，静置后取其清液保存于棕色瓶中
对氨基苯磺酸	0.34	0.5 g 氨基苯磺酸溶于 150 mL 2 mol/L HAc 溶液中
α-萘胺	—	0.3 g α-萘胺加 20 mL 水，加热煮沸，在所得溶液中加入 150 mL 2 mol/L HAc
钼酸铵	—	5 g 钼酸铵溶于 100 mL 水中，加入 35 mL HNO_3（密度 1.2 g/mg）
硫代硫酸钠	5	5 g 硫代硫酸钠溶于 100 mL 水中
钙指示剂	0.2	0.2 g 钙指示剂溶于 100 mL 水中
镁指示剂	0.0007	0.001 g 对硝基偶氮间苯二酚溶于 100 mL 2mol/L NaOH 中
铝试剂	1	1 g 铝试剂溶于 1 L 水中
二苯硫腙	0.01	10 mg 二苯硫腙溶于 100 mL CCl_4 中
丁二酮肟	1	1 g 丁二酮肟溶于 100 mL 95%乙醇中
醋酸铀酰锌	—	（1）10 g $UO_2(Ac)_2·2H_2O$ 和 6 mL 6mol/L HAc 溶于 50 mL 水中 （2）30 g $Zn(Ac)_2·2H_2O$ 和 3 mL 6mol/L HCl 溶于 50 mL 水中将（1）（2）两种溶液混合，24 h 后取清液使用
二苯碳酰二肼（二苯偕肼）	0.04	0.04 g 二苯碳酰二肼溶于 20 mL 95%乙醇中，边搅拌，边加入 80 mL(1∶9)H_2SO_4 中（存于冰箱中可用一个月）

续表

试剂名称	浓度/(mol/L)	配制方法
六亚硝酸合钴(Ⅲ)钠盐	—	$Na_3[Co(NO_2)_6]$和NaAc各20 g,溶解于20 mL冰醋酸和80 mL水混合溶液中,储于棕色瓶中备用(久置溶液,颜色由棕变红即失效)
2,4-二硝基肼	—	取2,4-二硝基肼3 g,溶于15 mL浓H_2SO_4,再加入75 mL 95%乙醇中,再加水稀释至100 mL
碘-碘化钾溶液	—	2 g碘和5 g碘化钾溶于100 mL水中
费林试剂	—	费林试剂A:取3.5g $CuSO_4·5H_2O$于100 mL水中,浑浊时过滤 费林试剂B:取酒石酸钾钠晶体17 g与15~20 mL热水中,加入20 mL 20%NaOH,释至100 mL 此两种溶液要分别贮存,使用时取等量的试剂A和试剂B混合即可
希夫试剂	—	取0.2 g对品红盐酸于100 mL热水中,冷却后,加入2 g亚硫酸钠和2 mL浓HCl,再用水稀释至200 mL
刚果红试密	—	取0.2 g刚果红溶于100 mL蒸馏水中,把滤纸放在刚果红溶液中浸泡,取出晾干,裁成纸条即可
氧化亚铜氨溶液	—	取1 g氯化亚铜加1~2 mL浓氨水和10 mL水,用力摇动,静置,倾出溶液,并投入一块铜片贮存备用
氯化锌-盐酸(Lucas)试剂	—	取34 g无水氯化锌在蒸皿中强热熔融,稍冷后,放在干燥器中冷室温。取出捣碎,溶于23 mL纯浓HCl(密度1.18)中,同时冷却
本尼迪克特试剂	—	取20 g柠檬酸和11.5 g无水$NaCO_3$于100 mL热水中,在不断搅拌下把2 g硫酸铜晶体加入其中
α-萘酚乙醇(Molish)试剂	—	取α-萘酚2 g溶于95%乙醇内,再用95%乙醇稀释至100 mL
托伦(Tallen)试剂	—	加1 mL 5%$AgNO_3$于一干净试管内,加入1滴10%NaOH,再滴加5%氨水沉淀消失为止
间苯二酚-盐酸试剂(Seliwanoff)	—	取间苯二酚0.05 g,溶于50 mL浓盐酸内,再加水稀释至100 mL
醋酸铜-联苯胺试剂	—	组分A:取150 g联苯胺溶于100 mL水及1 mL醋酸中,存放在棕色瓶中备用; 组分B:取286 g醋酸铜溶于100 mL水中,存放在棕色瓶中备用。使用前将两组分混合即可
二苯胺-硫酸溶液	—	取二苯胺0.5 g,溶于100 mL硫酸中
苯肼试剂	—	取5 g苯肼盐酸溶于100 mL水中,必要时可微助溶,然后加入9 g醋酸钠搅拌,使溶解。如溶液呈深色,加少许活性炭脱色,存于棕色瓶中。醋酸钠在此起缓冲作用,可调节pH在4~6范围内,这对成脎反应最为有利
蛋白质溶液	—	取新鲜鸡蛋清50 mL,加蒸馏水至100 mL,搅拌溶解,如果浑浊,加入5%氢氧化钠至刚清亮为止
0.1%茚三酮	—	将0.1 g茚三酮溶于124.9 mL 95%乙醇中,用时新配
无水乙醇	—	取一500 mL短颈圆底烧瓶,放入200 mL 95%乙醇,慢慢加入80 g小块的生石灰和约1 g氢氧化钠,再回流
高碘酸-硝酸银试剂	—	将25 g 12%高碘酸钾溶液与2 mL浓硝酸,2 mL 10%硝酸银溶液混合均匀,如有沉淀,过滤取透明液体备用
饱和亚硫酸氢钠溶液	—	在100 mL 40%亚硫酸氢钠溶液中,加入不含醛的无水乙醇25 mL
β-萘酚溶液	—	取4 g β-萘酚溶于40 mL 5%氢氧化钠溶液中

附录 19　常用二元体系展开剂的洗脱顺序一览表

1. 石油醚	15. 氯仿：乙醚 (8：2)	29. 苯：乙醚 (1：9)
2. 环己烷	16. 苯：丙酮 (8：2)	30. 乙醚
3. 二硫化碳	17. 氯仿：甲醇 (99：1)	31. 乙醚：甲醇 (99：1)
4. 四氯化碳	18. 苯：甲醇 (9：1)	32. 乙醚：二甲基甲酰胺 (99：1)
5. 苯：氯仿 (1：1)	19. 氯仿：丙酮 (85：15)	33. 乙酸乙酯
6. 苯：氯仿 (1：1)	20. 苯：醚 (4：6)	34. 乙酸乙酯：甲醇 (99：1)
7. 环己烷：乙酸乙酯 (8：2)	21. 苯：乙酸乙酯 (1：1)	35. 苯：酮 (1：1)
8. 氯仿：丙酮 (9：1)	22. 氯仿：乙醚 (6：4)	36. 氯仿：甲醇 (9：1)
9. 苯：丙酮 (9：1)	23. 环己烷：乙酸乙酯 (2：8)	37. 二氧六环
10. 苯：乙酸乙酯 (8：2)	24. 乙酸丁酯	38. 丙酮
11. 氯仿：乙醚 (9：1)	25. 氯仿：甲醇 (95：5)	39. 甲醇
12. 苯：甲醇 (95：5)	26. 氯仿：丙酮 (7：3)	40. 二氧六环：苯 (9：1)
13. 苯：乙醚 (6：4)	27. 苯：乙酸乙酯 (3：7)	41. 吡啶
14. 环己烷：乙酸乙酯 (1：1)	28. 乙酸丁酯：醇 (99：1)	42. 酸

参考文献

[1] 张国升，吴培云. 药用基础实验化学[M]. 北京：科学出版社，2006.
[2] 郭春. 药物合成反应实验[M]. 北京：中国医药科技出版社，2007.
[3] 查正根，郑小琦，汪志勇，等. 有机化学实验[M]. 合肥：中国科学技术大学出版社，2010.
[4] 尤启东. 药物化学实验与指导[M]. 北京：中国医药科技出版社，2010.
[5] 刘玮炜. 药物合成反应[M]. 北京：化学工业出版社，2012.
[6] 王玉良，陈华. 有机化学实验[M]. 北京：化学工业出版社，2012.
[7] 王绍杰，张星一，戚英波. 2,4-噻唑二酮的合成[J]. 中国药物化学杂志，2000，10(4)：89-90.
[8] 王福来. 有机化学实验[M]. 武汉：武汉大学出版社，2001.
[9] 李兆龙，阴金香，林天舒. 有机化学实验[M]. 北京：清华大学出版社，2001.
[10] 陈长水，刘汉兰，等. 微型有机化学实验[M]. 北京：化学工业出版社，1998.
[11] 吴苦峰，等. 微型有机化学实验[M]. 上海：上海大学出版社，1998.
[12] 顾可权，陈光沛，等. 半微量有机制备[M]. 北京：高等教育出版社，1990.
[13] 企钦汉，戴树珊，黄卡玛. 微波化学[M]. 北京：科学出版社，2001.
[14] 吴世晖，周景尧，林子森. 中级有机化学实验[M]. 北京：高等教育出版社，1986.
[15] 杜志强. 综合化学实验[M]. 北京：科学出版社，2005.
[16] 顾觉奋. 分离纯化工艺原理[M]. 北京：中国医药科技出版社，2002.
[17] 陆涛，陈继俊. 有机化学实验与指导[M]. 北京：中国医药科技出版社，2003.
[18] 赵临襄. 化学制药工艺学[M]. 北京：中国医药科技出版社，2003.
[19] 朱保泉. 新编药物合成手册(上、下册)[M]. 北京：化学工业出版社，2003.
[20] 王伯康. 综合化学实验[M]. 南京：南京大学出版社，2000.
[21] 张昭艾. 无机精细化工工艺学[M]. 北京：化学工业出版社，2002.